Lewis & Clark

Weather and Climate Data from the Expedition Journals

American Meteorological Society
Historical Monographs Series

* These Historical Monographs are in print and available to order from the AMS Order Department, 45 Beacon Street, Boston, MA 02108 • 617-227-2426, ext. 686 • amsorder@ametsoc.org

Lewis & Clark

Weather and Climate Data from the Expedition Journals

Edited by Vernon Preston

American Meteorological Society
Boston, Massachusetts

The most recent source for the expedition journals, which provided the most comprehensive data, is *The Definitive Journals of Lewis and Clark*, edited by Gary E. Moulton (University of Nebraska Press 2002). The editor of this monograph, Vernon Preston, is indebted to Moulton's work on the journals.

Photographs taken by and courtesy of Vernon Preston, Ed., unless otherwise noted.

Published by the American Meteorological Society
45 Beacon Street, Boston, Massachusetts 02108

Executive Director: Keith Seitter
Director of Publications: Ken Heideman
Books and Monographs Manager: Sarah Jane Shangraw
Historical Monographs Series Editor: John Perry

For a catalog of AMS Books, including meteorological and historical monographs, see www.ametsoc.org/pubs/books. To order AMS Books, call (617) 227-2426, extension 686,
or email amsorder@ametsoc.org.

ISBN 13: 978-1-878220-75-2
ISBN 10: 1-878220-75-6

Library of Congress Cataloging-in-Publication data is available at:
www.ametsoc.org/pubs/books/bookdesc.html#LWSCLARK.

Printed in the United States of America
by Capital Offset Company Inc.

To my honored parents, David and Ruth Mary, who believed
that history must be experienced, not just read;
who took our family to tour the West and its history;
who taught us to respect the land and its peoples;
and who encouraged my meteorological pursuits.

"This is our present Situation,! truly disagreeable. About 12 oClock the wind Shifted about to the NW and blew with great violence for the remainder of the day at maney times it blew for 15 or 20 minits with Such violence that I expected every moment to See trees taken up by the roots. Many were blown down. Those Squals were Suckceeded by rain, !O how Tremendious is the day. This dredfull wind and rain Continued with intervales of fair weather all the latter part of the night. O! How disagreeable is our Situation dureing this dreadfull weather."

—**Captain William Clark**
November 28, 1805
Mouth of the Columbia River

"About 12 oClock we arived in Site of St. Louis. Drew out the canoes then the party all considerable much rejoiced that we have the Expedition Completed and now we look for boarding in Town and wait for our Settlement and then we entend to return to our native homes to See our parents once more as we have been So long from them."

—**Sergeant John Ordway**
September 23, 1806
Return of Expedition to St. Louis

CONTENTS

ILLUSTRATIONS

FOREWORD

By Terry Nathans

The weather and climate of the trans-Mississippi west was virtually unknown at the beginning of the nineteenth century. This changed dramatically shortly after the Louisiana Purchase was signed in 1803, which set the stage for acquiring the first systematic weather measurements of the trans-Mississippi west. The framework for obtaining these measurements was outlined in the now famous June 20, 1803 letter from President Thomas Jefferson to his protégé and personal secretary, Captain Meriwether Lewis. In that letter, Jefferson instructed Lewis to plan and carry out an overland expedition to the Pacific Ocean for the purposes of commerce, and to observe and record a broad range of natural history subjects, including the

> ...climate, as characterised by the thermometer, by the proportion of rainy, cloudy & clear days, by lightning, hail, snow, ice, by the access & recess of frost, by the winds prevailing at different seasons, the dates at which particular plants put forth or lose their flower, or leaf... (Jackson 1978, p. 63).

Jefferson's instructions to Lewis, which were part of his decades-long ambition of launching an expedition to explore the interior of North America, were made at the threshold of what Fleming (1990) has called the "expanding horizons" in meteorology. During this period, more reliable meteorological instruments began to emerge allowing for a more comprehensive and systematic acquisition of weather data. Prior to Lewis and Clark's time, few advances were made in "meteorological science." Most of what was commonly known at the time about weather and climate was rooted in the theories first proposed by the philosophers of ancient Greece (Lamb 1977). Their theories, which related the character tendencies of the various peoples of the world to the warm, cold, and middle zones on Earth, were based on an axiomatic approach, wherein self-evident truths were postulated and built upon by deductive reasoning to reach conclusions about the natural world. It wasn't until the Enlightenment and the emergence of the scientific revolution in the seventeenth century that this Aristotelian philosophy was replaced with rationalism and the scientific method (Bowler 1993).

Jefferson was a product of the Enlightenment and, by the time the Louisiana Purchase was signed, he was uniquely positioned to propose and implement a broad-based exploratory expedition of the West. He was elected the third president of the United States and thus had political leverage as well as access to the resources of the government. He was a member of the American Philosophical Society (APS) in Philadelphia, the nexus of scientific thought at the time. And he had a unique and deep intellectual curiosity about ethnography, geography, science and climate, particularly of the land west of the Mississippi.

To lead his exploratory expedition of the West, Jefferson chose Meriwether Lewis, whom he believed to be the most qualified for the job, despite Lewis's lack of formal scientific training. Jefferson's rationale for choosing Lewis to lead the expedition is made clear in a March 2, 1803 letter to Robert Patterson, professor of mathematics at the University of Pennsylvania and fellow member of the APS. Jefferson wrote:

> I am now able to inform you..., that we...are likely to get the Missouri explored...I propose to send...a party...with Capt. Lewis, my secretary, at their head. If we could have got a person perfectly skilled in botany, natural history, mineralogy, astronomy, with at the same time the necessary firmness of body & mind, habits of living in the woods & familiarity with the Indian character, it would have been better. But I know of no such character who would undertake an enterprise so perilous. To all the latter qualities Capt. Lewis joins a great stock of *scientific* accurate observation on the subjects of the three kingdoms which are found in our own country but not according to their scientific nomenclatures...and I shall be particularly obliged to you for any advice or instruction you can give him (Jackson 1978, p. 21).

Lewis's "scientific training" for the expedition mostly occurred during May and June of 1803, when several of Jefferson's fellow APS members tutored Lewis on a variety of subjects, including surveying, botany, and medicine. Professor Robert Patterson provided instruction on various scientific instruments; Major Andrew Ellicott provided instruction on surveying; Benjamin Smith Barton, who wrote the first American textbook on botany, advised on the collection of scientific data; Casper Wistar, who wrote the first American textbook on anatomy, advised on zoology; and Dr. Benjamin Rush, the preeminent physician of the time, counseled him on medical issues. To round out his training, Lewis also purchased several scientific books that would serve as his traveling library. These books included *Elements of Botany* by Barton, *Elements of Mineralogy* by Kirwan, *A Practical Introduction to Spherics*, and *The Nautical Almanac and Astronomical Ephemeris* (Ambrose 1996).

There is no written evidence that has been found suggesting that Lewis's scientific training in Philadelphia formally included meteorology. This may at first be surprising since Philadelphia could arguably be viewed as the nation's center for meteorological thought during the eighteenth and early nineteenth centuries. For example, Philadelphia already had a long history of its citizenry keeping weather diaries (Ludlum 1966). Thomas Jefferson, in fact, made his first entry into his weather diary while in Philadelphia (Malone 1970). Benjamin Franklin, longtime resident of Philadelphia and founder of the American Philosophical Society, not only kept weather diaries but conducted the seminal experiments on lightning and planted the first intellectual seeds that eventually led to theories on the circulation of coastal storms, the connection between heated air and small-scale vortices such as waterspouts, and the equator-to-pole atmospheric circulation (Fleming 1990). One of the leading medical institutions at the time, Philadelphia's College of Physicians, included in its charter the importance of meteorological observations in linking diseases with weather (Hindle 1956), a link already hypothesized by Dr. Benjamin Rush (Rush 1789).

In view of the intellectual resources available in Philadelphia and Jefferson's longtime interest in recording weather data, why didn't Lewis receive formal meteorological training? First and foremost was that meteorology was not considered a formal science in the way chemistry and physics were. Broad, contemporaneous meteorological measurements were decades into the future and the laws governing atmospheric motion had yet to be established. For the learned men and the populace at large, weather was most important for the role it played in agriculture. Thus formal meteorological training of Lewis was neither possible nor essential; daily recordings of the weather—wind, temperature, cloud cover etc.—which required no special skill, would suffice.

Following Lewis's scientific training, there were two key communiqués. On June 19, Lewis wrote a letter to longtime friend and former military superior, William Clark, asking him to co-lead the expedition. Clark accepted several weeks later. On June 20, Jefferson wrote a letter to Lewis, which formally laid out the plans for conducting an exploration of the trans-Mississippi west. The letter is remarkable for its far-reaching vision, clarity of scientific goals, and ultimate impact on our nation's history. Jefferson wrote:

> ...The object of your mission is to explore the Missouri river, & such principal stream of it, as, by [its] course and communication with the waters of the Pacific ocean, ..., for the purposes of commerce...You will take, *careful* observations of latitude and longitude, at all remarkable points on the river...make yourself aquainted ...with the names of the nations and their numbers,...their language; their ordinary occupations in agriculture...Other objects worthy of notice will be the soil and face of the country...; the animals of the country; ...the mineral productions;...climate (Jackson 1978, pp. 61–63).

The letter formally mapped out a broad-based mission whose objectives were to provide new information on the economic, ethnographic, geographic, and scientific aspects of the West. Executing the mission required the cooperation of several branches of government, funding for supplies and equipment, and scientific training in several disciplines. As noted by historian James Ronda, this was the United States' first foray into what is now called "big science" (Ronda 1998).

By the time the Lewis and Clark expedition came to fruition in the early nineteenth century, several scientific instruments had been invented that became an integral part of the expedition. Among these instruments were an octant and sextant for determining latitude, a theodolite and chronometer for determining longitude, a telescope, a surveying compass, and a spirit level. Lewis and Clark also carried three thermometers, which had been invented more than 175 years before the expedition.

A barometer was not carried on the expedition, despite being more than a century and a half old by the time Lewis and Clark began their journey. At the time it was well known that the barometer measured the pressure of a column of air above the gauge and could also be used to measure elevation. The fragility of the barometer, however, made it impractical for transport. It was likely for these reasons that Lewis and Clark did not carry a barometer. Even if they did, the connection between barometric pressure and air masses and fronts was unknown. Such knowledge would have to wait for the development of the telegraph in 1843 when the contemporaneous acquisition of weather data over large areas began. The theory of air masses and weather fronts didn't occur until around 1920, more than a century after Lewis and Clark.

From late June until the end of August, Lewis and Clark spent most of their time obtaining additional supplies, maps, and books, commissioning men for the expedition, purchasing a pirogue, and having a keelboat made in Pittsburgh, which was completed on August 31. By eleven o'clock that morning the boats were loaded and Lewis and several recruits were headed down the Ohio River for their eventual mid-October rendezvous with Clark.

On August 31, Lewis makes the first entry in what is known as the Lewis and Clark Journals. September 1 marks the first entry about the weather, and on September 2, Lewis recorded the first temperature of the expedition. In the September 1 entry Lewis speculates about the origin of the fog on the Ohio River:

> …the Fog appears to owe [its] origin to the difference of temperature between the *air* and *water* the latter at this season being much warmer than the former; the water being heated by the summer's sun dose not undergo so rapid a change from the absence of the sun as the air dose consequently when the air becomes cool which is about sunrise the fogg is thickest and appear to rise from the face of the water like the steem from boiling water… (Moulton 1986, p. 67).

Lewis's speculation about the origin of the fog, which we now know as evaporation-steam fog, touches on two key scientific points, neither of which was fully understood by science at the time. One is related to condensation and evaporation and the other to the difference in specific heat capacity between water and air. First, Lewis states that the *fog "owe[s] [its] origin to the difference between the air and water."* Lewis's statement is indeed correct and is related to the role of temperature in the condensation and evaporation of water vapor. Although instruments to measure the amount of water vapor in the atmosphere were around since the fifteenth century, theories regarding the condensation and evaporation of water vapor were only beginning to emerge in the mid-eighteenth century (Middleton 1969). A full explanation of the processes involved in the formation of fog would have to wait until the late nineteenth century when the kinetic theory of gasses was well established and the seminal experiments on condensation were being carried out (Mason 1957).

Lewis's second point regards "...*the water being heated by the summer's sun dose not undergo so rapid a change from the absence of the sun as the air....*" Lewis is making a statement about how the difference in specific heat capacity between water and air favors fog formation in the morning. Like condensation and evaporation, the scientific foundation for explaining Lewis's observation on specific heat had yet to be fully developed.

Lewis and Clark rendezvoused in mid-October on the north side of the Ohio River in Clarksville, Indiana Territory. After a couple of weeks in Clarksville selecting enlisted men for the expedition, the party moved down the Ohio then up the Mississippi. On December 13, 1803, they arrived at their winter camp on the Wood River, near the mouth of the Missouri River. Lewis and Clark would spend five months at the Wood River camp. During their stay at the camp they were busy collecting and describing the local flora and fauna, practicing using their celestial instruments, and recording the daily weather.

On May 21, 1804, the Lewis and Clark expedition, which became known as the Corps of Discovery, was on the eve of its epic journey, a journey that would carry them through several climate zones. Based on the twentieth-century Köppen system for climate classification, the Corps of Discovery would pass through five major climate zones: humid subtropical, humid continental, mid latitude steppe, highland, and marine west coast. Lewis and Clark would formally document, for the first time, the continental climate of the Northern Plains, the highland climate of the Rocky Mountains, and the marine west coast climate of the Pacific Northwest, all of which were far outside their life experience and sharply counter to Jefferson's erroneous assumption that they would be traveling through a region of "moderate climate" (Jackson 1978, p. 12).

At 6:00 a.m. on May, 22, 1804, under cloudy skies, the Corps of Discovery began its first full day as a unit, moving upstream against the powerful Missouri River. Until the party reached the site of their winter quarters at Fort Mandan near today's Bismarck, North Dakota, in November, the party toiled against powerful river currents, recorded information on flora and fauna new to science, traded with the local Indian tribes, and made weather observations.

From St. Louis to Fort Mandan, the party traveled through what the Köppen system broadly classifies as a humid continental climate. This climate class is characterized by severe winters and no dry season. During this leg of the journey the expedition experienced the capriciousness of the continental summertime climate: oppressive humidity followed by dry, northerly winds; torrential downpours and high winds followed by calm; and dense fog followed by sunshine. Weather extremes were common. On July 29, while on the Missouri between today's South Dakota and Iowa, Lewis notes the destruction caused by an apparent tornado: "... *above this high land & on the S. S. passed much falling timber apparently the ravages of a Dreadfull harican which had passed obliquely across the river from N.W. to S.E. about twelve months Since, many trees were broken off near the ground the trunks of which were Sound and four feet in Diameter,...*" (Moulton 1986, p. 427).

By early November the party was at the site of their winter quarters, Fort Mandan, named after the local Mandan Indians. Located about 45 miles northwest of present-day Bismarck, North Dakota, the region is notorious for its severe winters. While at Fort Mandan, Lewis and Clark systematically recorded the daily weather, producing the first long-term, systematic tabulation of weather data west of the Mississippi. Comments on the weather were included in the regular journal entries as well as in a separate weather diary. The weather diary also contained tables that listed the weather observations. These tables, which were similar to Jefferson's weather tables, included the temperature, wind direction, and state of the river. The temperature and wind direction were recorded twice a day, at sunrise and at four o'clock, which Jefferson believed to be the coldest and hottest times of the day, respectively (Jefferson 1955, p. 78).

In comparing current weather observations with those tabulated by Lewis and Clark during their stay at Fort Mandan, Solomon and Daniel (2004) conclude that *"On balance, …the temperatures documented by Lewis and Clark…were not unusual compared to typical modern observations."*

In early April, when the ice had broken up on the Missouri, the thirty three-member party loaded their boats and proceeded on. The party was slowly entering terra incognita, today's Montana, where the geography, Indian cultures, and climate were virtually unknown. The arid climate of this region—the steppe climate of the Northern Plains—was outside of their life experience. Here, potential evapotranspiration exceeds precipitation, and the winters are severe and the summers hot and dry. Lewis found the low humidity noteworthy enough to carry out a crude experiment:

[May 30, 1805]… circumstances indicate our near approach to a country whos climate differs considerably from that in which we have been for many months. the air of the open country is asstonishingly dry as well as pure. I found by experiment that a table spoon full of water exposed to the air in a saucer would evaporate in 36 hours when the mercury did not stand higher than the temperate point at the greatest heat of the day; my inkstand so frequently becoming dry put me on this experiment (Moulton 1987, p. 221).

The periods of very low humidity were often punctuated by severe thunderstorms, which were accompanied by torrents of rain, hail, and flash floods.

By mid-August, the Corps of Discovery was making its arduous trek across the continental divide near today's Lehmi Pass in southwestern Montana. The corps encountered steep cliffs, confusing and narrow ravines, and lack of game, all of which were exacerbated by the cold temperatures of the highland climate. Traveling was dangerous. The horses frequently fell, and on September 3 the last of the three thermometers was broken. Clark summed up the situation on September 16: *"…began to Snow about 3 hours before Day and Continud all day…I have been wet and as cold in every part as I ever was in my life…"* (Moulton 1988, p. 209). The grueling journey across the mountains was completed by late September.

In preparation for their journey to the coast, dugout canoes were made, and on October 7, the party was on the Clearwater River heading for the junction with the Snake River. In three days the party reached the Snake River, and two weeks after that, the Columbia River. As the party traveled down the Columbia from the east to the west side of the Cascade Mountains, there was a dramatic change in climate. The steppe climate east of the cascades rapidly transitioned into a marine west coast climate. The arid, thinly timbered landscape of the steppe gave way to another that was moist and heavily timbered.

The marine west coast climate of the Pacific Northwest is distinguished by mild temperatures throughout the year, no dry season and a warm summer. The region is renowned for powerful, moisture-laden storms that buffet the region throughout the winter. Upon the party's arrival to the region in early November, rain, fog, and strong winds were becoming increasingly common. By mid to late November, powerful storms were occurring almost daily. The inclement weather and the lack of permanent shelter were proving extremely difficult to the party. Clark wrote: *"[November 22, 1805] …the wind increased to a Storm from the S.S.E. and blew with violence throwing the water of the river with emence waves out of its banks almost overwhelming us in water, O! how horriable is the day…"* (Moulton 1990, p. 79).

On December 8, 1804, the party began the first full day at the site that would become Fort Clatsop, named after the local Clatsop Indians. Although Lewis and Clark noted in October a change in climate as they paddled down the Columbia from the east to the west side of the Cascades, they now realized that they were in a winter climate that was foreign to them.

The nearly relentless rainfall and relatively mild temperatures of the Pacific Northwest contrasted sharply with the cold, snowy winters that Lewis and Clark experienced in the East. The distinct climate of the Pacific Northwest is noted by Lewis on January 3, when he writes, "*I am confident that the climate is much warmer than in the same parallel of Latitude on the Atlantic Ocean tho' how many degrees is now out of my power to determine*" (Moulton 1990, p. 259).

Lewis's inferred connection between climate and latitude was known since the time of the ancient Greeks. However, not so well known at the time was the role that ocean currents and land–sea heating contrasts played in climate, factors that largely account for his observed difference in climates between the Pacific Northwest and locations of similar latitude on the East Coast.

From the beginning of the party's stay at Fort Clatsop on December 8 until their departure on March 23, the weather remained stormy. Of the 106 days that the party stayed at Fort Clatsop, there were 90 days of precipitation, of which there were 17 days of snowfall. There were only 12 days without precipitation; the sun shone fair for only six days.

Based on comparisons with averaged conditions for today's Astoria, Oregon, which is near the site of Fort Clatsop, Nathan (2005) has shown that the winter of 1805–06 at Fort Clatsop was indeed atypical. The frequency of precipitation, the frequency of snowfall, and the persistent southwesterly winds were all dramatically different than present-day measurements. Nathan speculates that the winter of 1805–06 was a La Niña year.

Lewis and Clark departed Fort Clastop on March 23, 1806. They continued to make daily weather observations on their return trip. At twelve o'clock on September 23, 1806, the party arrived to St. Louis and "*...were met by all the village and received a harty welcome from [its] inhabitants...*" (Moulton 1993, pp. 370–371).

Pursuant to Jefferson's instructions, Meriwether Lewis and William Clark successfully completed an expedition that spanned more than 28 months and covered more than 8,000 miles. They recorded for science more than 170 new plant species and more than 120 new animal species, and they produced hundreds of pages of cartographic, ethnographic, and scientific information. Moreover, the Lewis and Clark journals provide the first systematic instrumental and proxy data of the trans-Mississippi west. The continuous temperature measurements at Fort Mandan and the keen descriptive comments on the weather throughout the journey, which are contained within this monograph, can be combined with documentary, dendroclimatic, and ice-core evidence to form a more complete picture of the regional and global weather patterns during the early nineteenth century. Lewis and Clark's pioneering weather observations add another small piece to the climate puzzle, serving as an overarching link between early nineteenth century climate data and our efforts to model climate change today.

References

Ambrose, S. E., 1996: *Undaunted Courage*. Simon and Schuster, 511 pp.

Bowler, P. J., 1993: *The Norton History of the Environmental Sciences*. Norton and Company, 634 pp.

Fleming, J. R., 1990: *Meteorology in America, 1800–1870*. Johns Hopkins University Press, 264 pp.

Hindle, B., 1956: The pursuit of science in Revolutionary America, 1735–1789. *The Charter, Constitution and Bye Laws of the College of Physicians of Philadelphia (1790)*. University of North Carolina Press, 410 pp.

Jackson, D., ed., 1978. *Letters of the Lewis and Clark Expedition, with Related Documents: 1783–1854*. 2d ed. Vol. I. University of Illinois Press, 728 pp.

Jefferson, T., 1955: *Notes on the State of Virginia*. Reprint. University of North Carolina Press, 315 pp.

Lamb, H. H., 1977: *Climate – Past, Present and Future*. Vol. 2, *Climate History and the Future*. William Clowes and Sons, 835 pp.

Ludlum, M., 1966: *Early American Winters 1604–1820*. American Meteorological Society, 285 pp.

Malone, D., 1970: *Jefferson and His Time*. Vol. 6, *The Sage of Monticello*. Little, Brown and Company, 551 pp.

Mason, B. J., 1957: *The Physics of Clouds*. Oxford University Press, 481 pp.

Middleton, W. E. K., 1969: *Invention of the Meteorological Instruments*. The Johns Hopkins Press, 362 pp.

Moulton, G., ed., 1986: *The Journals of the Lewis & Clark Expedition*. Vol. 2, *August 30, 1803 – August 24, 1804*. University of Nebraska Press, 621 pp.

_____, 1987: *The Journals of the Lewis & Clark Expedition*. Vol. 4, *April 7–July 27, 1805*. University of Nebraska Press, 464 pp.

_____, 1988: *The Journals of the Lewis & Clark Expedition*. Vol. 5, *July 28–November 1, 1805*. University of Nebraska Press, 415 pp.

_____, 1990: *The Journals of the Lewis & Clark Expedition*. Vol. 6, *November 2, 1805–March 22, 1806*. University of Nebraska Press, 531 pp.

_____, 1993: *The Journals of the Lewis & Clark Expedition*. Vol. 8, *June 10–September 26, 1806*. University of Nebraska Press, 456 pp.

Nathan, T. R., 2005: "O! How Horriable is the Day". *We Proceeded On*, **31** (4), 10–18.

Ronda, J., 1998: 'So Vast an Enterprise': Thoughts on the Lewis and Clark expedition. *Voyages of Discovery: Essays on the Lewis and Clark Expedition*, J. Ronda, ed., Montana Historical Society Press, 1–28.

Rush, B., 1789: Account of the climate of Pennsylvania, and its influence on the human body. *American Museum*, **26** (4).

Solomon, S., and J. S. Daniel, 2004: Lewis and Clark – Pioneering meteorological observers in the American West. *Bull. Amer. Meteor. Soc.*, **85**, 1273–1288.

PREFACE

This immence river so far as we have yet ascended, waters one of the fairest portions of the globe, nor do I believe that there is in the universe a similar extent of country, equally fertile, well watered, and intersected by such a number of navigable streams.

—Meriwether Lewis
March 31, 1805
Letter to his mother Lucy Marks while at
Fort Mandan, North Dakota

In the annals of American scientific exploration and discovery, one journey that stands the enduring test of time and leads us on to new discoveries is that of Meriwether Lewis and William Clark. The two captains, with more than three dozen participants, explored across the heart of the vastly uncharted North American continent, from St. Louis up the Missouri River across the Continental Divide and down the tributaries of the Columbia River to the mighty Pacific Ocean. Inspired by President Thomas Jefferson and following years of planning and failed attempts, the successful Lewis and Clark "Corps of Discovery" journey established a foundation of commerce, science, and knowledge in what would become the expanding domain of the United States of America (Appleman 1975; Ambrose 1996, 68–79; Hayes 2001; Ronda 2001, viii, 1–16; Wheeler 1904). Historian Roy Appleman (1975, 3) notes that "In its scope and achievements, the expedition towers among the major explorations of the North American Continent and the world." This expedition between 1803 and 1806 vastly increased the knowledge of flora, fauna, geography, geology, native peoples, commerce trade possibilities, and routes. This special edition of the journals describes the systematic climatological, hydrological, and meteorological events the explorers encountered during their journey.

The journals kept by members of the Corp of Discovery contain archaic spellings, misspellings, and inconsistencies in spacing. The editors of this volume have endeavored to edit the excerpts only slightly in order to clarify meaning. Brackets are used to set off text not contained in the original journals, i.e., text that has been added by the source for these excerpts (in these cases most often Moulton, 2002) or by this edition's editor to clarify meaning. Parentheses in journal excerpts were originally included by the journal writer or, in keeping with the convention used by Moulton, they indicate a word once there and erased by the journal writer or a word added by the journal writer after the fact.

This Countrey may with propriety I think be termed the Deserts of America, as I do not Conceive any part can ever be Settled, as it is deficent in water, Timber & too Steep to be tilled.

—William Clark
May 26, 1805
Central Montana

PART I Meteorology and the Corps of Discovery

CHAPTER 1 Meteorological Synopsis of the Expedition

"Saw a black cloud rise in the west which we looked for emediate rain we made all the haste possable but had not got half way before the Shower met us and our hind extletree broke in too we were obledged to leave the load Standing and ran in great confusion to Camp the hail being So large and the wind So high and violent in the plains, and we being naked we were much bruuzed by the large hail. Some nearly killed one knocked down three times, and others without hats or any thing about their heads bleading and complained verry much…The plains are so wet that we could doe nothing this evening."

—Sergeant John Ordway
June 29, 1805
Along the Portage Route around the
Great Falls of the Missouri, Montana

"In the afternoon, there arose a storm of hard wind & rain; accompanied with amazing large hail at the upper camp. We caught several of the hail Stones which was measured & weighed by us, there were 7 inches in Surcumference and weighed 3 ounces—Captain Lewis made a small bowl of punch out of one of them. As luck would have it, we were all…Safe…the party that was at the upper camp, were under a good shelter, but we feel concerned about the men on the road with the baggage from the lower Camp—"

—Private Joseph Whitehouse
June 29, 1805
White Bear Island, Upper Portage Camp
Southwest Great Falls, Montana

For decades, exploration of inland portions of the North American continent had been a goal of many governments worldwide, and lucrative trade with Indian nations led many countries to develop remote trading posts. Thomas Jefferson was intrigued by the idea of an expedition up the Missouri some twenty years prior to the Lewis and Clark Expedition. He tried to interest General George Rogers Clark in making the expedition in 1783, but lack of funding prevented an attempt. While serving in Paris, Jefferson tried to engage John Ledyard to cross Russia, enter North America by way of Alaska, and explore eastward to St. Louis. This too fell through as Ledyard was stopped by Russian officials while trekking through Siberia. Jefferson's concern over who would control interests in the Pacific Northwest was further aroused when he learned of overland journeys by British explorer Alexander Mackenzie. Mackenzie made two westward trips from northern Alberta's Lake Athabasca. During his first journey in 1789, Mackenzie led a small party northwest to the Arctic Ocean down a broad river (later named for Mackenzie). On his second journey in 1793, Mackenzie made a trek to the Pacific Ocean down the Peace River and later Fraser River. Arriving at the coast, he threw down the gauntlet to other countries by painting the rocks near the shore with the following inscription: "Alexander Mackenzie, from Canada, by land, the twenty-second of July, one thousand seven hundred and ninety-three" (Mackenzie 1801; Bakeless 1947; Salisbury 1950; DeVoto 1953; Gilbert 1973; Allen 1975; Appleman 1975; Wood and Thiessen 1985; Ambrose 1996; Ronda 2000;

1

Hayes 2001; Ronda 2001; Saindon 2003). As a twist of irony, the Lewis and Clark Expedition took liberties of a similar nature during their journey, and one of these markings still remains at Pompey's Pillar near Billings, Montana; it is the only remaining physical evidence of their journey on the landscape.

Undaunted by other setbacks, Jefferson tried once again to enlist an explorer to tour the Missouri River system. In 1793, backed by the American Philosophical Society, Jefferson tried to hire Andre Michaux, a French botanist, to make the journey. The plan failed when it was learned Michaux was a French spy attempting to stir up trouble between the Americans and Spaniards (Salisbury 1950; Steffen 1977).

Thomas Jefferson found himself in a better position to promote an expedition when he became president of the United States. On January 18, 1803, he submitted a confidential message to Congress (see Appendix A). Near the end of the message was a small paragraph requesting "an appropriation of $2,500 for the purpose of extending the external commerce of the United States" (Richardson 1897; Bruun and Crosby 1999). On February 28, 1803, Jefferson received word from Congress that they had approved the journey. Meriwether Lewis, President Jefferson's personal secretary, was selected to lead the expedition and spent the spring of 1803 in Philadelphia preparing for the journey. Lewis requested a co-leader for the journey and chose his former army captain, William Clark. While in Philadelphia, Lewis also completed training in astronomy, natural history and sciences, health and medicine, and ethnology. In addition to his studies, he spent time purchasing and obtaining a vast array of materials needed to complete the journey successfully (Biddle 1814; Jackson 1978; Botkin 1995; Burroughs 1997; Burns 1997; Chuinard 1998; Paton 2001; Peck 2002; Cutright 2003; Patient 2003). Included in his packing list were three thermometers (see Appendix B).

Meteorological Instruments

There is uncertainty as to the type of thermometers used on the Lewis and Clark Expedition. Although not discovered until the late 1600s, the basic principle behind thermometers was known as far back as the third century B.C.E. Galileo is credited with inventing the first thermometer sometime around 1593. By 1641, Ferdinand II, Grand Duke of Tuscany, developed a sealed thermometer. Other advancements were made by Robert Boyle, who recognized the need for a standard scale. Various trials took place using water, air, liquor, spirits, alcohol, linseed oil, and finally, mercury as the measuring element within a thermometer. As science advanced during the 1700s, Robert Hooke increased the accuracy and established a fixed measurement for the freezing point of water. Dutch mathematician Christian Huygens is credited with suggesting two fixed points, the second being that of boiling water. Sir Isaac Newton chose a scale using fixed points of melting snow and of the human body. In 1714, Gabriel Fahrenheit developed the first mercury thermometer with a reliable scale. He established the first point of his scale by dipping the thermometer into a solution of ice, water and sal ammoniac, and/or sea salt, and designated it zero. A second point was assigned at 32° F when the instrument was placed in a mixture of water and ice only. The third point of 96° F was based on the temperature reached when the thermometer was placed in the mouth or armpit of a healthy man. Swedish astronomer Anders Celsius provided another alternative in 1742. He used two fixed points: that of boiling water, which he assigned zero on his scale, and the temperature of melting ice, 100, with equal marks between. It was Jean Pierre Christin of Lyons, however, who inverted the scale as it appears today. In fact, by 1779 there were as many as nineteen temperature scales in use (Middleton 1969; Frisinger 1983; Middleton 2003).

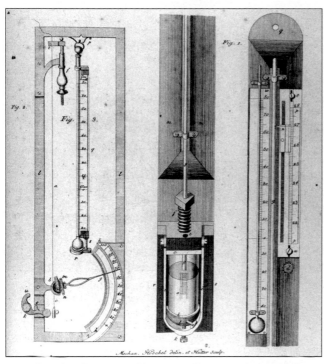

Examples of 18th-century thermometers, as seen in *Beschreiburg der meteorologischen Instumente* by Augusin Stark, published in 1815. Courtesy NOAA Photo Library.

A chronometer was an instrument used for determining time and noon-day sun. Lewis used this instrument for determining the two basic meteorological observation times each day. Photograph taken at Lewis and Clark National Historic Trail Interpretive Center, Great Falls Montana. (See May 17, 1806.)

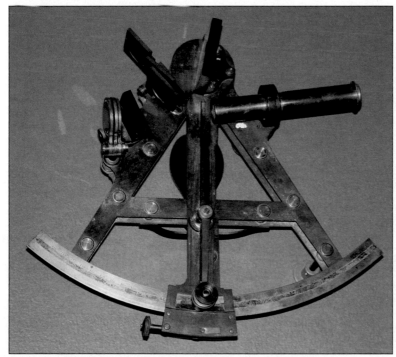

A sextant used for astronomical observations to determine latitude for navigation. Photograph taken at Jefferson National Expansion Memorial. (See April 5, 1806.)

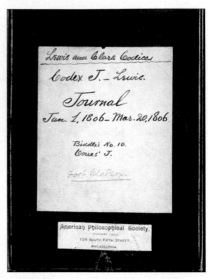

A portion of the famous map of the western portion of North America made by William Clark and formally published in 1814. Photograph by Vernon Preston, courtesy of the American Philosophical Society/U.S. National Park Service.

The front of a journal that was filled by Meriwether Lewis while the expedition spent the winter of 1805/06 at Fort Clatsop, near present day Astoria, Oregon. Photograph by Andrea Laliberte, USDA-ARS, and courtesy of the American Philosophical Society/U.S. National Park Service.

The Lewis and Clark Expedition Weather Diary entries for January 1, 1806 through January 12, 1806. During this time the Corps of Discovery was in winter quarters at Fort Clatsop, near present-day Astoria, Washington. Photograph by Vernon Preston, courtesy of the American Philosophical Society/ U.S. National Park Service.

Which scale was chosen for the Lewis and Clark Expedition? On their first entry (January 1, 1804) in what would become known as the "Weather Diary" (which was also referred to as the "Book of Thermometrical Observations" and the "Meteorological Register"), they performed experiments to determine errors in their thermometers by dipping them in a mixture of water and snow and marking the freezing point, and then making a similar mark when inserting them into boiling water. Lewis and Clark noted that they made these observations using "Ferenheit's Thermometer," which would imply they were using the Fahrenheit scale. It is not certain when they conducted these particular experiments, but during January 17–31, 1803, they made mention of current ambient air readings nearly hourly. Nowhere else during the journey did they write such entries. Clark at various times used words like "Ferenthiers" and "Ferents" thermometer and Lewis used "Ferrenheit," which help confirm what scale they were using (Clark, June 22, 1804; Lewis, September 3, 1803). Another mark they placed on the thermometers was an arbitrary summer temperature commonly known as "Summer heat," which was usually 75° or 76° F (Moulton 1986, 2: 316).

It is believed that the three thermometers noted on the requirements list were made in Philadelphia (see Appendix B) (Moulton 1986, 2: 69). In his January 3, 1804 entry, Clark makes mention of a particular company that made a thermometer: "John Donegan (or Denegan) and Joseph Donegany (Donegani) were making thermometers in Philadelphia in 1785. Although thermometers are among Lewis's list of requirements for the trip, there is no direct evidence that any were purchased" (Moulton 1986, 2: 146). Other stories abound as to their origin. Historian Donald Jackson (1978, 75) notes that "an undocumented family tradition," first related by Dye and renewed by Meany, declares that St. Louis physician Antoine Saugrain made thermometers for Lewis and Clark by scraping the mercury off the back of his wife's mirror. Saugrain had social contacts with the explorers before and after the expedition, but it is not likely that he made thermometers for them. Lewis kept temperature records on his way down the Ohio in the fall of 1803. Clark continued the practice at the Wood River (Dubois) Camp in the early months of 1804, and there is no evidence that the thermometers obtained in Philadelphia were not used. The last one was broken on September 3, 1805 when it was accidentally struck against a tree. The instruments must have been similar to that described by Jefferson in a request on June 5, 1804 to Isaac Briggs for two thermometers: "The kind preferred is that on a lackered plate slid into a mahogany case with a glass sliding cover, these being best exposed on exposure to the weather."

To further confirm this, the Weather Diary entry on September 6, 1805 notes *"Thermometer broke by the Box strikeing against a tree."* No other meteorological instruments are known to have been carried by the expedition. Except for temperature recordings, all other meteorological observations were taken using the natural senses or with other instruments used for navigation and measurement. To determine wind direction, they would stand facing the wind with a compass to determine a direction. For rise and fall of the river water, various marks were made on the bank and measured later with marked sticks, poles, or chains which used the English scale of inches and feet (Large 1986).

Final Instructions

Jefferson sent final instructions to Lewis in June 1803 giving specific directions on the scientific and commercial goals for the expedition (see Appendix C). As fortune would have it, Lewis returned to Washington, D.C. on July 4, 1803 to learn that Napoleon had decided to sell France's Louisiana territory to the United States. This changed the expedition's initial intent and expanded their commission to conduct diplomatic meetings with the various Indian nations and study the geography of the newly acquired landmass. Lewis went to Pittsburgh, Pennsylvania, via Harpers Ferry, Virginia, to load the many materials needed for the expedition as well as to obtain a keelboat and pirogues (large flat canoes). On August 31, 1803, he left Pittsburgh and headed down the Ohio River, moving slowly due to low water brought on by drought. Lewis noted the extremely low water of the Ohio as well as the perpetual morning fogs in his journal writings. In addition, he conducted mini-experiments by taking temperature readings of the ambient air and the surface water of the river and made interesting conclusions as to the cause of the fog: "*Fog appears to owe it's orrigin to the difference of temperature between the air and water the latter at this seson being much warmer than the former; the water being heated by the summer's sun does not undergo so rapid a change from the absence of the sun as the air dose consiquently when the air becomes most cool which is about sunrise the fogg is thickest and appears to rise from the face of the water like the steem from boiling water*" (Lewis, September 1, 1803).

Lewis arrived at the Falls of the Ohio near Clarksville, Indiana/Louisville, Kentucky on October 14, 1804. Here he met William Clark and more recruits. As acclaimed author and historian Stephen Ambrose (1996, 117) noted, "When they shook hands, the Lewis and Clark Expedition began." They set out from Louisville on October 26, arriving at the confluence of the Ohio and Mississippi rivers on November 14, 1803 and moved up the Mississippi near St. Louis by December 11. Heading up the Mississippi was made difficult by low water and strong currents, and became more burdensome with strong northwest winds from late-fall cold fronts pushing against the boats (Quaife 1916; Osgood 1964).

William Clark's drawing of the landmass around the mouth of the Columbia River and Pacific Ocean. Photograph by Vernon Preston, courtesy of the American Philosophical Society/U.S. National Park Service.

Segments of the Journey

CAMP DUBOIS — WINTER 1803/04

On December 12, 1803, Clark established Camp Dubois (Wood) at the mouth of the small Wood River on the east side of the Mississippi directly across from the confluence of the soon-to-be-explored Missouri River. The winter scene at Camp Dubois was fraught with boredom, endless drilling, and preparation of the boats for the journey. Clark kept a log during this time, known as the Camp Dubois Field Notes. However, it was not in the original manuscripts and was printed for the first time in the Moulton edition of the journals (1986). The weather that winter seems typical of the latitude. Their first recorded snows were on the day they established Camp Dubois. By December 22, ice was beginning to form in the rivers, and they settled in for the long winter. They had a white Christmas.

The new year started off with an inch of new snow. Entries began on January 1, 1804 in what today is known as the Weather Diary. This particular diary noted sunrise and 4 p.m. weather observations of temperature, wind direction, the state of weather, river rise and fall, and general remarks. The pattern of entries is very similar to President Jefferson's style of weather diary writings. It also follows the pattern Jefferson directed his friend James Madison to keep while conducting an experiment to debunk arguments by naturalist Comte de Buffon in the 1780s and 1790s. Buffon wrote scathing articles stating that the North American continent's "supposed" inferior weather patterns would lead to degenerated fauna (Druckenbrod 2003). In addition to the Weather Diary remarks, Clark routinely placed comments about weather or river conditions in his daily narrative journal entries. Ever vigilant to conduct quality scientific observations, Lewis and Clark seemed to take care in the placement of the thermometers to obtain the truest temperature readings possible. As many notes show, they found locations under trees out of direct sunlight. One such example contained in Lewis's notes was on July 22, 1805: *"I placed my thermometer in a good shade as was my custom about 4 PM."*

The winter season brought bouts of arctic outbreaks. Almost every day between January 17 and 31, Clark's narrative journals noted hourly temperature observations. No word is given as to why this was done, but it could be that they were conducting experiments to determine any errors on the scale that had been inscribed next to the thermometer. Late winter brought bouts of rain and snow, but fair weather set in by the middle of March, and the ice was off the rivers by the first day of spring. The horizon was occasionally obscured by smoke. A special treat greeted the camp on April 1, as red northern lights (aurora borealis) danced across the sky around 10 p.m. They watched with anticipation the river's daily rise and fall as the flood season progressed. Their first thunderstorm came on the leading edge of a cold front that passed by on April 5. Plants and trees were budding by the end of March, but Clark took special note on April 18, as he stated *"Vegetation appears to be Suppriseingly rapid for a fiew days past."*

STARTING UP THE MISSOURI — MAY 1804

After several days of falling river conditions and warming May days, the "Corps for North-western Discovery" left Camp Dubois. They pushed across the Mississippi, entered the mouth of the Missouri River, and moved upriver on a cloudy, showery May 14, 1804. At St. Charles (in today's Missouri) they stopped for a few days to readjust the loads in the keelboat and pirogues. Here they met Lewis, who was finishing diplomatic business in and around St. Louis.

Map showing the route of the Lewis and Clark Expedition from its origins in Virginia to St. Louis and on to the Pacific Coast. Image courtesy of NOAA Geodetic Survey.

The expedition finally set off from St. Charles on May 21. Going slowly upriver against the current, the party moved at 5 to 15 miles a day. Driftwood, snags, strong currents, and falling riverbanks from spring floods kept the Corps at a slow pace. Occasionally they would hoist a sail and use the wind to their advantage, but most of the time they poled, rowed, or towed by rope the boat and pirogues. Typical of the sultry Missouri climate, morning fogs were replaced by hot afternoon breezes. The occasional thunderstorm caused concern for the boats. Not much is known of the actual temperature readings during this part of the journey, as no record has been discovered. However, remarks from the daily narrative journal entries show it was hot and humid during the late spring and early summer on the Missouri. Some members had heat stroke, and many got sunburned. They passed present-day Kansas City, Missouri, on June 26. The highest temperature (96° F) the expedition would record for the entire journey occurred on June 30 as they neared present-day Leavenworth, Kansas.

On July 4, they stopped by a small stream and named it after the special day. Clark made special mention of the area: "*One of the most butifull Plains, I ever Saw, open & butifully diversified with hills & vallies all presenting themselves to the river covered with grass and a few scatttering trees. Nature appears to have exerted herself to butify the Senery by the variety of flours Delicately and highly flavered raised above the Grass, which Strickes & profumes the Sensation, and amuses the wind throws it into Conjectering the cause of So Magnificent a Senerey in a Country thus Situated far removed from the Sivilised world to be enjoyed by nothing bu the Buffalo Elk Deer & Bear in which it abounds & Savage Indians.*" A few days later, Clark and Sergeant Ordway noted the effects of the extreme heat on expedition members. "*Worthy of remark that the water of this river or Some other Cause, I think that the most*

Probable throws out a greater prepson. of Swet than I could Suppose Could pass thro: the humane body Those men that do no work at all will wet a Shirt in a Few minits & those who work, the Swet will run off in Streams" (Clark and Ordway, July 6, 1804).

On July 12, in what would become an expedition ritual, Clark climbed a nearby hill to take celestial observations and inscribe his name, day, and month near an Indian pictograph of animals and a boat. Summer thunderstorms brought relief from the excessive humidity but also brought perilous moments, as the boats were nearly capsized or run ashore by the strong winds and waves.

Passing north of present-day Omaha, Nebraska, the Corps found a swath of large- diameter trees twisted and mowed down. Clark's observation notes say "*passed much fallen timber apparently the ravages of a Dreadfull haricane.*" These trees were probably part of a recent tornado path. Farther up the river, they began to note how much of the prairie had been burned. As they would learn, starting a prairie fire was a way of notifying Indian nations that travelers were nearby or that their presence at a council was requested. Later, they would find bluffs with seams of coal on fire producing a sulphurous odor.

August continued sultry, with afternoon and evening showers and thunderstorms. Clark began to note changes in the air as they moved north: "*the air is pure and helthy So far as we can Judge—.*" As they entered the Great Plains and left the protection of the dense forests of Missouri, the strong daily breezes on the prairie acted as both a blessing and a curse to the expedition. On many days, the prevailing southeast winds assisted their upstream progress against the rapid current, but sudden changes as frontal boundaries pushed through in late summer delayed their departure many times until late in the afternoon. The loose sands near the river often acted as a "summer blizzard" by reducing visibility, irritating their eyes, and filling everything full of gritty menace. The party became concerned one August day as they rounded a bend and saw the river full of white feathers that continued for three miles. Tension mounted, as they were fearful of meeting hostile Indians. A few miles later they came upon a flock of pelicans who were shedding their feathers in preparation for new growth, something the Corps had never seen before. Throughout the summer months bugs plagued the party, most notably the mosquito. In fact, for nearly the entire journey, mosquitoes became a scourge and generated a plethora of comments.

By September 8, the expedition had pushed into South Dakota and was headed toward the "Great Detour of the Missouri." The first cool rains of the early fall season met the Corps just downriver from the Big Bend of the Missouri, or as they called it, the Grand Detour. A few days later, the Weather Diary mysteriously ends its silence as daily observations resumed on September 19 with no explanation as to their previous absence. These were the first entries since leaving Camp Dubois on May 14. They noted the change of seasons by remarking that the leaves of the cottonwood were fading and the brant and plover were starting to migrate southward. Clark further expounded on a remarkable change in the humidity. On September 23, the Weather Diary entry noted, "*aire remarkably dry-plumbs & grapes fully ripe— in 36 hours two Spoonfuls of water aveporated in a sauser.*" Lewis would again experiment with evaporation rates in Montana the following spring (Lewis, May 30, 1805) (Large 1986).

During what seemed like an early fall, the party began to note the changes in the flora and fauna and experienced their first frost of the season on October 5. Brisk north winds swept cold, dense gray clouds (black flying clouds, as Lewis called them), indicating the first clue of a harsh winter to come. The expedition pushed into North Dakota in mid-October and received freezing rain and their first snow of the season just north of Bismarck, the present-day capital of North Dakota, on October 21. Having decided to winter near the Hidatsa/Mandan Villages, they established Fort Mandan west of present-day Washburn, North Dakota, on November 2, 1804.

FORT MANDAN — WINTER OF 1804/05

Continuing to note special circumstances, the narrative journals and Weather Diary entries show that the winter of 1804/05 at Fort Mandan was somewhat colder and wetter than modern observations (Burnette 2002). Tragedy struck the Mandan villages on October 29, as wind-whipped fire swept across the prairie into the Indian village, killing a man and woman and severely burning others. The Corps barely escaped and noted the fire passed the camp "*with great rapidity and looked Tremendious.*" Other unique occurrences during the winter of 1804/05 included a spectacular appearance of the aurora borealis (November 6) and the onset of ice in the river (November 13). Repeated heavy frost was noted many times, and one event impressed Sergeant Ordway on November 16: "*the Trees were covered with frost which was verry course white & thick even on the Bows of the trees all this day. Such a frost I never Saw in the States.*" Several instances of frostbite and snow blindness were recorded, along with heavy snowstorms, such as the one on November 29 when 13 inches fell. Corps members experienced additional hazards as winds stirred up blizzard conditions and produced significant snowdrifts. Another white Christmas was noted and celebrated by cannon fire in the morning and feasting during the afternoon. Extreme cold, which none of them had ever experienced, caused them to change the guard as often as once every half hour during the coldest times. December 17 marked the coldest temperature recorded during the journey, as the mercury fell to −45° F. They also experienced many visual oddities during the winter, including parhelion (sun dogs, December 8 and 11); a halo around the moon (January 12, 1805); several mirages; and even an eclipse of the moon on January 14. As the first year of the expedition came to a close, Private Whitehouse gave his summation of events at Fort Mandan: "*nothing particular occured Since christmas but we live in peace and tranquillity in our fort. The weather continued pleasant & the Air Serene— .*"

The year 1805 began with a mild 34° F, but by early evening a light sprinkle of rain gave way to overnight snow. The Corps learned about various Indian customs during their stay with the Mandans. One of particular note was how they kept their horses during the winter. During the daytime, they let them roam about, but the horses would return to the large mound lodges and spend the night inside with the families. Sometimes, during hunting parties, the Indians were forced to stay out all night in the subfreezing temperatures. Sergeant Gass noted one particular incident when an Indian survived by cutting branches from trees and lying on them to keep his body off the snow while also covering himself with a buffalo robe. Clark was so impressed with their heartiness that he commented "*...a man Came in who had also Stayed our without fire, and verry thinly Clothed, this mans was not the least injured— Customs & the habits of those people has ancered [inured them] to bare more Cold than I thought it possible for man to indure— .*"

Realizing they needed to prepare for the spring thaw, the party attempted to remove the boats from the river ice during January and again in February. Many ingenious methods were employed to cut through or thaw the ice. They finally succeeded just a couple of weeks before the river started to break up in early March. The snow began to melt and spring was just around the corner. On March 3, they saw a large flock of ducks, and their first insect on March 27. With the snow nearly gone, they witnessed the Indian nations preparing for the return of the buffalo by setting the prairie on fire. They learned that this would stimulate early grass growth and lure the herds toward their village (Moulton 1987, 3: 309). By late March, the party was preparing to depart as they watched geese and swans return from their winter migrations. They experienced several ice jams near the end of the month as large chunks broke loose well above their location but became stuck in a river bend. To their amazement, the Indians began jumping from one cake of ice to another in an attempt to catch buffalo as they floated down.

Their first thunderstorm of the season occurred on April 1. Lewis provided some additional observations on this particular day: "*A fine refreshing shower of rain fell about 2 PM this was the first shower of rain that we had witnessed since the fifteenth of September 1804 tho' it several times has fallen in very small quantities, and was noticed in this diary of the weather. The cloud came from the west, and was attended by hard thunder and Lightning. I have observed that all thunderclouds in the Western part of the continent, proceed from the westerly quarter, as they do in the Atlantic States. The air is remarkably dry and pure in this open country, very little rain or snow ether winter or summer. The atmosphere is more transparent than I ever observed it in any country through which I have passed.*"

MOVING TOWARD THE ROCKIES — SPRING AND SUMMER 1805

On April 7, 1805, the permanent party of thirty-three left the Mandan villages for the unexplored land to the west. A smaller party took the keelboat packed with journals containing the expedition's discoveries and specimens collected during the previous year's travel up the Missouri back to St. Louis. Specimens including various plants, animals, and minerals were shipped to President Jefferson. Some of these items are still on display at Monticello, Virginia, and in Harvard's Peale Library. As they headed up the Missouri, Lewis had time to note the significance of the date: "*Our vessels consisted of six small canoes, and two large perogues. This little fleet altho' not quite so rispectable as those of Columbus or Capt. Cook were still viewed by us with as much pleasure as those deservedly famed adventures ever beheld theirs; and I dare say with quite as much anxiety for their safety and preservation. We were now about to penetrate a country at least two thousand miles in width, on which the foot of civillized man had never trodden.*"

Incessant winds racing across the prairie plagued the expedition during their ascent of the Missouri into present-day Montana. Many times in April and May, they had to stop by 9 a.m. and did not return to the river until 3 p.m. The strong winds caused several near-mishaps as canoes would tip and nearly sink with their valuable cargo. More importantly, a few Corps members barely escaped drowning. Ordway also noted how bad the winds were: "*The Sand blew off the sand bars & beaches so that we could hardly see, it was like a thick fogg.*" The sandstorms drew complaints of sore eyes from the men, and as Lewis noted, "*So penitrating is this sand that we cannot keep any article free from it; in short we are compelled to eat, drink, and breath it very freely.*" The dust rose to great heights in immense columns and was visible for several miles.

Just two days into the ascent, their old nemeses were back. As Ordway succinctly put it on April 9, "*The Musquetoes begin to Suck our blood this afternoon.*" As spring progressed, their flora and fauna notations in this newly explored land increased dramatically. By the end of April, they began to note the plants had scarcely changed in their growth patterns and may have been at even earlier stages than those at the Mandan villages. Although not perceptible to the Corps, they were experiencing the effects of the higher-elevation plains as they neared the Rocky Mountains, as well as the higher latitude, which has a shorter growing season.

The expedition reached the confluence of the Missouri and Yellowstone rivers near the end of April and pushed toward central Montana. Spring brought normal changes including many morning frosts, water freezing on their oars, and occasional bouts with fog and snow. A couple of late-spring snowstorms shocked the party as they remarked how hearty the plants must be to tolerate such drastic weather changes. Clark expounded, "*The Snow which fell to day was about 1 In deep, verry extroadernaley Climate, to behold the tree Green & flowers Spred on the plain, & Snow an inch deep*" (Clark, May 2, 1805). As they passed from the flat barren plains toward central Montana, navigation became more difficult as the river grew more

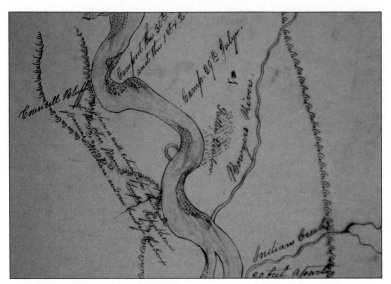

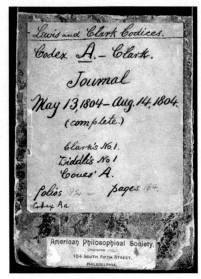

Clark drew exquisite maps of the river and its environs. In this image he noted the large areas of trees, which he described in the journal as *"much fallen timber apparently the ravages of a Dreadfull haricane."* The trees were likely affected by downburst winds or a tornado. (See July 29, 1805.) Photograph by Vernon Preston, courtesy of the American Philosophical Society/U.S. National Park Service.

The front of journal that was filled by William Clark while the expedition was traveling up the Missouri River from St. Louis toward present-day Omaha, Nebraska. Photograph by Andrea Laliberte, USDA-ARS and courtesy of the American Philosophical Society/U.S. National Park Service.

The Lewis and Clark Expedition Weather Diary entries for January 15, 1805 through February 2, 1805. During this time the Corps of Discovery was in winter quarters at Fort Mandan, near present-day Washburn, North Dakota. Photograph by Vernon Preston, courtesy of the American Philosophical Society/U.S. National Park Service.

narrow, with greater current and eddies from spring snowmelt flow. Dead and rotting buffalo, which they surmised had drowned when the ice gave way as they crossed frozen stretches of the upper Missouri, lined the banks.

Clark was the first to see the beginnings of *"those Shineing Mountains"* (The Little Rocky Mountains) on May 19, while walking on top of a ridge near the river. Lewis followed suit on May 26 with exuberance: *"while I viewed these mountains I felt a secret pleasure in finding myself so near the head of the heretofore conceived boundless Missouri."* The Corps then entered a scenic stretch of the river with spires and canyon walls that reminded the party of the marble buildings in the nation's capital. This area today is known as the Missouri Breaks National Monument. Lewis was so overtaken with the beauty he offered this emotional rendering of the area: *"As we passed on it seemed as if those seens of visionary inchantment would never have an end...So perfect indeed are those walls that I should have thought that nature had attempted here to rival the human art of masonary had I not recollected that she had first began her work"* (Lewis, May 31, 1805). The expedition members had lived in the lush green of the East Coast, so the treeless barren plains were of great concern as they were trying to provide images that would enhance the economic value of the Louisiana Purchase so hoped for by President Jefferson. So much to their astonishment was the appearance of Montana that Clark and Ordway bemoaned, *"this Countrey may with propriety I think be termed the Deserts of America, as I do not Conceive any part can ever be Settled, as it is deficent in water, Timber & too Steep to be tilled"* (Clark and Ordway, May 26, 1805).

As summer approached, the expedition came to a dilemma at the confluence of today's Marias and Missouri rivers. The captains could not tell which was the true Missouri due to spring flooding from the snowmelt off the distant mountains. After exploring both systems for many miles, the captains correctly choose the south fork. They were in search of what the Hidatsa Indians had called the "Great Falls of the Missouri." Lewis had speculated it would only require one day's portage before continuing. As the party proceeded, they began to comment nearly every day about the *"Snowclad"* mountains to the west. Snow in June caused more than a little anxiety, since they were accustomed to the Appalachian Mountains, which shed their snow by early spring. Lewis set out on the south fork with a small scouting party to locate the falls. Hearing a roar on June 13, Lewis noted its grandeur as *"to gaze on this sublimely grand specticle...at the cascade...on my right formes the grandest sight I ever beheld."* There were five falls of the Missouri, and the expedition would spend nearly a month portaging around them.

From June 13 to July 13, various members brought goods to the upper portage of the White Bear Islands at the southwest end of today's Great Falls, Montana. Many notable and significant weather events came to light during their monthlong stay, including numerous afternoon showers and thunderstorms — at least two of which were severe thunderstorms with large hail. One storm pummeled the men, knocking one down three times and leaving most of them bruised and bloodied. Another storm had hail the size of pigeon eggs and, as the Weather Diary noted, *"hail fell of an innomus size driven with violence almost incredible, when they struck the ground they would bound to the hight of ten to 12 feet and pass 20 or thirty before they touched again."* A flash flood nearly took the lives of Clark, Charbonneau, his wife Sacagawea, and their little son Jean Baptiste. Strange apparitions known as flying clouds were described several times. The resourceful portagers took advantage of strong winds as they hoisted the sails to assist them in moving the canoes overland to the upper portage. Strange noises that sounded like military cannon fire were heard by most of the Corps, but they could not understand what caused it. Clark discovered a large natural spring that produced enough water to be a river in itself. Another spring was laced with sulphur, which he used to treat Sacagawea, who was near death. The buffalo-trodden plains developed dried sharp points after summer rains, which cut into their feet. In addition to weather concerns, grizzly bear, prickly pear, and mosquitoes provided for lively discussions. They anxiously noted the looming snow-covered mountains to the west nearly every day.

Summer heat came late that year, as the wet June and early July temperatures struggled to get into the 70s. In fact, they did not record their first 80-degree temperature until July 16. The perpetual and persistent southwest winds continued through this entire time. Lewis and Clark finally had their fill and began to speculate on their origins. "*The winds has blown for Several days from the SW I think it possible that those almost perpetial SW winds, proceed from the agency of the Snowey mountains and the wide leavel and untimbered plains which Streach themselves along their borders for an emence distance, that the air comeing in Contact with Snow is Suddenly chilled and condensed, thus becoming heavyer than the air beneath in the plains, it glides down the Sides of those mountains and decends to the plains, where by the constant action of the Sun on the face of the untimbered country there is a partial vacuom formed for it's reception. I have observed that the winds from this quarter is always the Coaldest and most violent which we experience, yet I am far from giveing full credit to this hypothesis on this Subject; if I find however on the opposit Side of these mountains that the winds take a contrary direction I Shall then have full faith*" (Lewis, July 2, 1805; Clark, July 3, 1805).

The expedition was again moving up the Missouri and into the heart of the Rocky Mountains on July 13. They passed and named the "*Gates of the Rocky Mountains*" on July 19 and came to the Three Forks of the Missouri on July 25. To their astonishment, they found additional mountain ranges that reminded them of "*amphitheaters*," many of which were still covered with snow. After much scouting, the party correctly chose the southwest fork. Since all three forks were of nearly equal size, they named each of them for a prominent member of the Jefferson Administration — the Gallatin after the Secretary of the Treasury; the Madison after the Secretary of State; and the southwest fork "*in honor of that illustrious personage President Jefferson.*" The late-summer heat now began to affect the expedition, with temperatures rising into the 90s. They found some relief as they were constantly in the cool waters of the Jefferson trying to pull the canoes up the smaller and increasingly rapid waterway. Near Dillon, Montana, Lewis finally recognized that "*The mountains do not appear very high in any direction tho' the tops of some of them are partially covered with snow. This convinces me that we have ascended to a great hight since we have entered the rocky Mountains, yet the ascent has been so gradual along the vallies that it was scarcely perceptable by land*" (Lewis, August 10, 1805).

Over the Mountains — Late Summer and Fall 1805

Needing horses to cross the Rocky Mountains, Lewis set out with a small search party and reached the Continental Divide at Lemhi Pass, west of the present-day Clark Reservoir in southwest Montana, on August 12. Lemhi Pass is currently measured at 7,373 feet, and Lewis's group may have reached a couple of hundred feet higher depending on their exact route. This is the highest elevation the Corps would reach during the expedition. Since the time of Columbus, hopeful explorers had sought a mythical trade route known as the "Northwest Passage." The country that found and controlled this route would probably control the destiny of the continent (Duncan 1987). Lewis surely anticipated that he would find a gentle slope and river on the other side of this divide that would lead them toward the Pacific Ocean. Having the exultation of coming to the end of the Missouri River, Lewis instead must have felt centuries of hope come to a crushing defeat when he saw extensive snowcapped ranges in every direction. Lewis expounded, "*two miles below McNeal had exultingly stood with a foot on each side of this little rivulet and thanked his god that he had lived to bestride the mighty & heretofore deemed endless Missouri. After refreshing ourselves we proceeded on to the top of the dividing ridge from which I discovered immence ranges of high mountains still to the West of us with their tops partially covered with snow. Here I first tasted the water of the great Columbia river"* (Lewis, August 12, 1805).

They descended into the Lemhi Valley and found the Shoshone Indian Nation. Convincing Chief Cameahwait and a small party to return over the divide, they held council with Clark and the main party at Camp Fortunate in what is today Clark Reservoir. They experienced their first frost of the season while in the dry high mountain air (August 19). On August 21, they noted that the dry, cooler temperatures caused their writing ink and water in small vessels to freeze and that they needed two blankets or more to remain comfortable all night. Looking for a fast way to the ocean, Clark explored down Lewis's River (today known as the Salmon) to see if it could be navigated. He soon found, as the Shoshone chiefs had described, a river that had no shore, just tall rock cliffs and walls (still known today as the River of No Return). After acquiring the needed horses, the expedition then set out by land over the Lost Trail Pass back toward Montana. A horse carrying the last thermometer stumbled and broke the case and instrument while climbing rocky snow- and ice-covered terrain on September 3.

Fall came early to the Rocky Mountain region as they proceeded north down the Bitterroot Valley and toward an old Indian trail known today as the Lolo. Anxieties began to rise once again as they saw snowcapped rugged mountains over their left shoulders every day. On September 9, they arrived at Travelers Rest, near present-day Lolo, Montana, and on the 11th the Corps proceeded into the heart of the Bitterroot Mountains. They awoke one morning to nearly 10 inches of snow and many others to frost during the perilous eleven-day passage. They suffered from shortness of breath, great fatigue, and discomfort from extended time at the high elevations (6,000 to 7,000 feet) and from lack of food. The higher mountains stunned them with their perpetual snow cover that did not melt and their rugged appearance. Sergeant Gass, alarmed by the heights, wrote: *"proceed over the most terrible mountains I ever beheld."* They lost the trail at one point and had to climb nearly straight up a canyon side before returning to the correct course. The trail was littered with downed trees, and several areas suffered from the ravages of fire. Nearly starved to death, they staggered onto the Weippe Prairie and met the Nez Perce Indian Nation. They proceeded to near present-day Orofino, Idaho, and established Canoe Camp to build vessels for travel down the Clearwater, Snake, and Columbia rivers. The warm early-fall temperatures in the deep valley of the Clearwater, combined with a change in diet and water, made many of the expedition members ill.

TO THE PACIFIC — FALL 1805

On October 7, the Corps of Discovery was once again under way down the Clearwater River with the current to their backs and the river at seasonal low depths. On October 10, they passed the confluence of the Snake and Clearwater rivers and proceeded swiftly through various rapids toward the Columbia. By October 17, they were proceeding down the Columbia, and knew they were close to reaching their destination, as Clark noted a conical mountain to the southwest annotated on their maps as Mount Hood. They found the river crowded with salmon, both alive and dead. Cool, frosty mornings and pleasant daytime temperatures greeted the weary travelers as they moved through the Great Falls of the Columbia, to the long and short narrows, and on to the Cascades of the Columbia. Their weather fortune changed on October 27 as strong winds brought the first of what would be numerous winter storms into the Pacific Coast. Some suggest this was one of the stormiest winters in the Pacific Northwest (Lange 1979, 1986; Burnette 2002).

Knowing that they were nearing the Pacific Coast, they expectantly looked for the ebb and flow of the ocean tide at each major falls they came to on the Columbia. They thought they had found it on October 26, but were dismayed that it was water being backed up by yet another fall. On November 2, they passed beyond the Cascades of the Columbia and saw Beacon Rock, noted by Lieutenant William Broughton of George Vancouver's visit some thirteen years before (Ambrose 1996). They also noted the long-sought-after tide and reported it raising 9 inches. The once-constricted Columbia now widened as they rowed by the sites of present-day Vancouver, Washington, and Portland, Oregon, through thick and frequent fog. On the evening of November 4, the rains came. As the expedition would soon find out, the Pacific coastal area provided a harsh, wet, and windy climate. From this date until their return to the Columbia River Gorge the following spring, they would experience only twelve days without rain and only six with sunshine for half or more of the day (Gass, April 8, 1806).

As they moved into the estuary of the Columbia and neared the mouth, Clark made a premature revelation in his diary on November 7: "*We are in view of the opening of the Ocian, which Creates great joy.... Ocian in View! O! The joy. Great joy in camp we are in View of the Ocian, this great Pacific Octean which we been So long anxious to See. and the roreing or noise made by the waves brakeing on the rockey Showers (as I Suppose) may be heard distinctly.*" Sergeant Gass now noted the tidal fluctuations were over 4 feet and soon would reach 8 feet at the mouth of the Columbia River. After being pinned down in a cove for several days of stormy weather, Clark lamented, "*It would be distressing to a feeling person to See our Situation, at this time all wet and colde with our bedding &c. also wet, in a Cove Scercely large enough to Contain us, our Baggage in a Small holler about ? mile from us, and Canoes at the mercey of the waves & drift wood*" (Clark, November 12, 1805). He also noted, "*!O how Tremendious is the day. This dredfull wind and rain Continued with intervals of fair weather all the latter part of the night. O! How disagreeable is our Situation dureing this dreadfull weather. most tremendous and terrible winds*" (Clark, November 28, 1805). Members of the party now began to show concern about the rotting of their animal skin clothes in the wet conditions and the lack of drinking water in the salty estuary. In fact, the rain persisted for eleven straight days, never ceasing for more than two hours, with several thunderstorms and hail. As the party rowed toward the mouth of the Columbia River, many became seasick as the high waves and strong winds buffeted their canoes (Appleman 1975).

Finally, after a year and a half of journeying, the weather let up and the Corps of Discovery proceeded on to the mouth of the Columbia and their final destination, the Pacific Ocean. Sergeant Gass noted on November 16, 1805, "*We are now at the end of our voyage, which has been completely accomplished according to the intention of the expedition, the object of which was to discover a passage by the way of the Missouri and Columbia rivers to the Pacific Ocean; notwithstanding the difficulties, privations and dangers, which we had to encounter, endure and surmount.*" Private Whitehouse noted, "*We are now in plain view of the Pacific Ocean. The waves rolling, & the surf roaring very loud. We are now of opinion that we cannot go any further with our Canoes, & think that we are at an end of our Voyage to the Pacific Ocean.*" Mileage calculated by William Clark using dead reckoning placed the distance from St. Louis to the mouth of the Columbia River at 4,162 miles. He was within 40 miles of the actual distance (Ambrose 1998, 175; Duncan and Burns 1999). Clark led a small party up the Washington State coast and, remembering the indelible mark inscribed by Mackenzie in 1793, Clark inscribed on a tree, "*William Clark December 3rd 1805. By Land from the U. States in 1804 & 1805— .*" Most of the party followed suit during the coming weeks.

FORT CLATSOP — WINTER 1805/06

Needing to establish winter quarters, the members cast votes for their preferred winter location on the evening of November 24. They settled on exploring the south side in hopes that a ship might come by and allow them to obtain provisions and send some members and journals back via the sea. After traveling upriver, the party crossed and searched for a suitable location for winter quarters. The expedition established Fort Clatsop on December 7, 1805 and remained there until they left on March 23, 1806. This site is near present-day Astoria, Oregon.

They settled in for a long stay and set up a furnace near present-day Seaside, Oregon, to cook ocean water to obtain salt. They experienced numerous storms, including one about which Clark commented, "*a most dreadfull night the rain continues, with Tremendious gusts of wind. The winds violent from the SE. With some risque proceeded on thro high waves in the river, a tempestious disagreeable day. Trees falling in every direction, whorl winds, with gusts of rain Hail & Thunder, this kind of weather lasted all day, Certainly one of the worst days that ever was!*" (Clark, December 16, 1805). Christmas came and went as "*a warm, rainy, wet, showery, disagreeable day.*" The last year of their journey started like the first. Sergeant Gass remarked, "*The year commenced with a wet day; but the weather still continues warm; and the ticks, flies and other insects are in abundance, which appears to us very extraordinary at this season of the year, in a latitude so far north.*"

The mild Pacific air kept them quite comfortable, and Lewis pondered the possibility that this climate was warmer than that of the same latitude on the Atlantic Coast. He regretted he did not have a thermometer to verify his hypothesis (Weather Diary, January 3, 1806). They did not experience their first snows until January 25, 1806, and the cold weather remained for only a few weeks. But the rains of the winter of 1805/06 were incessant! They lamented nearly every day about the dreary, cloudy, rainy weather. Thunderstorms, sleet, hail, and winds buffeted the fort numerous times. Author Robert Lange noted, "In all the journals of the expedition, nowhere do we find any one word to be as repetitious as the word *rain*" (Lange 1979). The term "disagreeable weather" soon became their motto. The continual cloudy conditions kept them from taking observations. Lewis exclaimed, "*I am mortifyed at not having it in my power to make more celestial observations since we have been at Fort Clatsop, but such has been the state of the weather that I have found it utterly impracticable—*" (Lewis, February 25, 1806).

DEEP MOUNTAIN SNOW DELAYS RETURN — SPRING 1806

With enough food and salt for the journey to the Nez Perce Nation, they set out from Fort Clatsop on March 23, 1806 and started up the swollen Columbia River. Having labored against the strong current for seven days, they camped near present-day Portland, Oregon, between March 30 and April 6. Clark and a small party ventured up the Multnomah River (Willamette) on April 2–3. They had missed this tributary in the thick fog on the way down. Touring the river on April 2–3, he marked on his maps that this might still be a possible link to the mythical Northwest Passage. The weather in the tidal reaches of the Columbia was just as troublesome as what they had experienced the previous fall. Thick fog, drenching downpours, mist, and gray, overcast clouds kept them from taking celestial observations. Plants and flowers were coming to life in the temperate, moist Pacific air. As they proceeded into the Columbia Gorge, gale winds, channeled by the high timber-laden hills, caused several delays, as they were fearful of capsizing their canoes. Many members discussed the rising floodwaters, some 12 feet higher than the previous fall, as the spring thaw commenced in the Cascade Mountains. As the expedition moved past Dalles, Oregon, Lewis commented, "*The plain is covered with a rich virdure of grass and herbs from four to nine inches high and exhibits a beautifull seen particularly pleasing after having been so long imprisoned in mountains and those almost impenetrably thick forrests of the seacoast*" (Lewis, April 17, 1806).

The cold, damp, rainy, coastal weather gave way to leeside spring warmth and a break for the weary travelers, and they traveled by land from near Wishram, Washington, to the confluence of the Walla Walla and Columbia rivers, noting, "*there are now no dews in these plains, and from the appearance of the earth there appears to have been no rain for several weeks"* (Lewis and Clark, April 24, 1806). From here, they took an Indian shortcut through the steep rolling hills of southeast Washington, passing through present-day Waitsburg, Dayton, and Pomeroy to the Snake River near Lewiston, Idaho / Clarkston, Washington. Learning from the Nez Perce leaders that it would be late June before snow in the Bitterroots would melt enough to allow passage, the expedition set up camp near present-day Kamiah, Idaho (today known as Long Camp or Camp Chopunnish), between May 14 and June 10. During their stay, they learned the customs of the Nez Perce, increased their food stocks, and wrote more journal entries. The weather was typical for this time in spring, with warm temperatures in the Clearwater Valley and cool air in the higher plains. Journal entries describe their anxiety over whether they would make it through the mountains in time to return to the United States this season. Of particular note, they pondered the rain in the valley while snow would blanket the Weippe Prairie just a few hundred feet above. They observed the river daily and noted its variations hourly during the spring snowmelt, in hopes that when it fell and remained down for several days, it would mark the time to start into the Bitterroots. By the end of May, the river had reached flood levels *as high as any marks of [its] having been for several years past"* (Weather Diary, May 31, 1806). As they experienced the previous year, the growing season in the higher elevations had delayed plants, flowers, and—most important for the horses—the grass from reaching their spring maturity.

On June 10, after the Clearwater fell for five straight days, the expedition left the confines of the warm canyon and proceeded up the hill to the Weippe Prairie and toward the Lolo Trail. The Corps' attitude was still apprehensive as they viewed the white snow-covered mountains to the east. Near despair, Lewis wrote on June 14, *"every body seems anxious to be in motion, convinced that we have not now any time to delay if the calculation is to reach the United States this season; this I am detirmined to accomplish if within the compass of human power."* As they proceeded up the trail with their packhorses, mounds of snow increased steadily in depth. Compact and firm, the snow supported the weight of the Corps. Up they went, first on 4 feet of snow, then 8 feet, then 12 feet. Finally by June 17 they were on snows that reached a depth of 15 to 18 feet. The trail, which the Indians marked by scratching trees in higher snow seasons, was not visible due to the high snowpack. Due to the winter being extremely wet with a potential El Niño pattern, the excessive snows of this particular year forced the expedition to retreat for the first time (Ambrose 1978; Quinn et al. 1987; Quinn and Neal 1995). They returned down the mountain and spent the next week and a half asking the Nez Perce to provide a guide as they apprehensively waited for the snow to diminish.

They started a second attempt at the Lolo Trail on June 25 with the assistance of two Nez Perce guides, who entertained them the night before by lighting the undersides of fir trees on fire, a custom that they believed brought fair weather for the journey over the mountains. Lewis and Clark noted this spectacle: *"they have a great number of dry limbs near their bodies which when Set on fire creates a very sudden and emmence blaize from bottom to top of those tall trees. They are a beautifull object in this situation at night. this exhibition reminded me of a display of fireworks"* (Journal entry, June 25, 1806). As they proceeded with their newly acquired guides, the expedition reported that the snow where they had stopped had reduced to 10 feet 10 inches deep but generally was about 7 feet deep. Just as the Nez Perce guides had predicted with their tree ceremony, fair weather prevailed as they moved over the high terrain. On the Montana side, they stopped and refreshed themselves at a hot spring (Lolo Hot Springs), which they noted was as hot as any in Virginia. They arrived back at Travelers Rest on June 30 and rested for a couple of days.

EXPEDITION SEPARATES — SUMMER 1806

The expedition now split into two parties. Lewis took a volunteer contingent to the White Bear Islands Camp near the Great Falls of the Missouri via a trail the Indians had described as their path to the summer hunting grounds. Lewis left most of his group here under the direction of Sergeant Gass to prepare for a portage back below the Great Falls of the Missouri. Lewis, meanwhile, led a small exploratory party back to the Marias River to determine its northernmost extent. Clark, along with the remaining members, headed southeast through the Big Hole Valley and to Camp Fortunate to pick up cached supplies and specimens. From there, they took horses and canoes and floated down the Beaverhead and Jefferson rivers to the Three Forks. Here, Clark split his party again. Sergeant Ordway led a small party with the canoes and specimens down the Missouri to meet with Gass and portage the falls. Clark led his small contingent over present-day Bozeman Pass, recommended by Sacagawea, who guided them to the Yellowstone River, which they then descended by canoes. All expedition parties planned to meet again at the confluence of the Missouri and Yellowstone around August 5.

Weather remained cool and wet for the high country of Montana, sometimes with frosty mornings and many times with rain and afternoon thunderstorms. Clark and Ordway noted frost on July 10 at Camp Fortunate. The incessant prairie winds returned, while the mosquitoes redoubled their war on the Corps and became a daily theme in the journals. Lewis noted, *"My dog even howls with the torture he experiences from them"* (Lewis, July 15, 1806). Both Lewis and Clark continue to note the dry conditions across the prairie. On July 22, Lewis's party reached a point on the Marias where they realized that it would not carry them above 50° latitude and decided to spend the day taking celestial observations. However, due to the persistent clouds, they remained until the twenty-sixth. This would prove tragic, as a skirmish with a party of Blackfeet Piegan Indians broke out the next morning. Reuben Field and Lewis each killed an Indian who was attempting to steal the party's rifles and horses. On the night of the twenty-seventh, as they raced back to the Marias River, they experienced *"heavy thunderclouds lowering all around us and lightning on every quarter but that from which the moon gave us light"* (Lewis, August 27, 1806). The next day, Ordway noted hailstones from an afternoon thunderstorm larger than a "muskat ball." Meanwhile, on the Yellowstone River, Clark's party was building canoes and began their descent. On July 22, Clark noted a peculiar cloud formation as a thunderstorm approached. *"The Cloud appd. to hang to the SW, wind blew hard from different points from 5 to 8 PM which time it thundered and Lightened"* (Clark, July 22, 1806). Could this be a description of a wall cloud? Although they noted excessive heat a couple of times, it was the repeated rains and afternoon thunderstorms that filled their journals.

DOWN THE MISSOURI — EARLY FALL 1806

The weather continued unsettled for much of August, with many thunderstorms and daily rains noted on their parchment. They also commented on the continued coolness of the air. All members of the expedition finally reunited southeast of Williston, North Dakota, on August 12 and proceeded expeditiously toward the Mandan villages, where they parted company with Charbonneau, Sacagawea, little Pomp, and John Colter. Anxious to return home, they left the Mandan villages on August 17 and proceeded at a rapid pace, carried along by the current and their paddles. Indifferent to being rebuffed by the usual southerly winds, the Corps made tremendous progress, sometimes covering 50, 60, and even 70 miles a day. By September 1, the Corps was straddling the border of Nebraska and South Dakota, and the showers, thunderstorms, and winds persisted. On September 7, north of present-day Omaha, Clark again expounded on the excessive evaporation from his inkstand. He also made a startling revelation: "*The Missouri at this place does not appear to Contain (as much) more water than it did 1000 Miles above this place, the evaporation must be emence; in the last 1000 miles this river receives the water 20 rivers and maney Creeks Several of the Rivers large and the Size of this river or the quantity of water does not appear to increas any—*" (Clark, September 8, 1806). By the ninth, he mentioned that the nights were warm enough that the party was comfortable sleeping under a thin blanket. They were now returning to the humid climate of Missouri with its thick vegetation, trees, lakes and marshlands.

By the time the expedition reached Kansas City, they were complaining about the sultry heat and the weather being "*disagreeably [warm].*" Constant staring at the water and glare of the sun as they rowed brought complaints of near-blindness, although this may be attributed to bacterial infections (Chuinard 1998). As they rowed into St. Charles, Missouri, on September 21, the rains began again, and a final thunderstorm greeted them at daybreak on the twenty-second as they prepared for the final leg of their journey. Just as it bade them farewell in May of 1804, the rains greeted their return to the confluence of the Missouri and Mississippi and St. Louis on September 23, 1806. Ordway exultantly remarks on this momentous occasion: "*a wet disagreeable morning. Soon arived at the Mouth of the Missouri entered the Mississippi River and landed at River deboise where we wintered in 1804. About 12 oClock we arived in Site of St. Louis. Drew out the canoes then the party all considerable much rejoiced that we have the Expedition Completed and now we look for boarding in Town and wait for our Settlement and then we entend to return to our native homes to See our parents once more as we have been So long from them*" (September 23, 1806). Lewis and Clark hastily wrote letters— Lewis to President Jefferson and Clark to his brother in Kentucky—describing their safe return and findings (see Appendices D and E). The last entry in the Lewis and Clark Expedition journals was made on September 26 as Clark, still vigilant on reporting the weather, inscribed "*a fine morning we commenced wrightin &c.*" (Ordway, September 23, 1806).

Thus ended the Lewis and Clark Corps of Discovery Expedition. Their findings were numerous. Along the way, they described in scientific detail for the first time one hundred twenty-two new animal species and one hundred seventy-eight new plants, flowers, and trees. Clark recorded the journey through cartographic illustrations, and in 1814, produced the most advanced map of North America to that time. Although they failed to find the elusive Northwest Passage, they found the shortest route across the continent if you follow the Missouri River system. As fur traders, trailblazers, and immigrants later discovered, it was not the quickest way across the continent. Their extensive daily writings on weather conditions, though, are among the most detailed of their time. Nearly 200 years after they were recorded, the weather records have been awaiting an opportunity to burst forth and draw a new breath of research and debate. This publication hopes to establish this course and induce renewed enlightenment.

In the writings of Thomas Jefferson: "*The work we are now doing is, I trust, done for posterity, in such a way that they need not repeat it.... We shall delineate with correctness the great arteries of this great county; those who come after us will extend the ramifications as they become acquinted with them, and fill up the canvas we begin*" (Duncan and Burns 1997, 224; Ronda 2000, 50).

And so we learn from the national treasure of writings known as the Journals of the Lewis and Clark Expedition, and take up the scientific reins and continue to fill up the canvas that was begun so long ago. Of all the things that we can learn or discover about the Lewis and Clark Expedition, foremost is this: "At the core of discovery is the recognition that all human kind are but travelers making rough journeys, sustained now and again by the kindness of strangers" (Ronda 1984; Ronda 2001, 37).

CHAPTER 2   The Expedition Journals

"Came a Dredfulle hard Storme from the South which Lasted for about one ouer and half which Cosed us to Jump out and hold hir the wind fare Sailed"

—Sergeant Charles Floyd
July 14, 1804
North of St. Joseph, Missouri

"At day break it began to rain and continued until seven when it abated, and we set forward; but in a short time a gust of wind and rain came on so violent, that all hands had to leap into the water to save the boat. Fortunately this storm did not last long, and we went on to a convenient place and landed."

—Sergeant Patrick Gass
July 14, 1804
North of St. Joseph, Missouri

Recording of Weather and Climate in the Early 1800s

While several scientific books have been written describing the expedition's study of flora and fauna,[*] its advances in the fields of geology, geography, and cartography,[†] and its members' medical needs,[‡] the expedition's systematic daily observations of climate, water, and weather elements have largely been ignored. However, the daily observations en route represent the dawn of modern meteorology, when only a handful of scientists were noting the changing weather patterns. In general, regular daily observations, although noted back to the Greeks as early as the fifth century B.C.E., were not recorded until the late 1600s as instruments such as the thermometer, barometer, and hygrometer were developed (Frisinger 1983).

Weather diaries based on meteorological instruments were very uncommon in the United States in the late eighteenth century. Thomas Jefferson and James Madison in Virginia, John Winthrop in Massachusetts, and Dr. John Lining along with a few others in South Carolina were noted for taking and documenting daily observations (Druckenbrod 2003, 62). In recent writings, scientists have used this small record of historical diaries to reconstruct climate patterns.[§]

[*]Allen 1975; Cutright 1976; Bergon 1989; Burroughs 1995; Botkin 1996; Blume 1999; Moulton 1999, 12; Cutright 2001; Wells and Anzinger 2001; Cutright 2003; Patient 2003.

[†]Allen 1975; Moulton 1986, 1; Allen 1991; Plamondon II 1991; Goetzmann and Williams 1992; Ifland 1998; Schmidt 1998; Blume 1999; Cordes 1999; Hunt 1999; Schmidt 1999; Franslow 2000; Hayes 2000; Plamondon II 2000; Preston 2000; Plamondon II 2001; Space 2001; Bedini 2002; Cohen 2002; Eastman 2002; Hoganson and Murphy 2003; Plamondon II 2004.

[‡]Chuinard 1998; Paton 2001; and Peck 2002.

[§]Ludlam 1966; Middleton 1969; Ingram et al. 1978; Frisinger 1983; Bedini 1986; Bradley and Jones 1993; Baron 1995; Catchpole 1995; Pfister 1995; Quinn and Neal 1995; Glaser et al. 1999; Druckenbrod et al. 2003

No national approach to daily scientific observations was in place during this time. However, using the expedition's journals, we now have an opportunity to study three years of weather, water, and climate data from Pittsburgh, Pennsylvania, to the Pacific Ocean and back to St. Louis, Missouri. More than 60 years would pass before systematic scientific observations would be conducted on a daily basis in this region. Therefore, we have a true historical snapshot to compare today's weather, water, and climate records to those recorded by the Lewis and Clark Corps of Discovery Expedition.

Few selected writings can be found that describe limited aspects of the weather, water, or climate of the Lewis and Clark Expedition. Those that do include descriptions of snow conditions along the Lolo Trail (Ambrose 1978); the inclement weather at the mouth of the Columbia River (Large 1979); general weather observations (Large 1986); scientific instruments used on the Expedition (Plamondon II 1991); a master thesis on the weather conditions at Fort Mandan, North Dakota, and Fort Clatsop, Oregon (Burnette 2002); a discussion on temperature variations along the trail (Solomon and Daniel 2004); a discussion on climatic conditions during the Expedition (Knapp 2004); and 1805/06 winter conditions at Fort Clatsop (Miller 2004). No documentation of the Lewis and Clark Expedition has collected all weather-, water-, and climate-related records into one volume for use by the meteorological or hydrologic scientific communities.

Overview of the Journals

The journals of the Lewis and Clark Expedition are many and varied. Donald Jackson (1978, vii) noted that Lewis and Clark were "the writingest explorers of their time. They wrote constantly and abundantly, afloat and ashore, legibly and illegibly, and always with an urgent sense of purpose." The principal writers of the journals were Meriwether Lewis and William Clark. They recorded data into daily narrative diaries made from rough "Field Notes" that were translated at the end of the day by Clark. They maintained several other documents and booklets containing astronomy, botany, ethnology, geography, military orders, mineralogy, thermometrical and weather observations, and zoology. Clark produced numerous sketches and maps. Both wrote numerous letters before, during, and after the expedition. Much has been written discussing the methods used and history behind the journals (Biddle 1814; Coues 1893, 1: cvii–cxxxii; Thwaites 1904, 1: xvii–xciii; Bakeless 1964; Cutright 1976; Jackson 1978; Moulton 1986, 2: 8–48, 530–567; Bergon 1989; Beckham 2003; Saindon 2003). The reader is encouraged to view these source documents for further revelation on the journals.

President Thomas Jefferson did not order the keeping of separate journals by anyone other than the captains. His final instructions to Lewis did suggest that "several copies of these as well as of your other notes should be made at leisure times, & put into the care of the most trust-worthy of your attendants, to guard, by multiplying them, against the accidental losses to which they will be exposed" (Jackson 1978, 1: 62; Moulton 1996, 10: xi). In addition to those by the captains, three sergeants and a private are known to have written journals. Some of these journals have come to light only during the last 50 to 100 years. There may be up to three additional journals that have not been found. Lewis gives credence to this in his last communication to Jefferson from Fort Mandan in April 1805: "*We have encouraged our men to keep journals, and seven of them do so, to whom in this respect we give every assistance in our power*" (Jackson 1978, 2: 232).

The Journals of Lewis and Clark have been reproduced only a few times in the past 200 years. The first issuance was under the guidance of Clark, who after the untimely death of Lewis, took control of the known journals. Nicholas Biddle produced the first edition (1814), a two-volume set, using a general narrative paraphrase without much scientific content. Biddle turned the manuscript over to Paul Allen, whose name appears on the title page, for final revision. Biddle may have followed a literary custom of the time, which mandated that a gentleman did not publish under his own name (Moulton 1986, 2: 37; Ambrose 1996, 469–470). Elliott Coues (1893) produced the next edition of the journals, introducing numerous scientific discoveries that had been left silent for nearly a century. It is believed that it was Coues who rekindled the nation's interest in Lewis and Clark (Moulton 1986, 2: 39); however, Coues produced only a small subset of the full journal writings. Ruben Gold Thwaites published the first full edition of known journal writings for the centennial (1904); extensive research and discovery of new documents greatly enhanced his edition. Thwaites' eight-volume edition included copies of Clark's cartography in a special atlas. For the first time in history, the bulk of the Lewis and Clark journal writings were then available to the public.

Other valuable but small renderings were made by Milo Quaife (1916) and Earnest Osgood (1964); and concise, abridged editions such as those of John Bakeless (1947) and Bernard DeVoto (1953) were published as new materials became available after 1904. Discovery of missing journals and letters written by the captains and other expedition members led to the newest, most complete edition yet. Started in 1986 and completed in 2001 by University of Nebraska professor Dr. Gary Moulton, this compilation contains thirteen volumes with an atlas and journals by Lewis and Clark; Sergeants Floyd, Gass, and Ordway; and Private Whitehouse. Donald Jackson (1978) produced a two-volume set of letters that were written before, during, and after the expedition that complements the Moulton edition, tying together events outside of the journals. For further history about the journals, the reader is referred to the following source documents: Coues (1893, 1: cvii–cxxxii), Cutright (1976), Jackson (1978), Moulton (1986, 2: 8–48, 530–567), and Thwaites (1904, 1: xvii–xciii).

The Daily Narrative Journals

"A Cloudy fogey morning, a little rain. Ocian in View! O! The joy. Great joy in camp we are in View of the Ocian, this great Pacific Octean which we been So long anxious to See. and the roreing or noise made by the waves brakeing on the rockey Showers (as I Suppose) may be heard distinctly."

—William Clark
November 7, 1805
Columbia River Estuary, a few miles from the Pacific Ocean

This edition of content from the Journals of Lewis and Clark is designed to provide users a scientific approach to the meteorological, hydrological, and climatological information documented during the three-year time frame of known journal entries. Sources for this material include the first paraphrased edition by Biddle (1814); the second edition by Coues (1893); and the first full edition of known journals by Thwaites (1904). The most recent source is the thirteen-volume edition published between 1986 and 2001 by Dr. Gary Moulton through the University of Nebraska Press. In addition to original journal entries, Jackson (1978) contains the largest collection of known letters and other documents related to the Expedition.

The editorial procedures were designed to focus on the weather, water, and climate entries in the daily narratives of the two captains, the three sergeants, and one private, with an overall goal of incorporating all related entries into this edition. Redundant entries regarding strong river current have been omitted. Rather, selected notations are made to provide the reader a feel of the river course changes. Lewis and Clark provided entries about streams, creeks, and rivers as they traveled along or by them, and gave many of them names for the first time or noted names used by the Native Americans. They also scientifically described the river characteristics at main confluences, and many of these descriptions have been included in this edition. Since Moulton and Thwaites provide actual journey entries describing these in detail and distance from St. Louis or the Pacific Ocean, this edition does not try to re-create this information (Thwaites 1904, 6: 3–79; Moulton 1987, 3: 333–385; Moulton 1993, 8: 376–411). The reader is directed to these two authors when conducting extensive research on streams, creeks, and rivers.

Most general references to geography and its effect on climate are given by Jefferson's Instructions to Lewis (see Appendix C). "*By the access & recess of frost, by the winds, prevailing at different seasons, the dates at which particular plants put forth or lose their flowers, or leaf, times of appearance of particular birds, reptiles or insects*" were added to this edition. However, any general references made in Lewis's extensive botanical and zoological entries were not added in this edition, as other authors have made mention in their flora and/or fauna publications (Cutright 1976; Burroughs 1995; Botkin 1996; Wells and Anzinger 2001; Patient 2003).

When dual or parallel entries made by Meriwether Lewis and William Clark occur, Lewis was chosen as the first entry since he was given charge of organizing the Expedition under the direction of President Thomas Jefferson. Clark's entry is placed second. Each had a special technique for writing: Lewis was the more educated and verbose; Clark had a more frontiersman style, with numerous spelling and punctuation challenges. Clark kept field notes and transferred them the same day, or at times edited on later dates. When it came to the most diligent and faithful journal keeper, the honor went to Sergeant John Ordway. Not once during the entire 863-day journey did he fail to make an entry. Clark was a close second, with missing entries during a hunting trip from February 3–12, 1805, which was summarized when he returned. So for all practical purposes, his journal is complete (Cutright 1976). Lewis's journal entries begin in a field book started on the day he left Pittsburgh, Pennsylvania: August 31, 1803. His first reference to water was on the extreme low flow of the Ohio River, which he later referenced to drought conditions. He also noted the water being "*sufficiently temperate.*" His first meteorological entry was a discussion about a "*thick fogg on the face of the water that no object was visible 40 paces.*" Clark's first known travel entries were in some of his field notes taken near the confluence of the Ohio and Mississippi rivers, and his first shared journal entries were made on November 28, 1803 as Lewis left Clark in charge of the boats. Clark's first meteorological reference was "*This morning being verry Smokey prevents my being as acurate as I Could wish—.*" The same day he noted "*The horozon became darkened that I could not see across the River, which appeared to windened, the Current much Swifter than usial.*"

In order to preserve the record, at Fort Mandan and again at Fort Clatsop, duplicate journals were made by Lewis and Clark. Where duplicate entries are present, this edition uses a combination name "Lewis and Clark" without differentiation. On occasion, Clark would write two versions of his daily entries: an on-the-spot draft version in his field notes, and a much cleaner version for the final journal entry. When editing data into this edition, a combination of the two daily entries was used. This will allow the reviewer to see all the thoughts that were written for a particular day. Lewis's journal entries are much more sporadic, and many scholars have researched and discussed why Lewis's journal entries include large gaps. Some believe it was due to fits of depression, while others believe the journals may have been lost, destroyed, or misplaced (Appleman 1975; Cutright 1976; Moulton 1986, 2; Ambrose 1996). The only known reason for a particular gap—his last—was an accidental gunshot wound on August 11,

1806. Rejoining Clark and the rest of the party the next day, Lewis decided he would relinquish his journal entries to Clark and wrote his last words in the journals: "*I shall desist untill I recover and leave to my frind Capt. C. the continuation of our journal...This cherry...is now ripe...I have never seen it in blume.*" Clark's last reference to the topic of this edition, which is used as closing sentiments for many an author was, "*a fine morning we commenced wrightin &c,*" entered on September 26, 1806, three days after their arrival back in St. Louis.

After the entries of the two captains, those of three sergeants and one private have been placed in alphabetical order for this edition. Sergeant Charles Floyd's entries are the shortest since he was the only member of the expedition to die during the journey; he passed away near present-day Sioux City, Iowa, on August 20, 1804. His entries provided a more conscientious look at daily happenings. As Moulton (1995, 9:xviii) notes, "Floyd apparently had an eye for such details, which makes us regret all the more that he did not live to complete a record of the whole journey." He kept entries until two days before his death. Sergeant Patrick Gass and John Ordway wrote their journals from the day they left Camp Dubois on May 14, 1803 through September 23, 1806, the date of their arrival back in St. Louis.

Gass's (1807) journal is not from his original writings, as their whereabouts are not known. His journal was the first full account of the Expedition published. David McKeehan edited it and smoothed his rough frontier prose. Many editors have reprinted Gass' journal over the years. In this edition, Moulton (1996, 10) was the primary source. As a carpenter, he paid particular attention to details other journalist did not, such as times of particular rain events.

Ordway's entries are in his own hand and provided a substantial amount of data included in this edition on days when other journalists did not annotate data. Being an educated gentleman, his text is quite refined. Besides Clark's journal entries and the Weather Diary, Ordway provides the most useful weather, water, and climate information during the journey. In this edition, Moulton (1995, 9) was the primary source.

The final journalist used for this edition is Private Joseph Whitehouse. As with Gass, Whitehouse had an original version that was very rough and provides some distinct language about certain incidents. It dates from May 14, 1804 to November 6, 1805. A paraphrased journal discovered in a bookstore in Philadelphia, Pennsylvania, in 1966 provides entries from May 14, 1804 to March 23, 1806. There is speculation that he may have kept a journal through the end of the journey, but it has not been found. Just as with Clark and his rough and final journal versions, when two entries are found on a particular day, the data are blended in this edition to give the reviewer the context of what was written by Whitehouse. In this edition, Moulton (1997, 11) was the primary source. As with all of the sergeants' and privates' entries, the reader will note similar or duplicate observations at times. Various scholars have documented that the journalists would copy from one another at times, or fill out data for another member when they were away. Each duplicate entry is added in this edition except as highlighted.

Care was taken to keep as much of the original spelling and context as possible. Occasionally, punctuation or certain spelling corrections have been made for ease of reading. Many times the journalist did not place an end-of-sentence punctuation mark. Extra spacing has been used between sentences to help the reader differentiate a new line of thought. If the word was difficult to ascertain, [brackets] with a suggested word have been added. Since this edition uses Moulton as its primary source, his exhaustive research shows when certain words were changed or added by journalists or earlier editors. This edition does not make that differentiation. Footnotes have been added at the end of each month's/segment's entries when additional detail was necessary. To assist the reader, text has been added between daily entries to describe the location of the Expedition party with respect to well-known geographic and political boundaries.

Occasionally the journalists used abbreviations in their narrative entries. Some that have been included in this volume are:

d.	degree
do.	ditto
h.	hour
L. Larb, Ld. or Ldb S.	larboard (left) side
Latd.	latitude
Longtd.	longitude
m. mts.	minutes
mes. mls. ms.	miles
pt.	point
s.	second
St. Star. Starbd. S.Stb. or Stbd.	starboard (right) side

The Weather Diary

Lewis kept a Weather Diary from January 1, 1804 through May 1804. This "Diary" is also known as "The Book of Thermometrical Observations" (Thwaites 1904) and "The Meteorological Register" (Coues 1893). Half of it is comprised of the weather observation tables. The second half is narrative remarks related to daily weather, climate, and water changes, general characteristics of flora and fauna, and miscellaneous comments about events of the day. Both types of content are presented together under dated headings in Part II of this book. The tables represent a snapshot of what occurred during the Expedition. To fully comprehend the ramifications of weather, water, and climate on the Corps of Discovery, the reviewer is encouraged to read Lewis and Clark daily narrative journal entries[5], and this is why in this volume the two have been placed together by date.

Weather entries in the diary were in both Lewis's and Clark's hands. Clark repeated these entries in his Codex C journal. It is probable that Lewis copied the entries from Clark's journal, as Lewis was in St. Louis for an extensive amount of time while the party was at Camp Dubois. After May 14, there is a gap in the sunrise and 4 p.m. observations in both captains' books until September 19. During this time, general remarks on natural history, rather than meteorology or hydrology, were placed in Lewis's Weather Diary. Between September 19, 1804 and April 3, 1805, the twice-a-day observations and remarks resumed with few interruptions. After April 3, Lewis began placing weather data with his daily narrative journal entries, while Clark continued to make weather tables in his various journals. Clark kept the record going when gaps in Lewis diary entries are found. When the two parties split at Travelers Rest on their return trip, both Captains kept separate daily logs for July and portions of August. The weather notes indicate a substantial scientific record of atmospheric and hydrologic conditions,

---

[5]Coues, 3: 1,264; Moulton, 2: 168–169; Thwaites, 6, part 2: 165–166.

which included two entries for temperature readings, the general state of weather, the wind direction, and a single record of river rise and fall. In the remarks section, climatological references are made regarding flora and fauna, as well as daily weather and hydrologic occurrences. Various miscellaneous events were recorded occasionally, such as an Indian chief coming to visit. Although several methods were employed, it is certain that this portion of the journals was a collaborative effort of both captains (Moulton 1986, 2: 20, 30, 537).

The data in this edition follows much of what Coues (1893), Thwaites (1904), and Moulton (1986–1993) reproduced from the original journal entries, with a few variations. At the top of each table, notations were made to distinguish whether the data are a combination of various Lewis and Clark journals or entries that were made separately. Occasional capitalization of certain letters, such as certain column titles, T for Thunder, C for Cloudy, and L for Lightning, was omitted. The titles used at the top of each column varied during the journey. This edition generally follows what was entered but organizes them consistently from table to table to assist the reader. For various observations, the occasional period (.) next to numbers or letters has been omitted. Due to space limitations in the tables, the River Rise and Fall, and River Feet and Inches were combined into two columns and with an apostrophe (') used for feet and quotes (") used for inches. Also, to assist in the reading of the columns, the River Rise and Fall are placed in capitals. Lewis or Clark used either the word "ditto" or "do" to show a duplicate or identical journal entries, and the convention is continued.

The journals kept by members of the Corp of Discovery contain archaic spellings, misspellings, and inconsistencies in spacing. The editors of this volume have endeavored to edit the excerpts only slightly in order to clarify meaning. Brackets are used to set off text not contained in the original journals, i.e., text that has been added by the source for these excerpts (in these cases most often Moulton, 2002) or by this edition's editor to clarify meaning. Parentheses in journal excerpts were originally included by the journal writer or, in keeping with the convention used by Moulton, they indicate a word once there and erased by the journal writer or a word added by the journal writer after the fact.

Remarks that appear in the Weather Diary but have no relevance to the weather, waterways, or climate are not included in this edition but are replaced by four periods in a row (....). Additional journal entries from various manuscripts for a particular day have been delineated by parentheses in the remarks sections to identify the source. The original spelling of the words used by the authors of the journals has been preserved to maintain contextual significance.

Temperature observations in the table listed with "a0" "a" or "b0" "b" are in relation to the Fahrenheit temperature scale. For example, the December 17, 1804 journal entry of 43b0 translates to 43° below zero Fahrenheit. As noted in postscript discussions in Moulton's (1986) edition, the error of the thermometer readings may vary from 8 to 11 degrees.

The accuracy of the data is relevant to the Coues (1893), Thwaites (1904), and Moulton (1986–1993) editions of the journals, and they make extensive notes on the variations from the different journal entries. Many of these footnotes have been added to this compilation; however, it is recommended that researchers review the original journals or the three cited editions for additional postscript explanations, temperature corrections, and other entries.

The tables represent a snapshot of what occurred. To fully comprehend the ramifications of weather, water, and climate read Lewis and Clark's daily narratives.

LEWIS'S GENERAL COMMENTS ABOUT DIARY ENTRY PROCEDURES

THERMOMETRICAL OBSERVATION SHEWING ALSO THE RISE AND FALL OF THE MISSISSIPPI,* APPEARANCES OF WEATHER WINDS &C AT THE MOUTH OF THE RIVER DUBOIS COMMENCING
1ST JANY 1804. IN LONGITUDE 89° 57' 45" W. LATITUDE 38° 55' 19.6" N. THERMOMETER ON THE N. SIDE OF A LARGE TREE IN THE WOODS.

EXPLANATIONS.

IN THE MISCELLANIOUS COLUMN OR COLUMN OF REMARKS ARE NOTED, THE APPEARANCE QUANTITY AND THICKNESS OF THE FLOATING OR STATIONARY ICE, THE APPEARANCE AND QUANTITY OF DRIFT-WOOD, [THE RAPIDITY OF THE COURENT OF THE RIVER BELOW THE MOUTH
OF THE MISSOURI, THE FALLING OF THE BANKS —] THE APPEARANCE OF BIRDS, REPTILES AND INSECTS, IN THE SPRING DISAPPEARANCE IN THE FALL, LEAFING FLOWERING AND SEEDING OF PLANTS, FALL OF LEAF, ACCESS AND RECESS OF FROST, DEPTH OF SNOWS, THEIR DURATION OR DISAPPEARANCE.[LONGITUDE AND LATD.]

Notations of the weather

  f   means Fair
  c   —Cloudy
  r   —Rain
  s   —Snow
  h   —Hail
  t   —Thunder
  l   —Lightning
  a   —after— as f. a. r. means that it is *fair after rain* which has interveened since the last observation— c. a. s. *Cloudy after snow* intervening c.a.r.s. — cloudy after rain & snow—

Notations of the river

  R   means *risen* in the last 24 hours ending at sunrise
  F   —*fallen* in the same period

Notations of Thermometer

  a 0 means *above naught* (zero)
  b 0 means *below naught* (zero)

Remarks on the Thermometer

  —By two experiments made with Ferenheit's Thermometer which I used in these observations, I asscertained [its] error to be 11°[†] too low or additive + — I tested it with water and snow mixed for the friezing point, and boiling water for — the point marked boiling water.—

Note   when there is not room in the column for the necessary remarks it is transferred by the refference of Numbers to an adjoining part of this book.

---

*According to Moulton (1986, 2: 171) this is "Missouri" in Clark's Journal Codex C.

†According to Moulton (1986, 2: 169–172) Clark's Journal Codex C lists this temperature as "8°." See also Thwaites (1904, 6, part 2: 170).

PART II Excerpts from the Weather Diary and Narrative Journals

SECTION 1 East of the Mississippi

August 31 to December 11, 1803

On August 31, 1803, Meriwether Lewis left Pittsburgh, Pennsylvania, with a small party in a keelboat and canoes. They moved slowly down the Ohio River due to low water conditions brought on by drought. Lewis arrived at the Falls of the Ohio near Clarksville, Indiana / Louisville, Kentucky on October 14 and met William Clark with additional recruits. As author and historian Stephen Ambrose noted, "When they shook hands, the Lewis and Clark Expedition began" (Ambrose 1996, 117). They set out from Louisville on October 26, arriving at the confluence of the Ohio and Mississippi on November 14 and moved up the Mississippi through early December. Heading up the Mississippi was made difficult by low water and strong currents, and even more burdensome as late-fall cold fronts and accompanying strong northwest winds pushed against the boats. They arrived near St. Louis on December 12 and established winter quarters across from the confluence of the Missouri and Mississippi rivers at Camp Dubois near the mouth of the Wood River (see Section 2).

The journal entries in this section are known as the Eastern Field Notes and came to light when the grandsons of Nicholas Biddle (the first editor of the Lewis and Clark Journals) discovered new documents in their grandfather's personal notes. Milo Milton Quaife published these for the first time in 1916 and this section's weather data are excerpted from that work, with some bracketed clarifications and footnotes from Moulton (1986, 2: 65–143). To learn more about the history and publication of the journals, see Cutright (1976) and Moulton (1986, 2: 8–48, 530–567).

Note   *Meriwether Lewis prepares for the journey to St. Louis in Pittsburgh, Pennsylvania, from July 15, 1803 until his departure down the Ohio River with a keelboat, canoes, and a small party on August 31.*

WEDNESDAY, AUGUST 31

Lewis   Left Pittsburgh this day at 11 ock with a party of 11 hands.* The river is extreemly low; said to be more so than it has been known for four years. The water being sufficiently temperate was much in our favor.

THURSDAY, SEPTEMBER 1

Lewis   The Pilott informed me that we were not far from a ripple which was much worse than nay we had yet passed, and as there was so thick a fogg on the face of the water that no object was visible 40 paces he advised remaining untill the sun should acquire a greater altitude when the fogg would asscend and disappear. Remained untill eight Oclock...we set out again. These Foggs are very common on the Ohio at this season of the year as also the spring but do not think them as frequent or thick in the spring. Perhaps this may in some measure assist us to account for the heavy dues

---

*Although the original journals have August 30 written for this date, Lewis's letter to Thomas Jefferson on September 8 notes that he left on August 31 (Jackson 1978, 121). Thus begins the initial writings leading up to the expedition departure up the Missouri the following year from Camp Dubois.

which are mor remarkable for their frequency and quantity than in any country I was ever in—…They are so heavy the drops falling from the trees from about midknight untill sunrise gives you the eydea of a constant gentle rain, this continues untill the sun has acquired sufficient altitude to dessipate the fogg by [its] influence, and it then ceases. The dues are likewise more heavy during summer than elsewhere but not so much so as at this season—   the Fog appears to owe [its] orrigin to the difference of temperature between the air and water the latter at this seson being much warmer than the former; the water being heated by the summer's sun does not undergo so rapid a change from the absence of the sun as the air dose consiquently when the air becomes most cool which is about sunrise the fogg is thickest and appears to rise from the face of the water like the steem from boiling water—

FRIDAY, SEPTEMBER 2

Lewis   the weather is extreemly dry but there was some appearance of rain this morning which seems now to have blown over—   Thermometer stood at seventy six in the cabbin   the temperature of the water in the river when emersed about the same—   Observed today the leaves of the buckeye, Gum, and sausafras begin to face, or become red—

SATURDAY, SEPTEMBER 3

Lewis   Verry foggy this morning. Thermometer 63° Ferrenheit, immersed the Thermometer in the river, and the murcury arose immediately to 75° or summer heat so that there is 12° difference is sufficient to shew the vapor which arrises from the water; the fogg this prodused is impenetrably thick at this moment; we were in consequence obliged to ly by untill 9 this morning.

SUNDAY, SEPTEMBER 4

Lewis   Morning foggy, obliged to wait. Thermometer at 63° — temperature of the river-water 73° being a difference of ten degrees, but yesterday there was a difference of twelve degrees, so that the water must have changed [its] temperature 2d in twenty four hours, coalder; at 1/4 past 8 the murcury rose in the open air to 68°   the fogg dispeared and we set sail; the difference therefore of 5° in temperature between the warter and air is not sufficient to produce the appearance of fogg—   from the watermark we fixed last evening it appeared that the river during the night had fallen an inch perpendicularly—   The water is so low and clear that we see a great number of fish of different kinds

MONDAY, SEPTEMBER 5

Lewis   Again foggey, loaded both my canoes and waited till the fogg disappeared   set out at 8 Ock. Rained at six this evening and continued with some intervals through the night to rain pretty hard

TUESDAY, SEPTEMBER 6

Lewis   The fogg was as thick as usual this morning   detained us untill 1/2 past 7 O'C. when we set out— observed the Thermometer in the air to stand at 71° water 73° — the fogg continued even with small difference between the temperature of the air and water. Hoisted our fore sale. We run two miles in a few minutes when the wind becoming so strong we were obliged to hall it in lest it should carry away the mast, but the wind abating in some measure we again spread it; a sudan squal broke the sprete and had very nearly carried away the mast   The wind blew so heard as to break the spreat of it.

WEDNESDAY, SEPTEMBER 7

Lewis   Foggy this morning according to custom. Observed the Thermometer at sun rise in the air to stand at 47° the temperature of the river water being 68° — difference = 21° — got over the riffle   the water gets low as it most commonly is from the begining of July to the last of September; the water from hence being much deeper and the navigation better than it is from Pittsburgh or any point above it—

FRIDAY, SEPTEMBER 9

Lewis   in attending to the security of my goods I was exposed to the rain and got wet to the skin as I remained untill about twelve at night; when I wrung out my saturated clothes...the rain was excessively [cold] for the season of the year— about the time we landed it began to rain very [hard] and continued to rain most powerfully all night with small intervales.

SATURDAY, SEPTEMBER 10

Lewis   The rain ceased about day, the clouds had not dispersed, and looked very much like giving us repetition of the last evening's frallic, there was but little fogg

SUNDAY, SEPTEMBER 11

Lewis   observed a number of squirrels swiming in the Ohio and universally passing from the W to the East shore   they appear to be making to the south; perhaps it may be mast or food which they are in serch of but I should reather suppose that it is climate which is their object as I find no difference in the quantity of mast on both sides of this river

MONDAY, SEPTEMBER 12

Lewis   it began to rain and continued with some intervals untill three in the evening

Note   *Lewis's party passes through Marietta, Ohio.*

TUESDAY, SEPTEMBER 13

Lewis   this morning being clare we persued our journey at sunrise

WEDNESDAY, SEPTEMBER 14

Lewis   saw many squirrels this day swiming the river from NW to SE.

THURSDAY, SEPTEMBER 15

Lewis   it rained very hard on us from 7 this morning untill about three when it broke away and evening was clear with a few flying clouds.

FRIDAY, SEPTEMBER 16

Lewis   Thermometer this morning in the air 54° in the water 72° a thick fogg which continued so thick that we did not set out untill 8 oClock in the morning   the day was fair

SATURDAY, SEPTEMBER 17

Lewis   The morning was foggy but bing informed by my pilot that we had good water for several miles I ventured to set out before the fog disappeared. Came seven miles to the old Town Bar…I determined to spend the day…dry my goods…wet by the rain of the 15th. I found on opening the goods that many of the articles were much Injured; particularly the articles of iron, which wer rusted very much. The evening was calm tho' the wind had blown extreemly hard up the river all day—   It is somewhat remarkable that the wind on this river, from much observation of my own, and the concurrent observation of those who inhabit [its] banks, blows or sets up agains [its] courent four days out of five during the course of the whole year; it will readily be concieved how much this circumstance will aid the navigation of the river—   When the Ohio is in [its] present low state, between the riffles and in many places for several miles together there is no preseptable courent, the whole surface being perfectly dead or taking the direction only which the wind may chance to give it, this makes the passage down this stream more difficult than would at first view be immageoned, when it is remembered also that the wind so frequently sets up the river the way the traveler makes in descending therefore is by the dint of hard rowing— or force of the oar or pole.

SUNDAY, SEPTEMBER 18

Lewis   The morning was clear

Note   *Lewis stops writing in the Eastern Field Notes and no further journal entries are known to have been recorded until November 11, 1803. From September 28 through October 6, Lewis and a small party spend time in Cincinnati, Ohio. They then proceed down the Ohio River and arrive in Clarksville, Indiana (also near Louisville, Kentucky), on October 14, where they meet William Clark and his recruits. The larger party sets off down the Ohio River with the keelboat and two pirogues on October 26. They arrive at Fort Massac, Illinois (near present-day Paducah, Kentucky), on November 11 and leave the next day.*

SATURDAY, NOVEMBER 12

Lewis   remained, took equal altitudes AM but was prevented from compleating the observation by taking an observation in the evening by the clouds—*

SUNDAY, NOVEMBER 13

Lewis   raind very hard in the eving

---

*Lewis once again starts writing the Eastern Field Notes on November 11, 1804, although no meteorological or hydrological data was observed on that date.

Note   *The expedition arrives at the confluence of the Ohio and Mississippi rivers and stays until November 20.*

MONDAY, NOVEMBER 14

Lewis   this evening landed on the point at which the Ohio and Mississippi form there junchon.

TUESDAY, NOVEMBER 15

Lewis   took equal altitudes   lost the afternoon from clouds which interveened and prevented them.

THURSDAY, NOVEMBER 17

Lewis   the [wind] blew very [hard] last night from N and continued without intermission throughout the day   it became [cold] about twelve oclock— the canoes were driven by the violence of the waves against the shore and filled with water. measured the hight of the bank in the point and found it 36 F[eet] 8 I[inches] above the level of the water at thime which may with much propriety be deemed low water mark as neither the Ohio or Missippi wer ever known to be lower—

Note   *On November 20, the party proceeds up the Mississippi River.*

TUESDAY, NOVEMBER 22

Lewis   the current very rapid and difficult

THURSDAY, NOVEMBER 24

Lewis   I am not confident with respect to the accuracy of the observation of this day, in consequence of some flying clouds which frequently interveended and obscured his sun's disk about noon and obliged me frequently to change the coloured glasses of the Sextant in order to make the observation as complete as possible.

FRIDAY, NOVEMBER 25

Lewis   The Mississippi when full throws large quantities of mud into the mouths of these rivers whose courents not being equal to contend with it's power become still or eddy for many miles up them.

SATURDAY, NOVEMBER 26

Lewis   When the river is high the courent setts in with great violence on the W side of this rock...these strong courants thus meeting each other form an immence and dangerous whirlpool which no boat dare approach in that state of the water. In the present state of the water there no danger in approaching it.

Note   *Except for a few miscellaneous entries and astronomical readings, this marks the end of Lewis's known daily narrative journal entries until April 1805.*

MONDAY, NOVEMBER 28

Clark   Set out this morning at 8 oClock...this morning being verry Smokey prevents my being as acurate as I Could wish—   The horozon became darkened that I could not see across the River, which appeared to windened, the Current much Swifter than usial. 5, 6

Note   Although Clark made some navigational notes between November 15–19, 1803, this is the first of his daily narrative writings in the Eastern Field Notes of the Lewis and Clark Journals, and the first of his weather- and hydrology-related remarks.

TUESDAY, DECEMBER 6

Clark   A Dark wet morning. Current is verry Swift.

WEDNESDAY, DECEMBER 7

Clark   A Dark rainey morning with hard wind at NE, upon which point it blew all the last night accompanyd. with rain—   Set out a quarter past 7 oClock, the wing [wind] much against us. About 10 oClock the wind changed to the SE and gave us an oppertunity to Sailing. At 12 oClock the wind was So violent as to take off one of the Mast's   came to at 3 oClock...in view of St. Louis which is about 2 ? miles distant.

SATURDAY, DECEMBER 10

Clark   took Meridian Atld...the sun was reather dim, therefore it possible that this observation may have been liable to a small error—

SUNDAY, DECEMBER 11

Clark   a Verry rainey morning   the wind from the NE   The wind changed to NW about 3 oClock   the rain Continud untill 3 oClock to day. The Current of the water is against the Westerley Shore, and the banks are falling, where there is no Rock.

Note   The party moves up the Mississippi River and lands at Wood River, across from the confluence of the Mississippi and Missouri rivers in present-day Illinois, and establishes winter camp. Across the Mississippi River to the southeast is the city of St. Louis. They establish Camp Dubois on December 12 and remain here until May 14, 1804.

SECTION 2   Camp Dubois

December 12, 1803 to May 14, 1804

This section is comprised of entries from the expedition's Field Notes made while wintering at Camp Dubois, Illinois, on the Wood River. Apparently the captains did not regard this journal as an official document because they were not traveling and the Expedition had not actually begun; hence it is extremely sketchy and disorganized (Moulton 1986, 2: 133). Data were entered in these field notes until the day the Expedition commenced up the Missouri on May 14, 1804.

Many experiments were conducted during this time, including thermometer calibration as annotated in the Weather Diary, begun on January 1, 1804. The Weather Diary included Observation Tables with data and a column for general complementary remarks.

In this section, each day's entry begins with the data from the Observation Table (when available), followed by the remarks (when available), and finally excerpts from the Camp Dubois Field Notes that pertain to weather and climate.

While at Camp Dubois, river observations were taken at the mouth of the Mississippi and Wood rivers at sunrise for a 24-hour period and recorded in the Observation Tables. On some days, both Lewis and Clark entered remarks, and this is noted. If not indicated, the day's remarks were made by Clark, who kept the Weather Diary notebook at Camp Dubois through April. Starting in May, the Weather Diary entries come from Lewis's data. Some observations were actually made from across the Mississippi River in St. Louis instead at the encampment on the Dubois River. Weather data included in this section are from Coues (1893, 3: 1264–1267), Moulton (1986, 2: 169–217), and Thwaites (1904, 6, part 2: 166–174).

The Camp Dubois Field Notes were made primarily by Clark and are labeled as such in this section. There are no notes available for February 10 through March 20. As reference for the 1804 Camp Dubois Notes this volume uses Osgood (1964) with bracketed additions and footnoted information from Moulton (1986, 2: 130–226). To learn more about the history of the journals, the reader is directed to Cutright (1976) and Moulton (1986, 2: 8–48, 530–567).

Note   *The expedition begins its winter at Camp Dubois, at the mouth of the Wood and Mississippi rivers.*

MONDAY, DECEMBER 12

Clark   A hard NW wind all last night. I came to in the mouth of a little river called Wood River, about 2 oClock and imediately after I had landed the NW wind which had been blowing all day increased to a Storm which was accompanied by Hail & Snow, & the wind Continued to blow from the Same point with violence.

TUESDAY, DECEMBER 13

Clark   a hard wind all day— flying Clouds

WEDNESDAY, DECEMBER 14

Clark   wind Continu to blow hard river riseing—

THURSDAY, DECEMBER 15

    Clark   Snow

FRIDAY, DECEMBER 16

    Clark   the winds high to day— Cloudy—

SATURDAY, DECEMBER 17

    Clark   a Cold fine morning

SUNDAY, DECEMBER 18

    Clark   Clear morning

MONDAY, DECEMBER 19

    Clark   a hard frosty morning

WEDNESDAY, DECEMBER 21

    Clark   Cloudy Day water fall verry fast

THURSDAY, DECEMBER 22

    Clark   a verry great Sleat this morning, the river Coverd with running Ice, and falls verry fast 15 Inches last night. Mist of the rain, which prevents our doeing much to our huts to day

FRIDAY, DECEMBER 23

    Clark   a raney Day   a raney Desagreeable day   the Ice run to day   the water falls fast

SATURDAY, DECEMBER 24

    Clark   Cloudy morning

SUNDAY, DECEMBER 25

    Clark   Snow this morning, Ice run all day

MONDAY, DECEMBER 26

    Clark   a Cloudy day   The Ice run, this day is moderate

TUESDAY, DECEMBER 27

    Clark   a fair day.

WEDNESDAY, DECEMBER 28

    Clark   a Cloudy day   No Ice in the river

THURSDAY, DECEMBER 29

    Clark   Snow this morning   Cloudey & wet all day. Rain at night

FRIDAY, DECEMBER 30

Clark   Snow in the morning   Cloudy morning

SATURDAY, DECEMBER 31

Clark   began to snow at Dark and Continued untill 9 oClock   Cloudy to day

SUNDAY, JANUARY 1

	Sunrise			4 p.m.			Mississippi River	
Temp	Weather	Wind	Temp	Weather	Wind	Rise/Fall	Feet	Inches
	Cloudy			*Cloudy*				

Weather Diary   Snow 1 Inch Deep

Clark   Snow about an inch deep. Cloudy to day

MONDAY, JANUARY 2

	Sunrise			4 p.m.			Mississippi River	
Temp	Weather	Wind	Temp	Weather	Wind	Rise/Fall	Feet	Inches
	c a s			*c*				

Weather Diary   Snow last night inconsiderable

Clark   Snow last night (rain). a mist to day

TUESDAY, JANUARY 3

	Sunrise			4 p.m.			Mississippi River	
Temp	Weather	Wind	Temp	Weather	Wind	Rise/Fall	Feet	Inches
			2 ½a0	*f*	*NWW*			

Weather Diary   wind blew hard

Clark   a Verry Cold blustering day Thermometer one oClock in the open air the (quick-silver) mercuria fell to 21 D. below the freezing point [11°F], I took the altitude of the suns…all the after part of the Day the wind so high that the View up the Missouris appeared Dredfull, as the wind blew off the Sand with fury as to Almost darken that part of the

atmespear   this added to agutation of the water apd. truly gloomy. The wind violent all Day
from NW & NW Excessive cold after Sunset

WEDNESDAY, JANUARY 4

	Sunrise			4 p.m.			Mississippi River		
Temp	Weather	Wind	Temp	Weather	Wind	Rise/Fall	Feet	Inches
11a0	f	W		c	W			

Weather Diary   river Covered with ice out of the Missouires

Clark   a Cold Clear morning, the river Covered with Ice from the Missouri, the
Massissippi above frosed across, the Wind from the West, the Thermometer
this morning at 19° below freezing [13° F], Continued Cold & Clear all day.
At 4 oClock the murcuria of the Thmtr in a corner of a warm room was
20 D. above 0—

THURSDAY, JANUARY 5

	Sunrise			4 p.m.			Mississippi River		
Temp	Weather	Wind	Temp	Weather	Wind	Rise/Fall	Feet	Inches
	f	W		f	W			

Weather Diary   the River a Dubois rise a little

Clark   the Creek rose Considerably last night   the river full of Ice, and the wind which
blows from the West blows it to this Shore, a madderate day

FRIDAY, JANUARY 6

	Sunrise			4 p.m.			Mississippi River		
Temp	Weather	Wind	Temp	Weather	Wind	Rise/Fall	Feet	Inches
	f	WNW	30a0	f	WNW			

Weather Diary   the River a Dubois rise a little

Clark   Thermometer at 12 oClock 31 above 0   at 4 oCk at 30° above 0

SATURDAY, JANUARY 7

	Sunrise			4 p.m.		Mississippi River		
Temp	Weather	Wind	Temp	Weather	Wind	Rise/Fall	Feet	Inches
	h	*SW*		*c & r**	*SW*			

Weather Diary   the River a Dubois rise a little

Clark   Some rain last night, a thow [thaw?] and Some rain to day. Up last night and frequent thro the rain to day attending the boat.

SUNDAY, JANUARY 8

	Sunrise			4 p.m.		Mississippi River		
Temp	Weather	Wind	Temp	Weather	Wind	Rise/Fall	Feet	Inches
	f	*SW*		*f*	*NW*			

Weather Diary   Ice run down the little river [Dubois River]

Clark   Rained moderately all last night, a butifull morning a few large sheets of thin Ice running this morning. Clouded up at ½ past 2 oClock the wind chifted to the NW moderate.

MONDAY, JANUARY 9

	Sunrise			4 p.m.		Mississippi River		
Temp	Weather	Wind	Temp	Weather	Wind	Rise/Fall	Feet	Inches
	f	*WNW*	*1b0*	*c*	*WNW*			

Weather Diary   Snow last night

Clark   Some Snow last night, a hard wind this morning from WNW   river Rises with large Sheets of Ice out of Mississippi, the morning is fair. I returned before Sun Set,

---

*According to Moulton (1986, 2: 172), Clark's Journal Codex C lists this weather data as "c a r h."

and found that my feet, which were wet had frozed to my Shoes, which rendered precaution necessary to prevent a frost bite, the Wind from the W, across the Sand Islands in the Mouth of the Missouries, raised Such a dust that I could not See in that derection, the Ice Continue to run & river rise Slowly— exceeding Cold day

TUESDAY, JANUARY 10

Sunrise			4 p.m.			Mississippi River		
Temp	Weather	Wind	Temp	Weather	Wind	Rise/Fall	Feet	Inches
	f			*f*				6

Weather Diary   Missouri rise

Clark   a fine day, the river rose 6 Inches last night. Joseph Fields returned & crossed the River between the Sheets of floating Ice with Some risque, his excuse for staying so long on the Mississippi were that the Ice run so thick in the Missourie where he was 20 miles up that there was no crossing

WEDNESDAY, JANUARY 11

Clark   a fine morning, the river Still riseing, the Missouries run with fine Ice   at 1 oclock the wind blew strong from the west and turned Couled & Cloudy this afternoon

THURSDAY, JANUARY 12

Clark   a fair morning, the wind from the S West, the river Continue to rise moderately… with large Sheets of ice running against Ice attached to the bank with great force

FRIDAY, JANUARY 13

Sunrise			4 p.m.			Mississippi River		
Temp	Weather	Wind	Temp	Weather	Wind	Rise/Fall	Feet	Inches
	c & s	SW	*0*	*r & s*	SW			

Weather Diary   Snow'd last night

Clark   the river rise, a fall of Snow last night, the Missouris is riseing and runs with Ice, a Cloudy & warm day. A fine rain in the evening.

SATURDAY, JANUARY 14

Sunrise			4 p.m.			Mississippi River		
Temp	Weather	Wind	Temp	Weather	Wind	Rise/Fall	Feet	Inches
	f a s			*f*				

Weather Diary Snow'd last night

Clark a Snow fall last night of about an Inch and one half   the River (Still riseing) falling and running with Ice, a fair Sun shineing morning— The Mississippi, is Closed with Ice.

SUNDAY, JANUARY 15

Camp Dubois Field Notes

Clark river falling & runs still with Ice. A Cold night.

MONDAY, JANUARY 16

[No data was entered for this day.]

TUESDAY, JANUARY 17

Sunrise			4 p.m.			Mississippi River		
Temp	Weather	Wind	Temp	Weather	Wind	Rise/Fall	Feet	Inches
8b0	*f*	NW	*1 ½b0*	*f*	NW	*F*		*6*

Weather Diary   river falls & full of Ice 5 ½ In. thick

Clark a verry Cold morning, at 7 oClock the Thermometer in the air fell 8° below 0, the wind from the NW, a Stiff Breeze, Ice run greatly out of the Missouries— at 9oClock the Thermometer 6d below 0— at 10 oClock 3d below 0— at 12 oClock at 0— at 1 oClock 1° above 0— at 2 oClock 1 ½° above 0— at 3 oClock at 0— at 4 oClock the Thermometer was 1 ½° below 0 at— at 5 the Ther: was at 3° below 0— at 9 oClock 6° below 0— a verry Cold night; the Missouris has fallen to day about 6 Inches, runs with Ice, Ice from Shore 20 yds in the river is 5 ½ Inches Thick—

WEDNESDAY, JANUARY 18

Weather Diary   river falls & full of Ice 5 ½ In. thick

	Sunrise			4 p.m.		Mississippi River		
Temp	Weather	Wind	Temp	Weather	Wind	Rise/Fall	Feet	Inches
1b0	c	NWW	1a0	s & f	NNW	F		

Clark a Cloudy morning with moderate breaze from the NW by W. The river run with Ice, at 8 oClock the Thromtr. Stood at 1° below 0— 9 oClock 1° above 0— at 10 oClock 2° abov 0— at 11 oClock rose to 4° above 0— , and began to Snow, at 12 oClock the Thermt. at 2 above 0— . Snow above 1 inch   at 1 oClock 2 abov 0— , at 2 oClock 1 abov 0— , and left off Snowing—

THURSDAY, JANUARY 19

	Sunrise			4 p.m.		Mississippi River		
Temp	Weather	Wind	Temp	Weather	Wind	Rise/Fall	Feet	Inches
13a0	c	NW	11a0	c	NW	F		

Weather Diary   no ice running

Clark   Som Snow fell last night, a Cloudy morning, the river continues to fall, & Some Ice running, at 8 oClock this morning the Thermormeter Stood at 13° above 0—, the wind moderate from the NW, at 9 oClock 15° above 0— at 10 oClock 16° above 0— at 11 oClock 16° above 0— at 12 oClock 19° above 0— at 1 oClock 17° above 0— at 2 oClock 15 1/2° above 0— at 3 oClock 13° above 0— at 4 oClock 11° above 0— at [5?] oClock 10 1/2° above 0—

FRIDAY, JANUARY 20

	Sunrise			4 p.m.		Mississippi River		
Temp	Weather	Wind	Temp	Weather	Wind	Rise/Fall	Feet	Inches
5b0	f	NW	8a0	c	NW	F		

Weather Diary   Ice running out of the Missippi 9 In thick (Lewis)   No ice passing to day (Clark)

Clark   a verry Cold night, river Still falling [some] no Ice running out of the Missouries, the wind this morning from NW— The Thermometer at 7 oClock 5° below 0— at

8 oClock 7° below 0— at 9 oClock 4° below 0— at 10 oClock 2° below 0— at 11 oClock 2° above 0— at 12 oClock 4° above 0— at 1 oClock the Thmtr. Stood at 6° above 0— at 2 oClock the Them at 80 above 0—, the river Mississippi raised & some [of] the Ice is 9 Inches Thick (now ice flotes down the) at 3 oClock 11° above 0—, Cloudy at 4 oClock 8° above 0—, at 5 oClock 7 ½° above 0— Cloudy

SATURDAY, JANUARY 21

	Sunrise			4 p.m.			Mississippi River		
Temp	Weather	Wind	Temp	Weather	Wind	Rise/Fall	Feet	Inches
7a0	c & s	NE	17a0	s & h	NE	F		

Weather Diary Snow 2 ½ In Deep (Lewis) Ice running out of the Missoury, 9 In thick Snow 2 ½ In Deep (Clark)

Clark The Snow this morning is about 2 ½ Inches Deep, & Snowing fast, the Thermometer Stood at 7d above 0—, at 8 oClock, & wind from the NE, the river running with Ice and falling a little, at 9 oClock the Thermtr. at 7° above 0—, at 10 oClock 7d above 0—, at 11 oClock 10° above 0— & Snows— at 12 oClock 10° above 0— [Snow] or fine Haile increas, at 1 oClock 10° above 0—, at 5 oClock 17° above 0— haileing fine hail.

SUNDAY, JANUARY 22

	Sunrise			4 p.m.			Mississippi River		
Temp	Weather	Wind	Temp	Weather	Wind	Rise/Fall	Feet	Inches
11a0	s	shify*	13a0	s	NW	F		

Weather Diary Snow 5 ¾ In Deep Ice running down the Missouri (Lewis) Ice running out of the Missouri, Snow 5 ¾ In Deep. (Clark)

Clark Snow all the last night, and Snows this morning, the debth is 5 ¾ Inches, the Thermometer Stands at 9 oClock this morning in the Open air at 11° above 0— at 12 oClock rose to 14° above 0— and Stoped Snowing. wind Easterly at 3 oClock 13d abv 0—, riv nearly Clear of Ice.

---

*According to Moulton (1986, 2: 172), Clark's Journal Codex C lists this wind direction as "S."

MONDAY, JANUARY 23

Sunrise			4 p.m.			Mississippi River		
Temp	Weather	Wind	Temp	Weather	Wind	Rise/Fall	Feet	Inches
11a0	c	NE	17a0	c	N	F		

Weather Diary   Ice Stoped

Clark   a Cloudy morning, but little Ice runig to day   The Thermtr. at 8 oClock 11° above
0— at 12 oClock Stood at 10° above 0— at 3 oClock 17° above 0—   in the evening
the wind raised and Shifted to the North

TUESDAY, JANUARY 24

Sunrise			4 p.m.			Mississippi River		
Temp	Weather	Wind	Temp	Weather	Wind	Rise/Fall	Feet	Inches
4a0	c	NW	11a0	c	W	F		

Weather Diary   The Trees covered with ice to day

Clark   a Butifull morning Clear Sunshine the winds light from the NW, the Thermometer
at 9 oClock 4° above 0— at 10 oClock 8° above 0— wind West, at 12 oClock
14° above 0— at 3 oClock 11° above 0— [Some] Small pieces of Ice running

WEDNESDAY, JANUARY 25

Sunrise			4 p.m.			Mississippi River		
Temp	Weather	Wind	Temp	Weather	Wind	Rise/Fall	Feet	Inches
2b0	f	WNW	16a0	f	W	F		

Weather Diary   Some ice

Clark   a verry Clear [moon] Shiney night a fair morning, last night was a verry Cold one,
The branches of Trees and the Small broth are gilded with Ice from the frost of last
night which affords one of the most magnificent appearances in nature, the river

began to Smoke at 8 oClock and the Thermometer Stood at 2° below 0— aat 9 oClock at 0— at 10 oClock 5° above 0— at 11 oClock 12° above 0— at 12 oClock 16d above 0— at 1 oClock 16° above 0— at 2 oClock 19° above 0— at 3 oClock 16° above 0—, wind from the WNW

THURSDAY, JANUARY 26

Sunrise			4 p.m.			Mississippi River		
Temp	Weather	Wind	Temp	Weather	Wind	Rise/Fall	Feet	Inches
	c	SW		c	SW	F		

Weather Diary   warm Day

Clark   a Cloudy warm Day. Verry little Ice running to day

FRIDAY, JANUARY 27

Sunrise			4 p.m.			Mississippi River		
Temp	Weather	Wind	Temp	Weather	Wind	Rise/Fall	Feet	Inches
	f			f				

Weather Diary   warm Day

Clark   a Cloudy morning Some Snow   at 1 oClock 28° above 0—

SATURDAY, JANUARY 28

Sunrise			4 p.m.			Mississippi River		
Temp	Weather	Wind	Temp	Weather	Wind	Rise/Fall	Feet	Inches
5a0	c s	NW	18a0	c a s	NW	R		

Weather Diary   cold & Ice runing

Clark   a Cloudy morning   verry cold wind from the NW Some floating Ice in the River at 9 oClock 5° above 0— , Snows at 10 oClock 8° above 0— at 11 oClock 10°

above 0— , Sun Shines, at 12 oClock 12° above 0— at 1 oClock 14° above 0—
at 2 oClock 18° above 0— at 3 oClock 20° above 0— at 4 oClock 18°
above 0— at 6 oClock 14° above 0— , Porter [beer] all frosed & several
bottles broke

SUNDAY, JANUARY 29

	Sunrise			4 p.m.			Mississippi River		
Temp	Weather	Wind	Temp	Weather	Wind	Rise/Fall	Feet	Inches
16a0	f	W	23a0	f		R		

Weather Diary   no Ice running

Clark   a butifull morning, the river rise a little, no Ice, The Thermometer at 9 oClock Stood
at 16° above 0— at 11 oClock 22d above 0— at 12 oClock 24° above 0— Took the
alltiude of Suns, at 3 oClock 28° above 0— at 4 oClock 26° above 0— at 5 oClock
23° above 0—

MONDAY, JANUARY 30

	Sunrise			4 p.m.			Mississippi River		
Temp	Weather	Wind	Temp	Weather	Wind	Rise/Fall	Feet	Inches
22a0	c & s	N	16a0	f a s		R		

Clark   a Cloudy morning, Some Snow at 9 oClock The Thermotr. Stood at 22d abov 0— ,
a little wind from N. at 10 oClock 24° above 0— , cleared up & Sun Shown,
Stoped Snowing, but little Ice running this morng— at 11 oClock 25° above 0—
at 12 oClock 25° above 0— at 1 oClock 25° above 0— at 2 oClock 26° above
0— Cloudy. at 3 oClock 28° above 0— at 4 oClock 27° above 0— at 8 oClock
16° above 0—

TUESDAY, JANUARY 31

	Sunrise			4 p.m.			Mississippi River		
Temp	Weather	Wind	Temp	Weather	Wind	Rise/Fall	Feet	Inches
10a0	f	SW by W	15a0	f	W	R		

Weather Diary   Ice run a little

Clark   a fair morning, the Trees guilded with ice, at 7 oClock the Thermometer Stood at 7 oClock 10° below 0— . At 9 oClock some Ice running this morning. at 12 oClock 24° above 0— at 2 oClock 28° above 0— at 4 oClock 28° above 0— at 9 oClock 15° above 0— , wind SW by W

WEDNESDAY, FEBRUARY 1

	Sunrise			4 p.m.			Mississippi River		
Temp	Weather	Wind	Temp	Weather	Wind	Rise/Fall	Feet	Inches
10a0	f	SW	20a0	f	SWS	R		

Weather Diary   the wind blew very hard, no frost, snow disapearing fast

Clark   a Cloudy morning & warm wind from the SW, a warm Day

THURSDAY, FEBRUARY 2

	Sunrise			4 p.m.			Mississippi River		
Temp	Weather	Wind	Temp	Weather	Wind	Rise/Fall	Feet	Inches
12a0	f	NW	10a0	f	NW	R		1/2*

Weather Diary   frost this morning, the snow has disapeared in spots

Clark   wind high from SW

FRIDAY, FEBRUARY 3

	Sunrise			4 p.m.			Mississippi River		
Temp	Weather	Wind	Temp	Weather	Wind	Rise/Fall	Feet	Inches
12a0	f	SW	19a0	f	W			

Weather Diary   frost this morning, the snow thawed considerably, raisd boat

Clark   fair Thawing Day

---

*According to Moulton (1986, 2: 178), Clark's Journal Codex C lists the river rise as "1 1/2 inches."

SATURDAY, FEBRUARY 4

	Sunrise			4 p.m.		Mississippi River		
Temp	Weather	Wind	Temp	Weather	Wind	Rise/Fall	Feet	Inches
17a0	f	SW	28a0	f	S	R		½

Weather Diary   frost, considerable number of sawn & Geese from N & S.

Clark   a warm Day, Some rain last night, in the Evening the River Covered with large Sheetes of Ice from both rivers, the River & Creek rised Suffecent to take the boat up the Creek some distance, moderate day. Wild fowl pass.

SUNDAY, FEBRUARY 5

	Sunrise			4 p.m.		Mississippi River		
Temp	Weather	Wind	Temp	Weather	Wind	Rise/Fall	Feet	Inches
18a0	f	SE	31a0	c a f	SES	R	2	6 ½

Weather Diary   emmence quantities of ice runing some of which 11 inches

Clark   river rising & Covered with Small Ice. Fowl pass.

MONDAY, FEBRUARY 6

	Sunrise			4 p.m.		Mississippi River		
Temp	Weather	Wind	Temp	Weather	Wind	Rise/Fall	Feet	Inches
19a0	f	NW ·	15a0	c	S			

Weather Diary   a small white frost, the snow disappeared a small snow storm (Lewis)   a quantity of Soft ice running   Swans passing (Clark)

Clark   a fair day Snow nearly gone, Some Ice Still runing. River began to fall

TUESDAY, FEBRUARY 7

	Sunrise			4 p.m.		Mississippi River		
Temp	Weather	Wind	Temp	Weather	Wind	Rise/Fall	Feet	Inches
29a0	r a c	SE	30a0	r & c	SE	F		8

Weather Diary   a small quantity of soft ice runing, Swans passing

Clark   Some rain last night, Rain this morning. The [river] falling 8 inches. Rain Incres a little, the Creek or River a Dubois rasin fast

WEDNESDAY, FEBRUARY 8

	Sunrise			4 p.m.		Mississippi River		
Temp	Weather	Wind	Temp	Weather	Wind	Rise/Fall	Feet	Inches
22a0	c a r	NW	20a0	c a s	N	R		1

Weather Diary   many swans from N.W. creek rose & took off my water mark,

Clark   a Cloudy morning   Some rain, and Snow a Great raft of Ice Come Down the Creek to day, the river rises & some running Ice.

THURSDAY, FEBRUARY 9

	Sunrise			4 p.m.		Mississippi River		
Temp	Weather	Wind	Temp	Weather	Wind	Rise/Fall	Feet	Inches
10a0	f a s	NNE	12a0	c	NE	R		2

Weather Diary   the river raised 2 feet, large quantity of drift ice from Misso[uri]

Clark   a fine morning. River Still rise & Ice pass down the greater part out of the Missouries

Note   *After February 9, there are no further known Field Note entries by Clark until March 21, 1804.*

FRIDAY, FEBRUARY 10

	Sunrise			4 p.m.			Mississippi River		
Temp	Weather	Wind	Temp	Weather	Wind	Rise/Fall	Feet	Inches
3a0	f	NE	17a0	f	SW	R	1	4

Weather Diary   ice still drifting in considerable quantities, some geese passed fr S

SATURDAY, FEBRUARY 11

	Sunrise			4 p.m.			Mississippi River		
Temp	Weather	Wind	Temp	Weather	Wind	Rise/Fall	Feet	Inches
18a0	h a c	SE	31a0	s a h f	SE	R		1

Weather Diary   Swans from the N. The sugar maple runs freely,

SUNDAY, FEBRUARY 12

	Sunrise			4 p.m.			Mississippi River		
Temp	Weather	Wind	Temp	Weather	Wind	Rise/Fall	Feet	Inches
15a0	f	SSE	25a0	f	SW	F		s

Weather Diary   Pigeons, ducks of varis kinds, and gese have returned

MONDAY, FEBRUARY 13

	Sunrise			4 p.m.			Mississippi River		
Temp	Weather	Wind	Temp	Weather	Wind	Rise/Fall	Feet	Inches
12a0	f	NW	20a0	f	W	R & F		1

Weather Diary   the fist appearance of the blue crain, sugar trees run

TUESDAY, FEBRUARY 14

Weather Diary   but little drift ice, the Misipi is not broken up (Lewis)   Sugar trees run
(Clark)

	Sunrise			4 p.m.			Mississippi River	
Temp	Weather	Wind	Temp	Weather	Wind	Rise/Fall	Feet	Inches
15a0	f	SW	32a0	f	SW	F		

WEDNESDAY, FEBRUARY 15

	Sunrise			4 p.m.			Mississippi River	
Temp	Weather	Wind	Temp	Weather	Wind	Rise/Fall	Feet	Inches
18a0	f	SW	32a0	f	W	F		

Weather Diary   immence quantities of Swan, in the [marsh]

THURSDAY, FEBRUARY 16

	Sunrise			4 p.m.			Mississippi River	
Temp	Weather	Wind	Temp	Weather	Wind	Rise/Fall	Feet	Inches
28a0	c	SE	30a0	r a c	SSE	R		2 $\frac{1}{2}$

FRIDAY, FEBRUARY 17

	Sunrise			4 p.m.			Mississippi River	
Temp	Weather	Wind	Temp	Weather	Wind	Rise/Fall	Feet	Inches
15a0	c a r	SW	32a0	f	W	R		2

SATURDAY, FEBRUARY 18

	Sunrise			4 p.m.			Mississippi River	
Temp	Weather	Wind	Temp	Weather	Wind	Rise/Fall	Feet	Inches
10a0	f	NW				R		7 $\frac{1}{2}$

SUNDAY, FEBRUARY 19

	Sunrise			4 p.m.			Mississippi River		
Temp	Weather	Wind	Temp	Weather	Wind	Rise/Fall	Feet	Inches
10	f	NW						

MONDAY, FEBRUARY 20

	Sunrise			4 p.m.			Mississippi River		
Temp	Weather	Wind	Temp	Weather	Wind	Rise/Fall	Feet	Inches
10a0	f	NW	28a0		SSW	F		2 $\frac{1}{2}$

TUESDAY, FEBRUARY 21

	Sunrise			4 p.m.			Mississippi River		
Temp	Weather	Wind	Temp	Weather	Wind	Rise/Fall	Feet	Inches
20a0	f	NW	34a0		NW	F		$\frac{1}{2}$*

Weather Diary   in the evening the river began to rise $\frac{1}{2}$ inch

WEDNESDAY, FEBRUARY 22

	Sunrise			4 p.m.			Mississippi River		
Temp	Weather	Wind	Temp	Weather	Wind	Rise/Fall	Feet	Inches
14a0	f	NE	26a0		E	R		1 $\frac{1}{2}$

---

*According to Moulton (1986, 2: 178), Clark's Journal Codex C lists the river rise as "1 $\frac{1}{2}$ inches."

THURSDAY, FEBRUARY 23

	Sunrise			4 p.m.			Mississippi River	
Temp	Weather	Wind	Temp	Weather	Wind	Rise/Fall	Feet	Inches
6a0	f	NW	24a0		NW	R		1

Weather Diary   [river] fall in the evening ½ inch

FRIDAY, FEBRUARY 24

	Sunrise			4 p.m.			Mississippi River	
Temp	Weather	Wind	Temp	Weather	Wind	Rise/Fall	Feet	Inches
6a0	f	NE	26a0		NE	F		2

SATURDAY, FEBRUARY 25

	Sunrise			4 p.m.			Mississippi River	
Temp	Weather	Wind	Temp	Weather	Wind	Rise/Fall	Feet	Inches
20a0	f	NE	38a0*		SSW			

Weather Diary   River on a Stand

SUNDAY, FEBRUARY 26

	Sunrise			4 p.m.			Mississippi River	
Temp	Weather	Wind	Temp	Weather	Wind	Rise/Fall	Feet	Inches
16a0	f	NE	30a0		NE	F		½

---

*According to Moulton (1986, 2: 178), Clark's Journal Codex C lists this temperature as "28."

MONDAY, FEBRUARY 27

	Sunrise			4 p.m.		Mississippi River		
Temp	Weather	Wind	Temp	Weather	Wind	Rise/Fall	Feet	Inches
21a0*	c	NE	24a0	r & f s	NW	F		1

Weather Diary   River rose 3 Inches & fell immediately

TUESDAY, FEBRUARY 28

	Sunrise			4 p.m.		Mississippi River		
Temp	Weather	Wind	Temp	Weather	Wind	Rise/Fall	Feet	Inches
4a0	c & s	NW	6a0	c a s	NW	F		2

Weather Diary   began to Snow and Continued all day

WEDNESDAY, FEBRUARY 29

	Sunrise			4 p.m.		Mississippi River		
Temp	Weather	Wind	Temp	Weather	Wind	Rise/Fall	Feet	Inches
8a0	h & s	NW	12a0	c a s	NW	F		2 ½

Weather Diary   Snow all night & untill 11 oClock a.m. & Cleared away the weather had been Clear since [Capt. Lewis] [left] Camp untill this.

THURSDAY, MARCH 1

	Sunrise			4 p.m.		Mississippi River		
Temp	Weather	Wind	Temp	Weather	Wind	Rise/Fall	Feet	Inches
12b0	f	NW	4a0		NW	F		9

*According to Moulton (1986, 2: 178), Clark's Journal Codex C lists this temperature as "4."

FRIDAY, MARCH 2

	Sunrise			4 p.m.			Mississippi River	
Temp	Weather	Wind	Temp	Weather	Wind	Rise/Fall	Feet	Inches
11b0	f	NW	22a0		E	F		3

SATURDAY, MARCH 3

	Sunrise			4 p.m.			Mississippi River	
Temp	Weather	Wind	Temp	Weather	Wind	Rise/Fall	Feet	Inches
10a0	f	E	18a0		SW	F		6 ½

SUNDAY, MARCH 4

	Sunrise			4 p.m.			Mississippi River	
Temp	Weather	Wind	Temp	Weather	Wind	Rise/Fall	Feet	Inches
4a0	f	NE	20a0		E	F		5

MONDAY, MARCH 5

	Sunrise			4 p.m.			Mississippi River	
Temp	Weather	Wind	Temp	Weather	Wind	Rise/Fall	Feet	Inches
10a0	f	NW	20a0		NW	F		3

TUESDAY, MARCH 6

	Sunrise			4 p.m.			Mississippi River	
Temp	Weather	Wind	Temp	Weather	Wind	Rise/Fall	Feet	Inches
4a0	f	NW	10a0		NW	F		3

WEDNESDAY, MARCH 7

Sunrise			4 p.m.			Mississippi River		
Temp	Weather	Wind	Temp	Weather	Wind	Rise/Fall	Feet	Inches
8b0	c & s	NW	18a0	s	NW			

Weather Diary   Saw the first Brant return

Note   *According to Moulton (1986), "This table follows Lewis's Weather Diary, kept by Clark this month; its temperature readings are eight degrees higher than those in (Clark's Journal) Codex C, with exceptions noted within the table. In Codex C, Clark indicated that the thermometer registered eight degrees too low (see notes for the January 1804 Weather Diary); Lewis gave the error as eleven degrees, but in April and May Lewis applies the eight-degree correction. Since Clark compensated for the error in March in the Weather Diary but not in Codex C, he apparently did not keep the two tables simultaneously."*

THURSDAY, MARCH 8

Sunrise			4 p.m.			Mississippi River		
Temp	Weather	Wind	Temp	Weather	Wind	Rise/Fall	Feet	Inches
6a0	c & s	NW	20a0	s	NW	F		1/2*

Weather Diary   Rain Suceeded by Snow & hail

FRIDAY, MARCH 9

Sunrise			4 p.m.			Mississippi River		
Temp	Weather	Wind	Temp	Weather	Wind	Rise/Fall	Feet	Inches
18a0	c	NW	28a0†	c	NW	R		2

Weather Diary   The weather has been generally fair but verry Cold, the ice run for Several days in Such quantities that it was impossible to pass the River [Mississippi] Visited St. Charles Saw the 1st Snake which was the kind usially termed the Garter Snake, Saw also a Beatle of black Colour with two red Stipes on his back passing each other Crosswise, from the but of the wing towards the extremity of the Same.

---

*According to Moulton (1986, 2: 186), Clark's Journal Codex C lists the river fall as "1 1/2 inches."
† According to Moulton (1986, 2: 186), Clark's Journal Codex C lists this temperature as "10 a 0."

SATURDAY, MARCH 10

	Sunrise			4 p.m.			Mississippi River	
Temp	Weather	Wind	Temp	Weather	Wind	Rise/Fall	Feet	Inches
14a0	c & f	NW	32a0	f	NW	R		2 ¹/₂

SUNDAY, MARCH 11

	Sunrise			4 p.m.			Mississippi River	
Temp	Weather	Wind	Temp	Weather	Wind	Rise/Fall	Feet	Inches
20a0	f	E	38a0*	f	SW	F		2 ¹/₂

MONDAY, MARCH 12

	Sunrise			4 p.m.			Mississippi River	
Temp	Weather	Wind	Temp	Weather	Wind	Rise/Fall	Feet	Inches
22a0	f	NE	24a0	f	NE	R		1 ¹/₂

TUESDAY, MARCH 13

	Sunrise			4 p.m.			Mississippi River	
Temp	Weather	Wind	Temp	Weather	Wind	Rise/Fall	Feet	Inches
16a0	f	NW	20a0	f	NW	F		1 ¹/₂

WEDNESDAY, MARCH 14

	Sunrise			4 p.m.			Mississippi River	
Temp	Weather	Wind	Temp	Weather	Wind	Rise/Fall	Feet	Inches
12a0	f	NE	18a0	f	NE	F		4 ¹/₂

*According to Moulton (1986, 2: 186), Clark's Journal Codex C lists this temperature as "20 a 0."

THURSDAY, MARCH 15

	Sunrise			4 p.m.		Mississippi River		
Temp	Weather	Wind	Temp	Weather	Wind	Rise/Fall	Feet	Inches
2a0	c & s	NW	48a0	r a s	NE	R		5

FRIDAY, MARCH 16

	Sunrise			4 p.m.		Mississippi River		
Temp	Weather	Wind	Temp	Weather	Wind	Rise/Fall	Feet	Inches
6a0	f	E	48a0	f	SSW	R		11

SATURDAY, MARCH 17

	Sunrise			4 p.m.		Mississippi River		
Temp	Weather	Wind	Temp	Weather	Wind	Rise/Fall	Feet	Inches
20a0	f	NE	46a0	f	NE	R		7

SUNDAY, MARCH 18

	Sunrise			4 p.m.		Mississippi River		
Temp	Weather	Wind	Temp	Weather	Wind	Rise/Fall	Feet	Inches
10a0	f	E	52a0	f	NE	F		3

MONDAY, MARCH 19

	Sunrise			4 p.m.		Mississippi River		
Temp	Weather	Wind	Temp	Weather	Wind	Rise/Fall	Feet	Inches
10a0	f	NE	60a0	f	SSW	F		2 1/2

Note   *The Eastern Field Note remarks begin again on this date and continue until the book is completed on the day the Expedition leaves camp on May 14, 1804.*

TUESDAY, MARCH 20

	Sunrise			4 p.m.		Mississippi River		
Temp	Weather	Wind	Temp	Weather	Wind	Rise/Fall	Feet	Inches
12a0	f	E	68a0	f	SSW	F		1 ½

Weather Diary Heard the 1st frogs on my return from St. Charles after having arrested the progress of Kickapoo war party.

WEDNESDAY, MARCH 21

	Sunrise			4 p.m.		Mississippi River		
Temp	Weather	Wind	Temp	Weather	Wind	Rise/Fall	Feet	Inches
34a0	f	SSW	54a0*	f	NW	F		2

Clark good W[eather] river rise

THURSDAY, MARCH 22

	Sunrise			4 p.m.		Mississippi River		
Temp	Weather	Wind	Temp	Weather	Wind	Rise/Fall	Feet	Inches
30a0	f	NW	48a0	f	NW	F		2

Clark butifull weath[er] river Missouries rise

FRIDAY, MARCH 23

	Sunrise			4 p.m.		Mississippi River		
Temp	Weather	Wind	Temp	Weather	Wind	Rise/Fall	Feet	Inches
22a0	f	NE	52a0	f	NE	R		4

Clark good weath[er] the [river] continu to rise— 10 Inches to day & 8 last night.

---

*According to Moulton (1986, 2: 186), Clark's Journal Codex C lists this temperature as "36 a 0."

SATURDAY, MARCH 24

	Sunrise			4 p.m.		Mississippi River		
Temp	Weather	Wind	Temp	Weather	Wind	Rise/Fall	Feet	Inches
14a0	f	E	60a0	f	SSW	R	1	5 ½

Clark   fair weather, river rise fast.

SUNDAY, MARCH 25

	Sunrise			4 p.m.		Mississippi River		
Temp	Weather	Wind	Temp	Weather	Wind	Rise/Fall	Feet	Inches
24a0	f	SSW	54a0	f	E	R		2

Weather Diary   Saw the 1st White Crain return

Clark   a fair morning, river rose 14 Inches last night. The musquetors are verry bad this evening.

MONDAY, MARCH 26

	Sunrise			4 p.m.		Mississippi River		
Temp	Weather	Wind	Temp	Weather	Wind	Rise/Fall	Feet	Inches
36a0	f	E	52a0	f	E	R		10

Weather Diary   the weather w[a]rm and fair

Clark   a verry Smokey day. The Mississippi R Continu to rise & discharge great quantity of foam

TUESDAY, MARCH 27

	Sunrise			4 p.m.		Mississippi River		
Temp	Weather	Wind	Temp	Weather	Wind	Rise/Fall	Feet	Inches
42a0	r & t	E	50a0	f a r	NE	R		7

Weather Diary   The buds of Spicewood appeared, the tausels of the mail Cotton wood were larger than a large Mulberry, and which the Shape and Colour of that froot, some of them had fallen from trees. the grass begins to Spring. The weather has been warm, and no falling weather until this time tho the atmispere has been verry Smokey and thick, a heavy fall of rain commenced which continued untill 12 at night, attended with thunder, and lightning— Saw large insects which resembled Musquitors, but doubt whether they are really those insects or the fly which produces them, they attempted to bite my horse, but I could not observe that they made nay impression with their Beaks.

Clark   rain last night verry hard with thunder, a Cloudy morning. River continue to rise.

WEDNESDAY, MARCH 28

Sunrise			4 p.m.			Mississippi River		
Temp	Weather	Wind	Temp	Weather	Wind	Rise/Fall	Feet	Inches
42a0	c	NE	52a0	c	E	R		5 ½

Weather Diary   day cloudy & warm

Clark   a Cloudy morning.

THURSDAY, MARCH 29

Sunrise			4 p.m.			Mississippi River		
Temp	Weather	Wind	Temp	Weather	Wind	Rise/Fall	Feet	Inches
28a0	r a t	NE	38a0	h & r	NE	R		1

Clark   Rained last night, a violent wind from the N this morning with rain, Some hail. A blustering day all day. River Continue to rise, Cloudy Day.

FRIDAY, MARCH 30

Sunrise			4 p.m.			Mississippi River		
Temp	Weather	Wind	Temp	Weather	Wind	Rise/Fall	Feet	Inches
	c a r	NW		f	NW	R		2

Clark   a fair day

SATURDAY, MARCH 31

	Sunrise			4 p.m.		Mississippi River		
Temp	Weather	Wind	Temp	Weather	Wind	Rise/Fall	Feet	Inches
	f	NW		*f*	NWW*	R		2

Weather Diary   Windey

Clark   a fine morning

SUNDAY, APRIL 1

	Sunrise†			4 p.m.		Mississippi River		
Temp	Weather	Wind	Temp	Weather	Wind	Rise/Fall	Feet	Inches
	f	NE		*f*	NE	R		2 1/2

Weather Diary   The Spicewood is in full bloe, the dogs tooth violet, and may apple appeared above ground, a northern light appeared at 10 o C P.M. verry red.

Clark   a fair morning. A northern Light Seen at about 10 oClock, & frequently Changing Coler, appearing as in the atmusfier &c.

MONDAY, APRIL 2

	Sunrise			4 p.m.		Mississippi River		
Temp	Weather	Wind	Temp	Weather	Wind	Rise/Fall	Feet	Inches
16a0	*f*			*f*	NE	R		3 1/2

---

*According to Moulton (1986, 2: 209), Clark's Journal Codex C lists this wind direction as "NW."

†According to Moulton (1986, 2: 209), "The temperature readings in (Clark's Journal) Codex C are generally eight degrees below those in the Weather Diary (Lewis's Journal), as in March, but some discrepancies are noted within the table."

TUESDAY, APRIL 3

Sunrise			4 p.m.			Mississippi River		
Temp	Weather	Wind	Temp	Weather	Wind	Rise/Fall	Feet	Inches
50a0	f	NE		r	NE	R		3 ½

Weather Diary   a cloudy day.

Clark   wind blew verry hard all night   Some rain

WEDNESDAY, APRIL 4

Sunrise			4 p.m.			Mississippi River		
Temp	Weather	Wind	Temp	Weather	Wind	Rise/Fall	Feet	Inches
52a0	c a r	NW				R		11

Clark   hard wind and rain last night

THURSDAY, APRIL 5

Sunrise			4 p.m.			Mississippi River		
Temp	Weather	Wind	Temp	Weather	Wind	Rise/Fall	Feet	Inches
32a0	c a r	NE		t r		R		2

Weather Diary   the buds of the peaches, apples & Cherrys appear — wind high

Clark   Thunder & lightning last night. The wind is violintly hard from the WNW all day. River Still rise. The [wind] Shift to the North at Sun Set & Cold. Banks falls in.

FRIDAY, APRIL 6

	Sunrise			4 p.m.		Mississippi River		
Temp	Weather	Wind	Temp	Weather	Wind	Rise/Fall	Feet	Inches
26a0	c r	NW		s a r		F		4 ¹/₂

Weather Diary   a large flock of Pellicans appear.

Clark   a Cloudy day river fall 10 Inches. At one oClock the wind bley [blew] hard from the NW, in this Countrey the windy points for rain & Snow is from SE to NE, the fair weather winds SW & West, Clear and Cold from the NW & N, wind Seldom blows from the South— at about 9 oClock PM began to Snow and Continued a Short time, wind blew hard from the N West.

SATURDAY, APRIL 7

	Sunrise			4 p.m.		Mississippi River		
Temp	Weather	Wind	Temp	Weather	Wind	Rise/Fall	Feet	Inches
18a0	f a c	NW		c		F		2

Weather Diary   the leaves of Some of the Apple trees have burst their coverts and put foth —,. maney of the wild plants have Sprung up and appear above ground. cold air.

SUNDAY, APRIL 8

	Sunrise			4 p.m.		Mississippi River		
Temp	Weather	Wind	Temp	Weather	Wind	Rise/Fall	Feet	Inches
18a0*	c	NE		c r		F		2 ¹/₂

---

*According to Moulton (1986, 2: 209), Clark's Journal Codex C lists this temperature as "10 a 0."

MONDAY, APRIL 9

	Sunrise			4 p.m.			Mississippi River		
Temp	Weather	Wind	Temp	Weather	Wind	Rise/Fall	Feet	Inches
26a0	f a c	NE		c		F		2

Weather Diary   windey

TUESDAY, APRIL 10

	Sunrise			4 p.m.			Mississippi River		
Temp	Weather	Wind	Temp	Weather	Wind	Rise/Fall	Feet	Inches
18a0	f	NW		f		F		6 1/2

Weather Diary   no appearance of the buds of the Osage Aple, the Osage Plumb has put forth their leaves and flower buds. tho it is not completely in bloe. windey

WEDNESDAY, APRIL 11

	Sunrise			4 p.m.			Mississippi River		
Temp	Weather	Wind	Temp	Weather	Wind	Rise/Fall	Feet	Inches
18a0	f	NE		f		F		7 1/2

Weather Diary   windey

THURSDAY, APRIL 12

	Sunrise			4 p.m.			Mississippi River		
Temp	Weather	Wind	Temp	Weather	Wind	Rise/Fall	Feet	Inches
24a0	c	NW		f a c		F		7

Weather Diary   windey

FRIDAY, APRIL 13

	Sunrise			4 p.m.			Mississippi River		
Temp	Weather	Wind	Temp	Weather	Wind	Rise/Fall	Feet	Inches
34a0	c	NE		c		F		6 ½

Weather Diary   The peach trees are partly in blume   the brant, Geese, Duck, Swan, Crain and other aquatic birds have disappeared verry much, within a fiew days and have gorn further North I pr[e]sume. the Summer duck raise their young in this neighborhood and are now here in great numbers. windey

Clark   a Cloudy day. After part of the day fair. River falling

SATURDAY, APRIL 14

	Sunrise			4 p.m.			Mississippi River		
Temp	Weather	Wind	Temp	Weather	Wind	Rise/Fall	Feet	Inches
30a0	f	SW		f		F		5

Weather Diary   windey

Clark   a fair day, wind high from the [blank]

SUNDAY, APRIL 15

	Sunrise			4 p.m.			Mississippi River		
Temp	Weather	Wind	Temp	Weather	Wind	Rise/Fall	Feet	Inches
30a0	f	NW				F		6 ½

Weather Diary   windey

Clark   a fair morning. Clouded up at 12 oClock, the wind from the SW blew verey hard. A Boat pass up the Mississippi under Sail at 1 oClock—

MONDAY, APRIL 16

	Sunrise			4 p.m.		Mississippi River		
Temp	Weather	Wind	Temp	Weather	Wind	Rise/Fall	Feet	Inches
44a0	c	NW		f a c		F		5 1/2

Weather Diary   Windey

Clark   a fair morning, Some rain last night & hard wind from the SW by W. Wind verry hard.

TUESDAY, APRIL 17

	Sunrise			4 p.m.		Mississippi River		
Temp	Weather	Wind	Temp	Weather	Wind	Rise/Fall	Feet	Inches
34a0	f a c	NW		f		F		5

Weather Diary   wind verry high every day Since the 3rd instant   Some frost to day   Peach trees in full Bloome, the Weaping Willow has put forth its leaves and are 1/5 of their Sise, the Violet the Doves foot, & cowslip are in [bloom], the dogs tooth violet is not yet in blume. The trees of the forest particularly the Cotton wood begin to obtain from their Size of their buds a Greenish Cast at a distance—   the Gooseberry which is also in this country and lilak have put forth their leaves—   frost

Clark   The after part of this day cool.

WEDNESDAY, APRIL 18

	Sunrise			4 p.m.		Mississippi River		
Temp	Weather	Wind	Temp	Weather	Wind	Rise/Fall	Feet	Inches
26a0*	f a c	NNW†		c		F		3

Weather Diary   Windy Day   at St. Louis

Clark   a fair morning. Vegetation appears to be Suppriseingly rapid for a fiew days past. The wind from the SE,   rained the greater Part of this night.

---

* According to Moulton (1986, 2: 209), Clark's Journal Codex C lists this temperature as "16 a 0."

† According to Moulton (1986, 2: 209), Clark's Journal Codex C lists this wind direction as "NNW."

THURSDAY, APRIL 19

	Sunrise			4 p.m.		Mississippi River		
Temp	Weather	Wind	Temp	Weather	Wind	Rise/Fall	Feet	Inches
42a0	r	SSE				F		4

Clark   a rainy morning, Thunder and lightning at 1 oClock. Rain at 2, rain Continue.

FRIDAY, APRIL 20

	Sunrise			4 p.m.		Mississippi River		
Temp	Weather	Wind	Temp	Weather	Wind	Rise/Fall	Feet	Inches
42a0	c r	SE	45a0	r	SE	F		3 1/2

Clark   rain last night & this morning. The river fall Sloly. Rain all day. Dark Sultrey weather, Some Thunder

SATURDAY, APRIL 21

	Sunrise			4 p.m.		Mississippi River		
Temp	Weather	Wind	Temp	Weather	Wind	Rise/Fall	Feet	Inches
39a0	r	SW	50a0	f a r	W	R	1	2

Clark   rain all last night Slowley, a Cloudy morning, some rain. River raised last night 12 Inches.

SUNDAY, APRIL 22

	Sunrise			4 p.m.		Mississippi River		
Temp	Weather	Wind	Temp	Weather	Wind	Rise/Fall	Feet	Inches
36a0	c	NW	42a0	c	NW	R	1	6

MONDAY, APRIL 23

	Sunrise			4 p.m.			Mississippi River		
Temp	Weather	Wind	Temp	Weather	Wind	Rise/Fall	Feet	Inches
30a	f	NW	72a	f	W	F		1

TUESDAY, APRIL 24

	Sunrise			4 p.m.			Mississippi River		
Temp	Weather	Wind	Temp	Weather	Wind	Rise/Fall	Feet	Inches
44a	f	NW	52a	f	NW	R		8

WEDNESDAY, APRIL 25

	Sunrise			4 p.m.			Mississippi River		
Temp	Weather	Wind	Temp	Weather	Wind	Rise/Fall	Feet	Inches
34a	f	NW	46a	c	NW	R		2 $^{1}/_{2}$

THURSDAY, APRIL 26

	Sunrise			4 p.m.			Mississippi River		
Temp	Weather	Wind	Temp	Weather	Wind	Rise/Fall	Feet	Inches
24a	f	NW	66a	f	NW	F		6

Weather Diary   The white frost Killed much froot near Kahokia, while that at St. Louis escaped with little injurey—

Clark   river falls

FRIDAY, APRIL 27

Sunrise			4 p.m.			Mississippi River		
Temp	Weather	Wind	Temp	Weather	Wind	Rise/Fall	Feet	Inches
36a	t l r	W	70a	f	SW	F		8

SATURDAY, APRIL 28

Sunrise			4 p.m.			Mississippi River		
Temp	Weather	Wind	Temp	Weather	Wind	Rise/Fall	Feet	Inches
38a	f	NW	72a	f	NW	F		7

Clark   river fall

SUNDAY, APRIL 29

Sunrise			4 p.m.			Mississippi River		
Temp	Weather	Wind	Temp	Weather	Wind	Rise/Fall	Feet	Inches
40a	f	NW	60a	f	SE	F		7

Clark   river Still fall

MONDAY, APRIL 30

Sunrise			4 p.m.			Mississippi River		
Temp	Weather	Wind	Temp	Weather	Wind	Rise/Fall	Feet	Inches
26a	f	SE	64a	f	NE	F		6

Weather Diary   white frost, Slight did but little injurey—.
Clark   a fair day. River Still fall.

Note *In early May, the expedition prepares to leave Camp Dubois. According to Moulton (1986, 2: 216–217), the tabular data for May "is based on Lewis's Weather Diary, kept by Lewis. Its temperature readings are consistently 8 degrees above those in Clark's Codex C."*

TUESDAY, MAY 1

	Sunrise			4 p.m.			Mississippi River	
Temp	Weather	Wind	Temp	Weather	Wind	Rise/Fall	Feet	Inches
28a	f	SE	62a	f	NE	F		4 ½

Clark   Some fog this morning

WEDNESDAY, MAY 2

	Sunrise			4 p.m.			Mississippi River	
Temp	Weather	Wind	Temp	Weather	Wind	Rise/Fall	Feet	Inches
27a	f	SE	76a	f	SSE	F		6

THURSDAY, MAY 3

	Sunrise			4 p.m.			Mississippi River	
Temp	Weather	Wind	Temp	Weather	Wind	Rise/Fall	Feet	Inches
32a	f	SSE	80a	f	SSW	F		4 ½

Clark   Some wind. River falling

FRIDAY, MAY 4

	Sunrise			4 p.m.			Mississippi River	
Temp	Weather	Wind	Temp	Weather	Wind	Rise/Fall	Feet	Inches
48a	t l c r	S	64a	c a r	S	R		2

Clark   a rainey Day. The river riseing a little.

SATURDAY, MAY 5

Sunrise			4 p.m.			Mississippi River		
Temp	Weather	Wind	Temp	Weather	Wind	Rise/Fall	Feet	Inches
50a	t l r	W	66a	c a r	W	R		2 1/2

Weather Diary The thunder and lightning excessively [hard] this morning
Clark a Cloudy day   rains at different times.

SUNDAY, MAY 6

Sunrise			4 p.m.			Mississippi River		
Temp	Weather	Wind	Temp	Weather	Wind	Rise/Fall	Feet	Inches
42a	f	SW	78a	f	SW	F		2 1/2

Clark a fair day

MONDAY, MAY 7

Sunrise			4 p.m.			Mississippi River		
Temp	Weather	Wind	Temp	Weather	Wind	Rise/Fall	Feet	Inches
46a	f	SE	60a	f	SSW	F		4 1/2

Clark a fair day.

TUESDAY, MAY 8

Sunrise			4 p.m.			Mississippi River		
Temp	Weather	Wind	Temp	Weather	Wind	Rise/Fall	Feet	Inches
52a	f	NE	70a	f	SW	F		4

Clark Verry hot day

WEDNESDAY, MAY 9

	Sunrise			4 p.m.			Mississippi River		
Temp	Weather	Wind	Temp	Weather	Wind	Rise/Fall	Feet	Inches
50a	f	E	84a	f	SW	F		2

Clark   a fair day, warm. I send to the Missouries water for drinking water, it being much Cooler than the Mississippi.

THURSDAY, MAY 10

	Sunrise			4 p.m.			Mississippi River		
Temp	Weather	Wind	Temp	Weather	Wind	Rise/Fall	Feet	Inches
54a	c	NE	75a	f	NW	F		3 1/2

Weather Diary   distant thunder, sutery this evening
Clark   Some rain last night, Cloudy morning verry hot, in the after part of the day.

FRIDAY, MAY 11

	Sunrise			4 p.m.			Mississippi River		
Temp	Weather	Wind	Temp	Weather	Wind	Rise/Fall	Feet	Inches
48a	f	E	78a	f	SW	F		2 1/2

Clark   a warm morning. In the evengn. about 4 oClock a violent gust from the NW by W

SATURDAY, MAY 12

	Sunrise			4 p.m.			Mississippi River		
Temp	Weather	Wind	Temp	Weather	Wind	Rise/Fall	Feet	Inches
44a	f	E	80a	f	W	F		3

Weather Diary   the wind at 4 was uncomly hard.
Clark   rain all the evening

SUNDAY, MAY 13

Sunrise			4 p.m.			Mississippi River		
Temp	Weather	Wind	Temp	Weather	Wind	Rise/Fall	Feet	Inches
50	c a r	W	48a	c a r	NW	F		2

Clark   a rainey Day.

Note   *May 14 marks the last observation of the Mississippi River at the mouth of the Wood River. No weather observations in tabular form have been found for the rest of the month, as the expedition commences its journey up the Missouri River (Moulton 1986, 2: 216–217).*

MONDAY, MAY 14

Sunrise			4 p.m.			Mississippi River		
Temp	Weather	Wind	Temp	Weather	Wind	Rise/Fall	Feet	Inches
42a	c	SE	64	f	N	F		0

Clark   a Cloudy morning. Rain at 9 oClock. Rained the greater part of the day. Rained. I refur to the Comsmt. [commencement] of my Journal No. 1.

Note   *Thus Clark ends his recording of weather in the Camp Dubois Field Notes. This marks the end of the "Dubois Journal."*

Section 3 Ascending the Missouri River

May 14 to November 1, 1804

Most of the expedition (between 45 and 50 members) started out on a rainy May 14, 1804, crossing the Mississippi to begin the ascent of the Missouri River. Lewis joined the party in St. Charles, Missouri, where the party remained until May 21. The expedition moved slowly against the current, passing present-day Kansas City, Missouri, in late June and Omaha, Nebraska, by late July under the sweltering summer heat. They traveled westward parallel to the Nebraska–South Dakota border for 10 days in late August through early September, and then turned north into central South Dakota, constantly buffeted by prairie winds. Their first frost blanketed them on October 5 as they neared the South Dakota–North Dakota state line. By mid-October they experienced their first snows near present-day Bismarck, North Dakota, and established winter quarters at Camp Mandan on November 2. Here they would endure the rigors of prairie winter, including the coldest temperature of the year.

As in the previous section, each day's entry begins with the data from the Observation Table (when available), followed by the remarks (when available). These are followed by excerpts from expedition members' daily narrative journals. The systematic daily entries of daily narrative journals by Lewis, Clark, and the army sergeants and privates began at this point in the expedition; however, not every journalist noted weather, water, or climate data each day.

Weather data included in this section are from Coues (1893, 3: 1267–1285), Moulton (1986, 2: 169–217 and 1987, 3: 223), and Thwaites (1904, 6, part 2: 173–177). Note that no weather observations in table form have been found for May 14 through September 18, 1804. Where there are no data, there are no tables in this volume. Remarks were entered sporadically during this time, however, and are included.

Different journals and notebooks were used during the expedition. For a more detailed explanation on the journals and entry practices consult Cutright (1976) and Moulton (1986, 2: 8–48, 530–567).

Note *In late May, the expedition leaves Camp Dubois, crosses the Mississippi River, enters the mouth of the Missouri River, and proceeds a couple of miles upriver. Clark's entry on May 14 is the first entry of the "Lewis and Clark Journals." These data are listed in Clark's Journal No. 1, and is now designated Codex A. Lewis makes his first entry related to weather and hydrology in the travel diaries the next day.*

Monday, May 14

Clark Set out from Camp River a Dubois at 4 oClock PM and proceeded up the Missouris under Sail. Cloudy rainey day. Wind from the NE. Rained the fore part of the day proceeded under a jentle brease up the Missouri a heavy rain this after-noon. Wind from NE.

Floyd Showery Day

Whitehouse hard Shower of rain. Hoisted sail having a fair wind from the SE and rain. The river Missouri is about one Mile wide…and its waters are always muddy… the current Runs at about five Miles & a half p hour.

TUESDAY, MAY 15

Lewis   It rained during the greater part of last night and continued untill 7 oCk. AM. The evening was fair.

Clark   rained all last night and this morning untill 7 oClock, all our fire extinguished. a fair after noon   Wind from the NE   the water excessively rapid & banks falling in.

Floyd   Rainey mornig   fair wind the Later part of the day   Sailed som and encamped on the N side.

Gass   It rained in the morning; but in the afternoon we had clear weather

Ordway   rainy morning. fair wind later part of the day

Whitehouse   This morning early we set sail, in a hard rain. The current Swift & water muddy. The latter part of the day proved clear.

Note   *The expedition camps near St. Charles, Missouri.*

WEDNESDAY, MAY 16

Weather Diary   arrived at St. Charles

Clark   a fair morning. A fine day.

Gass   We had a fine pleasant morning   This evening was showery

Ordway   this morning plesant

Whitehouse   a clear morning

THURSDAY, MAY 17

Clark   a fine fair day. Measured the river found it to be 720 yards wide.

Floyd   a fair but Rainey Night.

Ordway   a fair day, but Rainy night, nothing occured worthy of notice this day

Whitehouse   a pleasant morning. In the evening we had some rain—

FRIDAY, MAY 18

Clark   a fine morning. The wind hard from the SW.

Whitehouse   This morning was fair and pleasant.

SATURDAY, MAY 19

Clark   a Violent Wind last night from the WSW accompanied with rain which lasted about three hours. A Cloudy morning, Cleared away this morn'g at 8 oClock.

Floyd   a Rainey day

Ordway   a Rainy day

Whitehouse   This morning proved Rainey & wet

SUNDAY, MAY 20

Weather Diary   rained the after part of the Day

Lewis   The morning was fair, the weather was p[l]easant. At half after one PM our progress was interrupted the near approach of a violent thunder storm from the NW and concluded to take shelter in a little cabbin hard by untill the rain should be over; accordingly we alighted and remained about an hour and a half and regailed ourselves with a [cold] collation which we had taken the precaution to bring us from St. Louis. The Clouds continued to follow each other in rapid succession, insomuch that there was but little prospect of [its] ceasing to rain this evening; I set forward in the rain...to St. Charles. During the fore part of this day it rained excessively hard.

Clark   a Cloudy morning rained and a hard wind last night. At 3 oClock Capt. Lewis & [party] of St. Louis arrived thro a violent Shoure of rain. Rained the greater part of this evening

Note   *The expedition leaves St. Charles, Missouri, and proceeds up the Missouri River.*

MONDAY, MAY 21

Weather Diary   leave St. Charles   heavy rain in the evening with wind. great number of muscators

Clark   after 3 oClock...proceed on under a jentle Breese, at one mile a Violent rain with Wind from the SW. Camped, Soon after it commenced raining & continued the greater part of the night...which lasted with short intervales all night. Rains powerfully.

Floyd   Showery

Ordway   at 4 oClock PM Showery

Whitehouse   This morning we had some rain. We found the current of the River very rapid, the Banks steep, & the bottom very miry.

TUESDAY, MAY 22

Clark   a Cloudy morning. Rained Violently hard last night.

Floyd   Set out after a verry hard Rain   plesent day

Gass   we proceeded on our voyage with pleasant weather

Whitehouse   a fair morning. The current rapid.

WEDNESDAY, MAY 23

Clark   the water excessively Swift to day. A fair evening.

Whitehouse   a fair weather pleasant morning

THURSDAY, MAY 24

Clark   ...bank which was falling in so fast that the evident danger obliged us to cross between the starbd side and a sand bar in the middle of the river. The Swiftness of the Current wheeled the boat, Broke our Toe rope.

Whitehouse   a fair weather morning. Current of the river Swift and strong

Note   *The expedition passes La Charette, in Warren County, the westernmost "white" settlement on the Missouri River.*

FRIDAY, MAY 25

Weather Diary   strawbury in the praries ripe and abundant

Clark   rain last night. River fall Several inches. Some hard rain this evening.

SATURDAY, MAY 26

Clark Set out at 7 oClock after a hard rain & Wind (after a heavy Shour of rain), & proceed on verry well under Sale. Wind from the ENE. The wind favourable to day we made 18 miles a Cloud rais & wind & rain Closed the Day.

Floyd hard thunder and Rain this morning

Gass At seven we embarked and had loud thunder and heavy rain

Ordway hard thunder & rain this morning

Whitehouse a fair fine morning, clear. towards evening we had some rain and Thunder. The current still running rapid—

SUNDAY, MAY 27

Weather Diary serviceburries or wild Courants, ripe and abundant

Clark …a gentle Breese from the SE. Gasconade River…is 157 yard wide, a butifull stream of clear water. 19 feet Deep.

Whitehouse a fine fair weather morning

MONDAY, MAY 28

Clark rained hard all the last night   Some wind from the SW. Some thunder & lightning hard wind in the forepart of the night from the S.W. River begin to rise. this day So Cloudy that no observations could be taken

Whitehouse a fair pleasant morning. There was a Small Spring in a cave. It is the most remarkable cave I ever Saw, in my travels.

TUESDAY, MAY 29

Clark rained last night, the river rises fast. Cloudy morning. The Musquetors are verry bad. Rain all night. river still rised, water verry muddey

Floyd Rain Last night

Whitehouse this morning being clear

WEDNESDAY, MAY 30

Weather Diary Mulburies being to ripen, very abundant in the bottom

Clark Set out between 6 and 7 oClock after a heavy Shower of rain, rained all last night. Camped. A heavy wind accompanied with rain & hail. The river Continue to rise.

Floyd Set out after verry hard rain Last night   Rained all the with thunder and hail   Set out 7 ock after a very hard Rain and thunder   it Rained During the Gratiest part of the day with hail

Gass After experiencing a very disagreeable night, on account of the rain, we continued our voyage at 7 o'clock AM   At twelve we had a heavy shower of rain, accompanied with hail

Orwday we Set out at 7 oClock AM after a hard rain, rained all last night. At 12 oClock a hard Shower of rain & hail

Whitehouse a fair morning. About noon it began to rain.

THURSDAY, MAY 31

Clark rained the greater part of the last night, the wind from the West raised and blew with great force untill 5 oClock PM which obliged us to lay by.

Floyd   it Rained and Cleard up

Gass   We were obliged to remain at this encampment all day, on account of a strong wind from the west.

Ordway   we lay at panther creek on acct. of a hard wind from N. West

Whitehouse   a fair morning. The wind blowing hard, high wind.

Note   *On June 1–2, the expedition camps near the confluence of the Missouri and Osage rivers, several miles east of Jefferson City, the present-day capital of Missouri. According to Moulton (1986, 2: 336) "No tables of weather observations for June 1804 have been found, but both captains entered a few remarks, Clark in Codex C and Lewis in his Weather Diary. These are observations of natural phenomena related to seasonal change and climate such as plant ripening and animal migrations, rather than weather data as such."*

FRIDAY, JUNE 1

Clark   a fair morning. The wind a head from the West. Current exceeding rapid and banks falling in. A fair afternoon. Fell a number of trees in the Point to take observation, it up untill 1 oClock to take Som observations &c.

Floyd   the wind from the west   the day Clear   the day Clear wind from the west   water strong

Whitehouse   the weather being pleasant this morning

SATURDAY, JUNE 2

Clark   the Osage R 397 yds wide.

Floyd   the day was Clear wind from the South

Ordway   the width of the Missouri at this place is 875 yards wide, the Osage River 397 yards

Whitehouse   the Missourie is at this place 875 yards wide

Note   *On June 3–4, the expedition passes the Jefferson City area.*

SUNDAY, JUNE 3

Clark   the fore part of the day fair. I attempted to take equal alltitudes…but was disapointed, the Clouds obsured the Sun. We set out at 5 oClock PM   Cloudy & rain.

Floyd   the Latter part Clouday with thunder and Rain wind from Est.

Whitehouse   a fair clear morning

MONDAY, JUNE 4

Clark   a fair Day.

Floyd   the Clear morning   Strong water

Whitehouse   a fair weather morning

TUESDAY, JUNE 5

Clark   …we had a fine fair wind, but could not make use of it, our Mast being broke

Floyd   fair day   Water strong

Whitehouse   a fair morning.

WEDNESDAY, JUNE 6

Clark   Set out at 7 oClock, under a Jentle Braise from SE by S. River rose last night a foot. Some wind in the after part of to day from the SE. The Banks are falling in greatly in this part of the river.

THURSDAY, JUNE 7

Ordway   fair day & f. wind

FRIDAY, JUNE 8

Clark   Camped…Commenced raining Soon after we Came to which prevented th party Cooking their provisions—   rained this night

Floyd   the day Clear wind from the west   Land well timberd

Ordway   some rain   the Country on the right is verry fine—

Whitehouse   had strong Watter to Goe throug

SATURDAY, JUNE 9

Clark   a fair morning. The river rise a little. Water verry swift. A Wind from the S at 4 oClock. Some dift & Snags…this was a disagreeable and Dangerous Situation, particularly as immense large trees were Drifting down. I can Say with Confidence that our party is not inferior to any that was ever on the waters of the Missoppie. Camped…the wind from the SW. The river continue to rise Slowly,   Current excessive rapid. Comminced raining at 5 oClock and continued by intervales the greater part of the night.

Floyd   Set out after a verry hard Rain Last night the morning Clear wind from the Est. The Latter part of the day Clouday with rain

Ordway   Rain last night

Whitehouse   The day proving stormy

SUNDAY, JUNE 10

Weather Diary   rasberreis perple, ripe and abundant

Clark   Some hard rain last night. Passed a part of the River that the banks are falling in takeing with them large trees of Cotton woods which is the Common groth in the Bottoms Subject to the flud. The current is excessively Swift. Wind from the NW. The evening is Cloudy, our party in high Spirits.

Floyd   the Current is Strong about this place   day Clear wind from the NW

Gass   and remained there the whole of the next day, the wind blowing too violent for us to proceed

Ordway   we Set eairly after Some rain, a fair day

Whitehouse   the wind blowing hard

MONDAY, JUNE 11

Weather Diary   many small birds are now setting   some have young, the whiperwill setting

Clark   as the wind blew all this day hard & cold from the NW which was imedeately a head we Could not Stur, but took the advantage of the Dealy and Dried our wet articles. The river beginning to fall

Floyd   Day Clear wind from the N West   Lay by all Day on account of the wind   the
Latter part of the day Clouday

Ordway   we lay by on acct. of the wind Blowing hard from the NW

Whitehouse   the wind blew So strong in the morning that the Commanding Officer halted
there that day.

TUESDAY, JUNE 12

Floyd   the day Clear wind from the west

Ordway   a fair (day) morning

Whitehouse   with a fair wind and fine weather, all hands being well and in high spirits; we
found the current of the River still running strong.

WEDNESDAY, JUNE 13

Clark   a fair Day. Took Some Looner Observations which Kept Cap L. & me Self up untill
half past 11 oClock.

Ordway   fair morning

Whitehouse   the Currant Being Rappid Neerly Swept the men of their legs while Bearing her
up

THURSDAY, JUNE 14

Clark   We set out at 6 oClock after a thick fog

Floyd   day Clear water strong

Gass   The river having risen during the night was difficult to ascend.

Ordway   foggy but fair day

Whitehouse   the river rose

FRIDAY, JUNE 15

Clark   the river is riseing fast & the water exceedingly Swift. The Current was So Strong
that we Could not Stem it with our Sales under a Stiff breese in addition to our ores.
This evening the river beginning to fall

Floyd   Strong water

Ordway   a fair day

Whitehouse   a fair fresh wind from the SE. We Crouded Sail and Saild 16 miles

SATURDAY, JUNE 16

Weather Diary   the wood duck now has [its] young, this duck is abundant, and except one
Solatary Pelican and a few gees these ducks were the only aquatic fowls we
have yet seen

Clark   some rain this morning. Heavy rain came on & lasted a Short time

Floyd   day Clouday with rain   water verry Strong

Gass   had cloudy weather and rapid water all day

Ordway   the Current is verry Strong all this day, So that we were obledged to waid & Toe
the boat over sand bars

Whitehouse   The wind rose from the South East; and we set all our Sails, we found the
current running very Strong   towards the evening, the wind lull'd & died
away.

SUNDAY, JUNE 17

Clark   Cloudy morning   Wind from the SE
Gass   This morning was clear

MONDAY, JUNE 18

Clark   Some rain last night, and some Showers this morning which delay our work verry
much. Heavy rain all the fore part of the day. The misquiter verry bad.
Floyd   Clouday with Rain and Thunder and wind from the Est.
Ordway   hard Rain this morning
Whitehouse   We remain'd here this day, in the forenoon of which we had severe thunder &
lightning after which succeeded a Violent rain, in the afternoon it cleared up.

TUESDAY, JUNE 19

Clark   rain last night. Arrange everry thing and Set out 8 oCk wind in favor with a jentle
breese from the SE
Floyd   day Clouday   wind from the Est. Strong water
Ordway   we set out at 9 oC with a fair wind   the water So Swift that we were obledged to
hole the Boat by a Rope
Whitehouse   the weather being fine & clear, about 9 oClock AM a good Brees sprung up
from the South East, We set sail

WEDNESDAY, JUNE 20

Clark   Set out after a heavy Shower of rain and proceeded   A gentle breese from the SW.
Passed Som verry Swift water to day. We took Some Loner [luner] observations,
which detained us untill 1 oClock a butifull night but the air exceedingly Damp &
the mosquiters verry troublesom
Floyd   Clouday day Rain Strong water
Ordway   we Set out at 5 oC and after some rain passed Tiger Creek on the N. side
Whitehouse   This morning as we were preparing to start a Rain came on which detained us,
for some time, in about an hour the weather got clear. The Currant was Strong

THURSDAY, JUNE 21

Clark   The river rose 3 Inches last night. Some wind from the SSE at 3 oClock. At Sun Set
the atmespier presented every appearance of wind, Blue & white Streeks Centering at
the Sun as She disappeared and the Clouds Situated to the SW, Guilded in the most
butifull manner. At Sun Set The ellement had every appearance of wind
Floyd   Clear day   strong water
Gass   we had rapid water
Whitehouse   the Current setting so strong against us

FRIDAY, JUNE 22

Clark   River rose 4 Inches last night. I was waken'd (at) before day light this morning by the
guard prepareing the boat to receive an apparent Storm which thretened violence
from the West. At day light a Violent gust of wind accompanied with rain cam from
the West and lasted about one hour, it Cleared away and we Set out and proceeded
on under a gentle breeze from the NW. Ferenthiers (Ferents) Thermometr at 3 oClock
PM 87 d which is 11 d above Summer heat. Passed Some verry Swift water Crouded
with Snags

Floyd   Set out at 7 oclock after a verry hard Storm   thunder (wind from the NE) and Rain wind [WC: from the West, proceeded on under a gentle breeze from the NW] Strong water

Gass   It rained hard from four to seven in the morning

Ordway   we Set out at 7 oC after a hard Shower of rain & high wind from NW Thunder and lightning &.C— the day fair

Whitehouse   This morning we were detained from starting by a heavy Rain which continued till 7 oClock AM. The weather proved excessive hot, as it was the two succeeding days, & being by far the warmest weather that we met with for a long time. The current running very strong against us, and having to tow the boat it can hardly be imagined the fatague that we underwent

SATURDAY, JUNE 23

Clark   Some wind this morning from the NW. The wind blew hard and down the river which prevented the Pty moveing from this Island the whole day. Evening…the wind continueing to blow prevented their moveing. The river fell 8 Inches last night.

Floyd   a Small Brese from the NW (Set out   day Clouday)

Gass   at 12, the wind blew so strong down the river that we were unable to proceed

Ordway   Some wind this morning from the NW   we Set out at 7 OC   the wind Raised

Whitehouse   the wind arose and blew ahead of us, which render'd our towing the boat extreme difficult & fataigueing;   It blew so hard that Captain Clark who was on shore could not come off to us—

SUNDAY, JUNE 24

Clark   good water. Wind blowing So hard Down the river. The Party in high Spirits.

Floyd   wind from the NE Sailed   Day Clear

Ordway   a fair day

Whitehouse   at 12 oClock AM we stopped to Jerk our meat, the weather being so warm that we were afraid it would spoil

MONDAY, JUNE 25

Clark   a heavy thick fog detained us about an hour untill 8 oClock. The wind from the NW. Hard water & logs, Bank falling in. The river falling fast about 8 Inches in 24 hours. The river is still falling.

Floyd   we Set out at 8 oClock after the Fogue was Gon.

Gass   The morning was foggy and at seven o'clock we pursued our voyage. The river here is narrow

Ordway   a foggy morning. It Detained us about an hour.

Whitehouse   the current still running strong

Note   *The expedition arrives at the confluence of the Kansas and Missouri rivers, at present-day Kansas City, Missouri.*

TUESDAY, JUNE 26

Clark   wind from the SW. The river falling a little. We Killed a large rattle Snake, Sunning himself in the bank

Ordway   Swift water this afternoon [near present-day Kansas City, Missouri]

Whitehouse   the morning was fine and clear. The water was strong at the head of the Island we Campd on   The day proving extreamly warm which still added to our fataigue

WEDNESDAY, JUNE 27

Clark   a fair warm morning. The river rose a little last night. The Kansas River by an angle and made it 230 yds wide, it is wider above the mouth   the Missouries at this place is about 500 yards wide.

Floyd   day Clouday

Ordway   The Kansas River is 230 yrds wide at the mouth

THURSDAY, JUNE 28

Clark   examining our Provisions we found Several articles Spoiled from the wet or dampness they had received, a verry warm Day, the wind from the South, the river Missourie has raised yesterday last night & to day about 2 foot. This evening it is on a Stand. The waters of the Kansas is verry disigreeably tasted to me.

Ordway   pleasant   The width of the M. here is 500 yd. wide.

FRIDAY, JUNE 29

Clark   obsvd. the distance of sun and moon, took Equal & maridinal atld. SW wind.

SATURDAY, JUNE 30

Clark   came to at 12 oClock & rested three hours, the [sun?] being hot the men becom verry feeble, Farnsts. Thermometer at 3 oClock Stood at 96° above 0

Gass   The day was clear

Whitehouse   the water was Strong

Note   *The expedition camps on islands near present-day Fort Leavenworth on the Kansas–Missouri border.*

SUNDAY, JULY 1

Clark   The river still falling a little, a verry warm Day. Capt. Lewis took Medn. Altitude & we delayed three hours, the day being excessively hot. The wind from SW.

Floyd   Clear Day

Ordway   the Day is exceding hot. So we Stoped at 12 oClock & Delayed about 3 hours to rest in the heat of the day

Whitehouse   the current set strong against us this day

MONDAY, JULY 2

Clark   a 12 oClock came to on the Island...detained four hours, exceedingly hot, wind in forepart of the day from the SE. The river not So wide. A verry hot day. I observed that the river was Crouded with drift wood, and dangerous to pass   as this dead timber Continued only about half an hour, I concluded that Some Island of Drift had given way.

Ordway   we Delayed at 12 oC for to put up a Temperary mast as the wind was fair

Whitehouse   Found the water to run very strong against us

TUESDAY, JULY 3

Clark   Set out verry early this morning and proceeded on under a gentle Breeze from the
South

Floyd   water verry strong   the Land is verry mirey

Ordway   we Set out eairly & proceeded on under a gentle Breese from the South.

Whitehouse   The wind being in our favour, and blowing a good breeze

Note   *The expedition passes present-day Atchison, Kansas, and names the river after this special day in United States history.*

WEDNESDAY, JULY 4

Weather Diary   a great number of young geese and swan in a lake oposit to the mouth of
the 4th of July Creek, in this lake are also abundances of fish of various
species. The pike [hap?] catt, sunfish &c perch Carp, or buffaloe fish

Clark   passed the mouth of a Beyeue...this Lake is large and was once the bend of the River.
As this Creek has no name, and this day is the 4th of July, we name this
Independence us. Creek. One of the most butifull Plains, I ever Saw, open & butifully
diversified with hills & vallies all presenting themselves to the river covered with
grass and a few scatttering trees. Nature appears to have exerted herself to butify the
Senery by the variety of flours Delicately and highly flavered raised above the Grass,
which Strickes & profumes the Sensation, and amuses the wind   throws it into
Conjectering the cause of So Magnificent a Senerey in a Country thus Situated far
removed from the Sivilised world to be enjoyed by nothing bu the Buffalo Elk Deer
& Bear in which it abounds & Savage Indians

Whitehouse   the wind being favourable and the water being good Roed on Successfully. The
day proved mighty hot very warm.

THURSDAY, JULY 5

Clark   The river Continue to fall a little—   Wind from SE

FRIDAY, JULY 6

Clark   the river falls Slowly. Wind from the SW. A verry warm day. Worthy of remark that
the water of this river or Some other Cause, I think that the most Probable throws
out a greater prepson. of Swet than I could Suppose Could pass thro: the humane
body   Those men that do no work at all will wet a Shirt in a Few minits & those
who work, the Swet will run off in Streams.

Ordway   we Set out eairly this morning   proceeded on (the river falls Slowly) the weather is
verry warm, Several day's, the Sweet pores off the men in Streams

Whitehouse   the water was tolarably Good. Had good Sailing

Note   *The expedition passes present-day St. Joseph, Missouri.*

SATURDAY, JULY 7

Clark   some verry Swift water. A verry warm morning. At 4 oClock pass a Verry narrow part of the river water Confd. in a bead not more than 200 yards wide at this place. A hard wind (a Violent Ghust of wind) with Some rain from the NE at Dark - 7 oClock which lasted half an hour, with thunder & lighting. The river fall a little. Once man sick (Frasure) Stuck with the Sun, Capt. Lewis bled him & gave Niter which has revived him much.

Floyd   passed Some Strong water on the South Side   Clear morning   verry warm

Ordway   verry warm morning

SUNDAY, JULY 8

Clark   the Sick man [Frazer] much better. Passed...Nadawa [Nodaway] River is abt. 70 yards wide at its mouth...jentl Current

Floyd   Rain last night with wind from the E.

Gass   The river here is crooked and narrow

Whitehouse   This morning we embark'd early with a fair wind, and sail'd for 8 hours

MONDAY, JULY 9

Clark   at 8 oClock [AM] it commenced raining, the wind changed from NE to SW wind Shifted to the NW in the evening. Saw a fire on the S.S.

Floyd   Rain to day   Sailed the Gratist part of the day   Saw a fire on the N Side thougt it was (Indians) ouer flanken partey

Gass   it rained hard till 12 o'clock

Ordway   Rainy. The wind changed from the NE to the SW   the wind Shifted to the NW in the evening

Whitehouse   shortly after we had started a Rain came on, which continued most of the day—

TUESDAY, JULY 10

Clark   Set out this morning with a view to Land near the fire Seen last night...discover our men were at the fire, they were a Sleep early last evening, and from the Course of the Wind, which blew hard, their yells were not heard by party in the perogue. The river is on a Stand.

Gass   had a fair day and a fair wind

Whitehouse   the water was Strong the Morning was Clear. the current still strong against us, & very hard rowing to stem it.

Note   *The expedition crosses the present-day Kansas–Nebraska border.*

WEDNESDAY, JULY 11

Clark   made Some Luner observations this evening.

Whitehouse   This morning having cloudy weather & appearance of Rain, at 10 oClock AM the Clouds dispers'd and the weather became clear

THURSDAY, JULY 12

Weather Diary   the deer and bear begin to get scarce and the Elk begin to appear

Lewis This is a mean of four observations which were not so perfect as I could have wished them, in consequence of the moon being obscured in some measure by the clouds, which soon became so general as to put an end to my observations during this evening—

Clark I assended a hill....I had an extensive view of the Serounding Plains, which afforded one of the most pleasing prospects I ever beheld. on a clift Sand Stone $^1/_2$ me. up & on Lower Side I marked my name & day of the month near an Indian mark or image of animals & a boat. Took Some Luner Observations.

FRIDAY, JULY 13

Clark My notes of the 13th of July by a Most unfortunate accident blew over Board in a Storm in the morning of the 14th Obliges me to refur to the Journals of Serjeants. Last night a violent Storm from the NNE— Last night about 10 oClock a violent Storm of wind from the NNE which lasted with Great violence for about one hour, at which time a Shower of rain Succeeded . Set out at Sun rise...under a gentle Breeze. Sailed under a Wind from the South all day. The Clouds appear to geather from the NW. A most agreeable Breeze from the South. Great appearance of a Storm from the North W this evening verry agreeable the wind Still from the South. Showers of rain all night.

Floyd a verry hard Storm Last night from the NE which Lasted for about one ouer proseded with a Small Souer of Rain   wind fare   Sailed all day   a Small Shouer of Rain.

Gass We were early under way this morning with a fair wind. The day was fine.

Ordway last night at 10 oClock a violent Storm from the NNE which lasted for one hour. a small Shower succeded the wind. The wind favourable from the South

Whitehouse The wind Rose. Encamped on an Island called little sandy Island, opposite the hurricane Priari—*

SATURDAY, JULY 14

Clark Some hard showers of rain accompanied with Some wind this morning prevented our setting out until 7 oclock, at half past seven, the atmisperr Became Suddenly darkened by a black and dismal looking cloud, at the time we were in a Situation...near an Island...the bank was falling in and lined with snags...in this Situation a Violent Storm of wind which passd over an open plain from the N.E. struck the our boast on the starbd. quarter, and would have thrown her up on the sand island dashed to pieces in an instant, had not the party leeped out on the leeward side and kepth her off with the assistance of the anker & cable untill the storm was over. In this situation we continued about 40 Minits. when the storm Sundenly Seased and the river became Instancetaniously as Smoth as glass in 1 minit. The two perogus dureing this Storm was in Similar Situation with the boast about half a mile above—   The wind Shifted to the SE and we Set Sail. River falls a little.

Floyd Came a Dredfulle hard Storme from the South which Lasted for about one ouer and half which Cosed us to Jump out and hold hir   the wind fare Sailed,

---

*Whitehouse uses a name for this prairie which no one else uses in the Journals. This reference may by to the violent wind of the next day, which Floyd calls "a Dredfulle hard Storme" (Moulton 1997, 11: 39).

Gass   At day break it began to rain and continued until seven when it abated, and we set forward; but in a short time a gust of wind and rain came on so violent, that all hands had to leap into the water to save the boat. Fortunately this storm did not last long, and we went on to a convenient place and landed.

Ordway   Some hard showers of rain acocmpanied with some wind which detained us untill about 7 oClock   Then their came up a violent Storm from the NE of wind & rain which passed through an open prarie, it came So Suddenly by a black cloud & dismal looking. We were in a Situation near the upper point of a Sand Island & on the opposite Shoer falling in, the boat nearly quartering & blowing down the current. The Boat was in danger of being thrown up on the Sand but the men were all out in an Instant holding hir out Stemming the wind   the anchor was immediately carried out. So by all exertion we could make we kept the boat from filling or takeing injury. This Storm Suddenly Seased, and in one minute the River was as Smooth as it was before, the wind Shifted to the SE and we Set Sail & proceded on   Capt. Clark notes & Reamks of 2 days blew Overboard this morning in the Storm, and he was much put to it to Recolect the courses &.C—

Whitehouse   This morning before we embarked a heavy Rain cam on, with a hard Smart wind that Inraged the watter and occasioned the River to run in high Waves to Such a degree that all hands had to Get in the Watter to keep up the boat— The Storm abated about 10 oClock AM and we proceeded

SUNDAY, JULY 15

Clark   a heavy fog this morning which Detained us untill 7 oClock

Floyd   water verry Strong

Ordway   a foggy morning which Detained us untill 7 oClock

Whitehouse   The morning being foggy, we had to waite 'till it clear away, at 9 oClock AM we set off

MONDAY, JULY 16

Clark   The morning was cloudy; proceeded on under a gentle breeze from the S. The river falling

Floyd   the wind from the South Sailed ouer boat   Sailed all day

Gass   had a fine day and fair wind

Ordway   the wind from the South

Whitehouse   the morning being fine and weather clear we started early, the wind in our favour we set all our Sails and had Good Sailing. The water Strong.

TUESDAY, JULY 17

Clark   wind from the SE. Camped…a puff of wind brought Swarms of Misquitors, which disapeared in two hours, blown off by a Continuation of the Same brees.

Ordway   a pleasant warm day

Whitehouse   This day being fine and clear, we staid at the Maha priari

Note   *The expedition crosses the present-day Missouri–Iowa border.*

WEDNESDAY, JULY 18

Clark   a fair morning. The river falling fast. Set out at Sunrise under a gentle Breeze from SE by S. Wind from the SW hard

Floyd   the day Clear wind fair   passed a verry Strong pace of Water

Gass   our voyage fair wind and pleasant weather

Ordway   we Set out at Sun rise under a gentle Breeze from the SE by S   a fair morning

Whitehouse   This morning being clear, we got under way at day light, with a fair wind, and sailed

THURSDAY, JULY 19

Clark   this prarie was Covered with grass about 18 Inches or 2 feat high and contained little of any thing else. The river Still falling a little

Whitehouse   the morning was clear

FRIDAY, JULY 20

Clark   a fog this morning and verry Cool. The Soil of Those Praries appears rich but much Parched with frequent fires— a verry Pleasant Breeze from the NW all night— river falling a little.

Ordway   a heavy Deaw last night. Some foggy this morning.

Whitehouse   a freash Bres of wind

Note   *The expedition arrives at the confluence of the Platte and Missouri rivers, in present-day Nebraska.*

SATURDAY, JULY 21

Clark   Set out verry early under a Gentle Breeze from the S.E   at 7 oClock [AM] the wind Seased and it Commenced raining. Passed…Mouth of the Great River Plate   this river which is much more rapid than the Missouri has thrown out imence quantities of Sand…found it shallow. Camped…a verry hard wind from the NW.

Floyd   Rain this morning   wind fair   Sailed passed the mouth of the Grait River Plate on the South Side   it is much more Rappided than the missorea   this river is not navigable for Boats to Go up it

Gass   I rained this morning but we had a fine breeze of wind. At nine the wind fell, and at one we came to the great river Platte, or shallow river.

Ordway   Some Rain this morning   We Set out at Sun rise under a gentle Breese from the South or SE   A hard (head) wind from the NW   praries in pt between the Missouris & the Great R. Platt but flat Subject to overflow.

Whitehouse   got underway, the wind became fair and Seased Bloing. The River Plate…runs strong at its mouth.

SUNDAY, JULY 22

Clark   Set out early with a view of Getting to Some Situation above in time to take equal altitudes and take Observations, as well as one Calculated to make our party Comfortabl in a Situation where they Could recive the benifit of a Shade— Camped. Wind hard from NW. Cold. The river rise a little.

Gass   with fair weather

MONDAY, JULY 23

Weather Diary   Cat fish is verry Common and easy taken in any part of this river. Some are nearly white perticulary above the Platte River.

Clark   a fair morning. Praries being on fire in the direction of the Village. Wind hard this afternoon from the NW.

Gass   clear weather

Ordway   clear morning

Whitehouse   We remained here this day in Order to get an Observation, the weather being fine & clear.

TUESDAY, JULY 24

Clark   a fair morning and day. The wind rose with the Sun & blows hard from the S. The breezes which are verry frequent on this part of the Missouri is cool and refreshing. Thos Southerly Breezes are dry Cool & refreshing. The Northerly Breezes which is more frequent is much Cooler and moist.

Floyd   Histed oeur Collars in the morning for the Reseptions of Indians who we expected Hear   when the Rain and wind Came so that we wase forst to take it down Continued Showery all day

Gass   some showers

Ordway   Some rain wind blew from NE

Whitehouse   We remained here this day, and had some Rain in the morning.

WEDNESDAY, JULY 25

Clark   a fair morning. Wind from the SE.

Gass   clear weather

Ordway   a pleasant morning

Whitehouse   a pleasant morning.

THURSDAY, JULY 26

Clark   the wind blustering and verry hard from the south all day which blowed the Clouds of Sand in such a manner that I could not complete my pan [map] in the tent, the Boat roled in Such a manner I could do nothing in that, I was obliged Combat with the Misqutr. under a Shade in the woods—   This evening we found verry pleasant—

Gass   clear weather

Ordway   pleasant morning. All the latter part of the day the wind blew hard from the S

Whitehouse   we had a pleasant morning. The latter part of the day, the wind blew hard from the South

Note   *On July 27 and 28, the expedition passes the location of present-day Omaha, Nebraska.*

FRIDAY, JULY 27

Lewis   I wished to have taken one or two sets more with the moon and Aquilae, but the clouds obscured the star. I was also anxious to have taken some sets with Aldeberan, then in reach of observation and East of the moon, but was prevented by the intervention of the clouds, which soon became so general as to obscure the whole horizon—

Clark   A Small Shower of rain this morning until near 10 oClock. At half past 1 oClock we Set Sale under a gentle breeze from the South. A butifull Breeze from the NW this evening which would have been verry agreeable, had the Misquiters been tolerably Pacifick, but thy were rageing all night, Some about the Sise of house Flais [flies].

Floyd   prossed on under a Jentil Brees from the SE   nothing worth Relating except the wind was verry villant from the South Est—

Gass   At 12 we proceeded with a fair wind, and pleasant weather

Ordway   cloud morning. We set out under Sail about one oClock proceeded very well   the River verry crooked.

Whitehouse   cloudy morning. The River is very Crooked in this days route

SATURDAY, JULY 28

Clark   Set out this morning early, the wind blou from the NW by N. A Dark Smokey Morning, some rain.

Floyd   Rain the fore part of the day   the Latter part Clear with wind from the North Est.

Gass   had a cloudy morning

Ordway   Cloudy morning. The wind hard from the NE Detained us Some time.

Whitehouse   The morning still continued Cloudy. The wind Blew hard from the NE—

SUNDAY, JULY 29

Clark   a Dark rainey morning   Wind from the WNW. Rained all last night— a Cold Day Wind from the NW   Some Rain the fore part of the Day. Above this high land & on the S.S...at the commencement of this course...passed much fallen timber apparently the ravages of a Dreadfull haricane which had passed obliquely across the river from NW to SE about twelve months since. Many trees were broken off near the ground the trunks which were Sound and four fect in diameter.

Gass   passed a bank, where there was a quantity of fallen timber

Ordway   Rain all last night. Cloudy morning.   The Missouri is much more crooked since we passed the Great River Platte than before but not so Rapid in general; & more praries, the Timber Scarser &C—

Whitehouse   This morning was rainey, we started at sunrise.

MONDAY, JULY 30

Clark   this Prarie is Covered with grass about 10 or 12 Inch high. a fair still evening and Cool. Great no. misquitors this evening.

TUESDAY, JULY 31

Clark   a fair Day. The evening verry Cool.

Ordway   pleasant & Cool this morning   the Missouri is verry crooked couses on one Side or the other all the way from the Great River Platte, but the current not So Swift as below.

Whitehouse   having fine clear weather.

Note   *From July 30 to August 2, the expedition camps at "Council Bluffs," north of present-day Omaha, Nebraska (not the site of present-day "Council Bluff," which is across the river from Omaha). This location would become Fort Calhoun.*

WEDNESDAY, AUGUST 1

Clark   a fair morning. The air is Cool and pleasing. Wind rose at 10 oClock and blowed Steedy from the WSW agreeable Breeze all Day. Very pleasant all day. Cool fine eveninge.

Ordway   a fair morning

Whitehouse   This morning was clear, we remain'd still at out encampment

THURSDAY, AUGUST 2

Clark   a verry pleasant Breeze from the SE. At Sunset...the wind Continue hard from the SE.

Ordway   Cool & pleasant this morning   The afternoon Cloudy. The wind Southerly appearence of rain.

FRIDAY, AUGUST 3

Clark   at 4 oClock [PM] Set our under a gentle Breeze from the SE. Set Sail. The Musquitors more numerous than I ever Saw them. Camped. Great appearance of wind and rain to the NW we prepare to rec've it—   the air is pure and helthy So far as we can Judge—

Floyd   embarked at 3 oclock PM under Jentell Brees from the South Est.

Gass   encamped....where we had a storm of wind and rain, which lasted two hours.

Ordway   a foggy morning. No Diew last night

Whitehouse   This morning was foggy

SATURDAY, AUGUST 4

Lewis   the suns disk was frequently obscured in the course of this observation

Clark   at 7 oClock last night the heavens darkened and we had a Violent wind from the NW Som little rain Succeeded, the wind lasted with violence for one hour   after the wind it was clear Sereen and Cool all night. Channel Confined within 200 yards....the Banks washing away & trees falling in constantly for 1 mile. The wind blew hard a head.

Floyd   Set out erly this morning after the Rain was over   it Rained Last night with wind and thunder from the NW   it Lasted about an [hour]   prossed on the morning Clear

Ordway   at 7 oClock last night, a violent wind from the NW & thunder & rain which lasted about an hour then ceased blowing but hard rain followed   proceeded on through a narrow part of the River which is filled with Snags & logs   the River in many places is confined within 200 yards.

Whitehouse   This morning we set off early, having fine Clear weather

SUNDAY, AUGUST 5

Clark   wind from the NE. Great appearance of Wind & rain. I have observed and remark that I have not heard much Thunder & lightning and is not common in this Countrey as it is in the atlantic States

Floyd   Cam 2 miles when a verry hard Storm of wind and Rain from the North (West) Est it Lasted about 2 ouers and Cleard up I have Remarked that   I have not heard much thunder in this Countrey Lightning is Common as in other Countreys

Gass   We set out early, but a storm of rain and wind obliged us to stop two hours. It then cleared and we continued   The river here is very crooked and winding

Ordway   a Shower came up from the NW Some wind attending it. which Detained us about 2 hours. A head wind. C. Clark was at the River below this point which is only 370 yards across

Monday, August 6

Clark   at 12 oClock last night a Violent Storm of wind from the NW. Some rain. One pr. of Colours lost in the Storm from the big Perogue.

Floyd   about 12 oclock Last night a villant Storm of Wind and Rain from the NW—

Gass   proceeded…after a stormy night of wind and rain

Ordway   a violent storm came up about 12 oClock last night of wind & rain from NW

Whitehouse   the morning was fair

Tuesday, August 7

Clark   last night about 8 oClock a Strom of wind from the NW lasted $3/4$ of an hour. Mosquitors more troublesom last night than I ever Saw them, Set out late this morning wind from the North—

Floyd   [proceeded] on   day Clear wind from the North west—   Water Good

Ordway   last night about 8 oClock a Storm from NW of wind and rain which lasted about $3/4$ of an hour. the wind N

Whitehouse   the morning was clear

Wednesday, August 8

Lewis   we had seen but a few aquatic fouls of any kind on the river since we commenced our journey up the Missouri. This day after we passed the river Souix…I saw a great number of feathers floating down the river   Those feathers had a very extraordinary appearance as they appeared in such quantities as to cover pretty generally sixty or seventy yards of the breadth of the river. For three miles after I saw those feathers continuing to run in that manner, we did not percieve from whence they came, at length we were surprised by the appearance of a flock of Pilican.

Clark   the wind as usial from the NW.

Floyd   Set out this morning at the usele time   day Clear wind from the NW

Ordway   the wind from NW   this River is about 80 yds wide & navagable for perogues for a considerable distance   it contains a Great quantity of fish common to the country.

Thursday, August 9

Clark   The fog of this morning being thick detained us untill $1/2$ passed 7 oClock at which time we Set out and proceeded on under Gentle Breeze from the SE. Musquetors worse this evening than ever I have Seen them.

Floyd   Set out at 7 oclocks AM (we could see about us the) after the fague was Gon which is verry thick in this Cuntrey   Passed a verry Bad place in the River whare the water is verry Shellow

Gass   The fog was so thick this morning, that we could not proceed before 7, when we went on under a gentle breeze

Ordway   a foggy morning, which detained us till past 7 oClock at which time we Set out under a gentle Breeze from SE

Whitehouse   This morning we set out in a fog which cleared up at 8 oClock AM. The Wind blew from the South; had Good Sailing. When the wind died away, we then took our Oars and rowed.

Friday, August 10

Clark   wind hard from the SW

Ordway   a fair day. The wind hard from the SW (Sailed Some.)

Whitehouse   The morning was Clear. The musquitoes was mighty troublesome Untill The Sun rose to Some hight. The wind being fair we Sail'd the greatest part of this day.

Saturday, August 11

Clark   about day this morning a hard wind from the NW followed by rain. The river is verry Crooked. A hard wind accompanied with rain from the SE after the rain was over Capt. Lewis myself & 10 men assended a Hill.

Floyd   Set out after a verry hard Storm this morning of wind and Rain continued untill 9 oclock AM and Cleared up   [proceeded] on

Gass   A storm came on at three o'clock this morning and continued till nine, notwithstanding which, we kept under way till ten

Ordway   hard Showers this morning   comenced at day break & lasted & detained us about an hour, hard wind from SW   we proceeded on   the wind hard   Some Thunder, the river verry crooked, after we passed this hill

Whitehouse   At 3 oClock AM this morning we had a rain, which was very heavy, which was immediately succeeded with a smart heavy wind from the South. At 6 oClock we set sail

Sunday, August 12

Clark   Set out early under a gentle Breeze from the South. The river wider than usial, and Shallow. The wind Comes round to the SE.   The wind for a few hours this evening was hard and from the SE. Musquitors verry troublesom untile the wind rose. At one or 2 oClock

Floyd   Set out at the usel time [proceeded] on under a Jentel Bres from the North Est. Sailed day Clear

Ordway   a fair morning. We Sailed on with a SE wind

Whitehouse   the morning was fair a Sharp Breese of wind Blew from the South Sailed

Monday, August 13

Clark   Set out this morning at Day light the usial time and proceeded on under a gentle Breeze from the SE. The SE wind Continues high, we took some Luner observations this evening. The air pleasant.

Floyd   [proceeded] on under a Jentel Brees from the South Est—   Sailed   (day C) morning Cloudy about 10 ock, it Cleared up

Gass   proceeded…with a fair wind

Ordway   proceeded on under a gentle Breese from Souhard

Tuesday, August 14

Clark   a fine morning wind from the SE

Whitehouse   they day was fair and pleasant

Wednesday, August 15

Clark   [party] to examine a fire which threw up an emence smoke from the praries on the N.E. side of the river and at no great distance from camp— in the evening this Party returned and informed, that the fire arose from Some high trees which had been left burning by a small Party of Seoux and hard wind of to day. the wind Setting from that point blew the smoke from that pt. over our Camp.

Gass There had been fire there some days, and the wind lately blowing had caused the fire to spread and smoke to rise.

Ordway a pleasant morning   hard wind F. NW

THURSDAY, AUGUST 16

Clark a Verry cool morning, the winds as usial from the NW. The wind Shifted around to the SE. Every evening a Breeze rises, Sometimes before night, which Blows off the Musquitors & Cools the atmispeire.

Ordway a pleasant morning.

Whitehouse This morning fine & clear

FRIDAY, AUGUST 17

Clark a fine morning, wind from the SE. Sent & fiered the Prarie near Camp to bring in the Mahars & Souex. A Cool evening

Ordway a clear morning, the wind from SE

Whitehouse the weather was fine

SATURDAY, AUGUST 18

Clark a fine morning, Wind from the SE in the after part of the day. A fine evening. Wind SE.

SUNDAY, AUGUST 19

Clark a fine morning. Wind from the SE.

Ordway pleasant, S wind

Note *The expedition is near present-day Sioux City, Iowa.*

MONDAY, AUGUST 20

Clark we Set our under a gentle breeze from the SE and proceeded on verry well— Sergt. Floyd [died]. Name Floyds River, the Bluffs Sergts. Floyds Bluff— Capt. Lewis read the funeral Service over him. Proceeded to the Mouth of a little river 30 yrd wide & Camped. a butifull evening.

Gass a fair wind and fine weather

Ordway pleasant, we Set of under a gentle Breeze from SE

Whitehouse we Set our eairly this morning under a gentle breeze from the SE. We continued sailing on very well 'till noon, when we landed to take dinner—

Note *The expedition travels parallel to the current day Nebraska–South Dakota border until September 8.*

TUESDAY, AUGUST 21

Clark   we Set out verry early this morning…under a gentle Breeze from the SE. Camped. Clouds appear to rise in the west & threten wind.

Ordway   we Set off eairly this morning under a hard Breeze from the S. we proceeded on verry well.   ….the wind blew so hard that we were obliged to take a reefe in our Sail & the Sand blew So thick from the Sand bars that we could not see the channel far ahead & it filled the air before us about a mile.

Whitehouse   we Set out eairly this morning under a hard stiff breeze from the South.

WEDNESDAY, AUGUST 22

Clark   set out early wind from the South. Sailed the greater part of this day with a hard wind from the SE great deel of Elk Sign, and a great appearance of wind from the NW. Is worth remark, that my Ink after Standing in the pot 3 or four days Soaks up & becons thick

Ordway   the current verry Swift. The wind hard from the South   we proceeded on under a fine Breeze from the South.

Whitehouse   This morning we set out early, and found the current running very Strong against us, the Wind blowing hard from the South.

THURSDAY, AUGUST 23

Clark   wind blew hard (I am obliged to make the next Corses Short on acknount) of the flying Sands which raise like a Cloud of Smoke from the Bars when the wind (rise) Blows, the Sand being fine and very light and containing a [great] perpotion of earth and when it lights it Sticks to every thing it touches. Such Clouds that we Could Scercely See.   At this time the grass is white— and in the Plain for a half a mile the distance I was out every Spire of Grass was covered with the Sand or Dust.

Gass   proceeded…with a fair wind. But the wind changed and we were obliged to halt for the present…at five in the evening we proceeded

Ordway   the wind favourable from the South   Buffelow But this was the first I ever Saw & as great a curioustiy to me. The wind blew So hard that it detained us the most of the afternoon, the Sand blew So thick from the Sand Island that we could not see across the River for a long time, towards evening the wind abated & we proceeded

Whitehouse   The wind blowing hard we Jerked out Meat & overhauled several articles on board the boat. Towards evening the wind died away, We Proceeded on 'till dark

Note   *The expedition passes present-day Vermillion, South Dakota.*

FRIDAY, AUGUST 24

Clark   Some rain last night & this morning. Those Bluffs appear to have been laterly on fire, and at this time is too hot for a man to bear his hand in the earth at any debth, gret appearance of Coal. All the after part of the day it rain. Verry wet, a Cloudey rainey night—

Gass   This morning was cloudy with some rain.

Ordway   Some Small Shower rain the latter part of last night. Rainy morning. We found also a burning bank or Bluff which was verry high & had fire in it. it had a

Sulpheras Smell   at the Same time we had a fine Shower of Rain which lasted abt. Half an hour

Whitehouse   We had some small showers of rain last night

Saturday, August 25

Clark   a Cloudy morning assended...the Mound which the Indians Call Mountain of little people or Spirits. We Concluded it was most probably the production of nature— The Surrounding Plains is open void of Timber and leavel to a great extent: hence the wind from whatever quarter it may blow, drives with unsisial force over the naked Plains and against this hill; the insects of various kinds are thus involuntaryly driven to the mound by the force of the wind. From the top of this Mound we behld a most butifull landscape. At two miles further up our Dog was So Heeted & fatigued we was obliged Send him back to the Creek. Capt. Lewis much fatigued from heat the day it being verry hot & he being in a debilitated state from the precautions he was obliged to take to prevent the effect of the cobalt, & minl substance which had like to have poisoned him two days ago. Several of the men complaining of Great thirst, deturmined us to make for the first water. We returned to the boat at Sunset, my Servant nearly exosted with heat thurst and fatique. We Set fire to the Praries in two Places to let the Sous know we were on the river. At 3 oClock Murcky. 86 abv. 0. We set the praries on fire as a signal for the Soues to come to the river. Verry dark...some rain this evening

Gass   we set sail with a gentle breeze from the SE

Ordway   a fair & pleasant morning.

Sunday, August 26

Clark   river wide. Misquetors bad to night.

Monday, August 27

Clark   This morning, the morning Star was observed to be very large...more than Common. We set out under a Gentle Breeze from the SE. Above this bluff we had the Prarie Set on fire to let the Souic See that we were on the river, & as a Signal for them to Come to it. Camped...the wind blew hard from the South. A Cool & Pleasant evening. The river has fallen verry Slowly and is now low.

Ordway   we Set off at Sun rise under a gentle Breeze from the SE

Whitehouse   We set out at sunrise under a gentle Breeze from the SE

Tuesday, August 28

Clark   The wind blew hard last night. Set out under a Stiff Breeze from S. Several Sand bars the river here is wide & Shallow full of Sand bars—   Wind blows hard this afternoon from the South.

Gass   The day was pleasant, and a fair wind from SE

Ordway   a pleasant morning   we Set off eairly under a fine Breeze from SE   at 2 oClock PM the wind Blew hard from Sw the large pearogue drove against the Shore on NS & hole got knocked in her So that it let the water in verry rapid they began to unload.

Whitehouse   Wet set off early this morning with a fine Breeze from the SE; and all Sails set. About 2 oClock PM the Wind blew hard from the SW.

WEDNESDAY, AUGUST 29

Clark   rained last night and Some this morning. Verry cloudy

Gass   At 8 o'clock last night a storm of wind and rain came on from the N. west, and the rain continued the greater part of the night. The morning was cloudy with some thunder.

Ordway   a hard Storm arose from the NW of wind & rain about 8 oC last night   rained considerable part of the Night. Cloudy morning. Some Thunder

Whitehouse   a hard Storm arose from the NW last night about 8 oClock PM with wind and rain, accompanied with Thunder lightning. Cloudy this morning, some Thunder.

THURSDAY, AUGUST 30

Clark   a Foggie morning. Wind hard from the South. I will here remark a Society which I had never before this day heard was in any nation of Indians— Those who become members of this Society must be brave active young men who take a Vow never to give back let the danger be what it may; in War Parties they always go forward without Screening themselves behind trees or any thing else   to this Vow they Strictly adheer their Lives— an instanc which happened not long Since, on a party in Crossing the R Missourie on the ice, a whole was in the ice imediately in their Course which might easily have been avoided by going around, the foremost man went on and was lost [drowned],   the others were caught by their party and draged around—

Gass   A foggy morning, and heavy dew.

Ordway   A foggy morning, a heavy diew last night. The fog remained on the River late this morning & So thick that we could not See the Indians camp on the opposite Shore—   at about 8 oClock the fog went away.

Whitehouse   the fog is So thick on the river this morning that we could not See across the river, untill late in the morning.

FRIDAY, AUGUST 31

Clark   a fair Day— the evening…Soon after a violent Wind from the NW accompanied with rain. The rain continued the greater part of the night. The river a riseing a little.

Gass   A clear morning.

Ordway   pleasant morning   directly after a hard Strom arose of the wind and rain from the NW which lasted 2 hours. Rained considerable part of the night

Whitehouse   This morning we had pleasant Weather.

Note   *The expedition camps just upriver from present-day Gavin Dam, along the Nebraska–South Dakota border, near future historical site of Fort Yankon.*

SATURDAY, SEPTEMBER 1

Clark   Set out under a gentle Breeze from the South (Raind half the last night), River rose 3 Inches last night. Some hard wind and rain, Cloudy all day—   the river wide

Gass   During last night we had hard wind and some rain, which continues to fall occasionally during the day—

Ordway   we proceeded on under an unsteady Breeze from SW

Whitehouse   the morning was rainy

SUNDAY, SEPTEMBER 2

Clark   The wind being hard a head from the NW. Verry Cold. Some rain all day much Thunder & lightning. Water riseing. At Sunset the [wind] luled and cleared up Cold—

Gass   At 1 o'clock last night we had hard thunder, lightning and rain, which continued about two hours. About twelve (noon), the wind blew so hard down the river, that we cold not proceed, and we landed on the north side   it was cloudy and rained till 4 when it cleared up.

Ordway   a hard Strom arose the latter part of last night 1/2 past 1 oC of wind & rain from NW which lasted abt. 2 hours. Cloudy this morning   we Set off eirily. Sailed a Short distance with a SE wind & in less than 2 hours the wind Shifted in to the NW which blew hard a head   The wind shifted into the North & Blew So heard that we were obliged to lay by at a high Bluff   The weather is Cool & rainy to day

Whitehouse   Last night we had a hard Storm of Wind, accompanied with Rain which was very heavy. It lasted for near 2 hours. Cloudy this morning.

MONDAY, SEPTEMBER 3

Clark   Set out at Sun rise, verry Cold morning clear and but little wind from the NW. The river wide and riseing a little. Scercely any timber in Countery except a little on the river.

Ordway   a Cool & pleasant morning   we Set off at Sun rise, the wind blew from the west a swift current

Whitehouse   cool and pleasant this morning.

TUESDAY, SEPTEMBER 4

Clark   a verry Cold wind from the SSE.

Ordway   the wind Shifted to the South & blew verry hard   we hoisted Sail   ran verry fast a Short time. Broke our mast, we [the] sand flew from the Sand bars verry thick   the water shoots in to the Missiouri verry Swift, & has thrown the Sand out, which makes a Sand bar & Sholes from the mouth a considerable distance

Whitehouse   the wind blowing fresh, we set all our Sails & proceeded on Sailing fast,

WEDNESDAY, SEPTEMBER 5

Clark   set our early, the wind blew hard from the South   Sailed.

Gass   We set sail early this morning with a fair wind, and had a clear day.

Ordway   the wind Blew hard from the South

Whitehouse   we Sailed on

THURSDAY, SEPTEMBER 6

Clark   a Storm this morning from the NW at day light which lasted a fiew minites, Set out after the Storm was over and proceeded on a hard wind from the NW a head. The morning and day verry Cold. The river riseing a little.

Gass   We set out early and had cloudy morning. About 9 o'clock it began to rain and we had strong wind ahead.

Ordway   a cloudy morning the wind from the NW   the current Swift & Shallow

Whitehouse   cloudy weather this morning

FRIDAY, SEPTEMBER 7

Clark   a verry Cold morning. Wind SE.
Gass   had a clear day
Ordway   a fair, cool morning   the wind from NW
Whitehouse   a clear morning

Note   The expedition leaves the north–central border of Nebraska and enters present-day South Dakota.

SATURDAY, SEPTEMBER 8

Lewis   the evening was cloudy, which prevented my taking the altitude of any fixed star.
Clark   Set out early and proceeded on under a Gentle breese from the SE. Hill on the SS recently burnt—
Gass   had a clear day and fair wind from the SE
Ordway   a pleasant morning. The wind from the SE we proceeded on under a gentle breeze Capt. Clark…travelled over a riged and mountanious Country without water & riseing 5 or 600 feet, where these hills had been lately burned over by the natives—
Whitehouse   This morning we had fine pleasant weather

SUNDAY, SEPTEMBER 9

Clark   a fair Day wind from the SE. The river shallow

MONDAY, SEPTEMBER 10

Clark   a Cloudy dark morning. Set out early under a Gentle Breeze from the SE. The river verry Shallow and falling. We proceeded on under a Stiff Breeze.
Gass   We had a foggy morning, but move on early
Ordway   a foggy morning   we Sailed on verry well
Whitehouse   a foggy morning

TUESDAY, SEPTEMBER 11

Clark   at Dark in a heavy Shower of rain, it Continued to rain the greater part of the night, with a hard wind from the NW   Cold—   a Cloudy morning. The river is verry Shallow & wide. At 12 oClock it became Cloudy and rained hard all the after noon, & most of the night, with a hard wind from the NW.
Gass   We set sail before day light with a fair wind   at 1 o'clock it began to rain   And we continued our voyage, until night, though it rained very hard
Ordway   Sailed on   The boat sailed on   Rained so hard my gun got wet loading
Whitehouse   We set out this morning at an early hour, with a fair wind & clear pleasant weather; and proceeded sailing on. We proceeded on & it began to rain very hard,   The Rain continued untill 7 oClock in the evening.

WEDNESDAY, SEPTEMBER 12

Clark   a Dark Cloudy Day the wind hard from the NW. The water Swift and Shallow, it took 3/4 of the day to make one mile. Rains a little all day.

Gass   had a cloudy day   the boat had much difficulty in passing on account of the sand bars and strong current

Ordway   the wind shifted Since last night in to the North.   The current swift & wind a head.   We had some difficulty owing to the river being Shallow.

Whitehouse   Clouday. The current running, so rapid against us

THURSDAY, SEPTEMBER 13

Clark   A Dark Drizzley Day. The winds from the NW. Verry Cold morning and day. Rains. Water is verry Shallow being Crouded with Sand Bars. Musquitors verry bad, wors than I have Seen them, qts [quantities] of mud wash into the rivr from a Small rain.

Gass   the morning was cloudy with some rain and wind ahead

Whitehouse   cloudy and had rain . High wind.

FRIDAY, SEPTEMBER 14

Clark   water wide & Shallow. Some drizzely rain in the forepart of this day. Cloudy & Disagreeable and Som hard heavy Showers.   All wett. The rain Continued the Greater part of the day. The Soil of those Plains washes down into the flats, with the Smallest rain & disolves & mixes with the water…what mud washed into the river within those few days has made it verry mudy. A rainy evening and night. In My ramble I observed, that all those parts of the hills which was Clear of Grass easily disolved and washed into the river and bottoms, and those hils under which the river run, Sliped into it and disolves and mixes with water of the river, the bottoms of the river was covered with the water and mud frome the hills about three Inches deep— those bottoms under the hils which is Covered with Grass also [receives?] a great quantity of mud.

Gass   We proceeded as yesterday, and with the same kind of weather. Had considerable difficulty in getting along, on account of the shallowness of the river

Ordway   a foggy morning. Cloudy. The water shallow. Some Rain.

Whitehouse   a great foggy morning, a cloudy day, some rain. Water is So Shallow that we had to waid

Note   *From September 15–17, the expedition camps near Chamberlain, South Dakota.*

SATURDAY, SEPTEMBER 15

Clark   the White River…is 400 yds wide….Current regularly Swift, much resembling the Missourie. Camped…This creek [American Crow Creek] raised 14 feet last rain. This evening is verry Cold. Great many wolves of different Sorts howling about us. The wind is hard from the NW this evening.

Gass   A cloudy morning. Passed White River…the current and colour of the water are much like those of the Missouri

Ordway   hard rain the greater part of last night.   We proceeded on till night with a head wind.

Whitehouse   the weather cloudy this morning.

SUNDAY, SEPTEMBER 16

Clark   we concluded to ly by at this place the ballance of this day and the next, in order to dry our baggage which was wet by the heavy showers of rain which had fallen within the last three days. The clouds during this day and night prevented my making any observations. These extensive planes had been lately birnt and the grass spring up and was about three inches high. Cloudy all day.

Ordway   Cool & Clear   we Camped on SS in a handsome bottom of thin Timbered land, lately burned over by the natives, it had grown up again with Green Grass which looked beautiful.

MONDAY, SEPTEMBER 17

Lewis   to amuse myself on shore with my gun and view the interior of the country…the shortness and virdue [verdure] of grass gave the plain the appearance throughout it's whole extent of beatifull bowling green in fine order. The surrounding country had been birnt about a month before and young grass had now sprung up to hight of 4 inches presenting the live green of the spring. This scenery already rich pleasing and beatiful, was still further hightened by immence herds of Buffaloe deer Elk and Antelopes…I do not think I exaggerate when I estimate the number of Buffaloe which could be compreed at one view to amount to 3000….drank of the water of a small pool which had collected on this plain from the rains which had fallen some days before.

Clark   Dried all those articles which had got wet by the last rain. A fine day. Wind from the SW.

Gass   As the weather was fair we remained here during the day.

Ordway   a pleasant day

TUESDAY, SEPTEMBER 18

Lewis   this day saw the first brant on their return from the north—

Clark   wind from the NW Modrt. The wind a head proceed verry Slowly. A Cole night for the Season. Before night the wind being verry hard & a head all day.

Gass   the day was clear and pleasant

Ordway   a fair morning   We camped on the South Side in a small grove of Timbers, 2 hours eairlier than usal the wind being a head,

Whitehouse   Clear & pleasant weather this morning

Note   *On September 19, the expedition begins the long trip (also known as the "Grand De Tour") around the Big Bend of the Missouri River in central South Dakota. On this day, both captains resume recording observations in the Weather Diary's Observation Tables, including two daily weather temperatures, state of weather, and wind direction readings. Neither recorded any information about the rise and fall of the river during the month. No explanation has ever been found from the captains as to the absence of weather observation data since May 14, or for the lack of remarks since July. For more information, see Moulton (1987, 3: 132).*

WEDNESDAY, SEPTEMBER 19

	Sunrise			4 p.m.		Mississippi River		
Temp	Weather	Wind	Temp	Weather	Wind	Rise/Fall	Feet	Inches
46a	f	SE	71a	f	SE			

Weather Diary   the leaves of some of the cottonwood begin to fade. Yesterday saw the first Brant passing from the NW to SE—

Clark   a Cool morning clear & Still. The wind from the SE. The wind being favourable for the boat all day. Here Commences a Butifull Countrey on both Sides of the Missourie. A fine evening.

Ordway   a pleasant morning   we proceeded on under a fine Sailing Breeze from ESE

Whitehouse   fine clear weather this morning

THURSDAY, SEPTEMBER 20

	Sunrise			4 p.m.		Mississippi River		
Temp	Weather	Wind	Temp	Weather	Wind	Rise/Fall	Feet	Inches
51a	f	SE	70a	f	SE			

Weather Diary   the antelope is now ruting, the swallow has disappeared 12 days

Clark   a fair morning wind from the SE

Gass   had a clear day and fair wind. From these and others of the same kind, the Missouri gets its muddy colour. The earth of which they are composed dissolves like sugar; every rain washes down great quantities of it, and the rapidity of the stream keeps it mixing and afloat in the water

Ordway   a fair morning   proceeded on under a gentle Breeze from the E   the current swift   the moon Shined pleasant all night—

Whitehouse   fine weather a clear day & fair wind. These bluffs were of a black Colour; from those and others of the same kind, it is supposed that the Water of the Mesouri river, derives its muddy colour; the black Mud lying on those black bluffs, melting like Snow at every Rain. At 10 oClock PM the Bank of the River on the side we were encamped began to fall in, It fell in so fast

FRIDAY, SEPTEMBER 21

Sunrise			4 p.m.			Mississippi River		
Temp	Weather	Wind	Temp	Weather	Wind	Rise/Fall	Feet	Inches
58a	f	SW	88a	f	SW			

Weather Diary Antilopes ruting, as are the Elk, the Buffaloe is nearly ceased— the latter commence the latter end of July or first of August

Clark the river here is wide nearly a mile in width— we observe an emence number of Plover of Different kind Collecting and takeing their flight Southerly, also Brants which appear to move in the same Direction. This day is worm, the wind which is not hard blows from the SE.

Ordway a clear & pleasant morning

Whitehouse fine pleasant weather   a clear day.

SATURDAY, SEPTEMBER 22

Sunrise			4 p.m.			Mississippi River		
Temp	Weather	Wind	Temp	Weather	Wind	Rise/Fall	Feet	Inches
52a	f	E	82a	f	SE			

Weather Diary a little foggy this morning, a great number of green leged plove passing down the river, also some geese & brant—

Clark a Thick fog this morning untill 7 oClock which detained us.

Gass We embarked early in a foggy morning

Ordway a foggy morning.

Whitehouse This morning we sett out early, having some fog, about 8 oClock AM the fog clear away

SUNDAY, SEPTEMBER 23

Sunrise			4 p.m.			Mississippi River		
Temp	Weather	Wind	Temp	Weather	Wind	Rise/Fall	Feet	Inches
50a	f	SE	86	f	SE			

Weather Diary aire remarkably dry-plumbs & grapes fully ripe— in 36 hours two Spoonfuls of water aveporated in a sauser

Clark Days & nights equal. Set out early under a gentle Breeze from the SE. We observed a great Smoke to the SW which is an Indian Signal of their haveing discovered us.

Ordway   a fair pleasant morning   the wind favourable from SE
Whitehouse   fine clear weather

Note   *The expedition passes the location of present-day capital of South Dakota, Pierre.*

MONDAY, SEPTEMBER 24

	Sunrise			4 p.m.			Mississippi River	
Temp	Weather	Wind	Temp	Weather	Wind	Rise/Fall	Feet	Inches
54a	f	E	82	f	W			

Clark   a fair morning and day. Set out early, wind from the East. The wind from the SE—
        Teton river [Bad River] is 70 yards wide at the mouth…has a considerable Current.
Gass   We set sail early with fair weather
Ordway   a Clear and pleasant morning. Proceeded on under a gentle breeze from SE

TUESDAY, SEPTEMBER 25

	Sunrise			4 p.m.			Mississippi River	
Temp	Weather	Wind	Temp	Weather	Wind	Rise/Fall	Feet	Inches
50	f	SW	79	f	W			

Clark   a fair morning   the wind from the SE.
Ordway   A clear and pleasant morning.

WEDNESDAY, SEPTEMBER 26

	Sunrise			4 p.m.			Mississippi River	
Temp	Weather	Wind	Temp	Weather	Wind	Rise/Fall	Feet	Inches
54	f	W	78	f	SW			

Clark   all in Spirits this evening   wind hard from the SE.
Ordway   a clear and pleasant morning

THURSDAY, SEPTEMBER 27

	Sunrise			4 p.m.		Mississippi River		
Temp	Weather	Wind	Temp	Weather	Wind	Rise/Fall	Feet	Inches
52	f	W	86	f	SW			

Weather Diary   Saw a large flock of white Gulls with wings tiped with black
Ordway   a clear and pleasant morning

FRIDAY, SEPTEMBER 28

	Sunrise			4 p.m.		Mississippi River		
Temp	Weather	Wind	Temp	Weather	Wind	Rise/Fall	Feet	Inches
45	f	SE	80	f	SE			

Clark   proceeded on under a Breeze from the SE.
Gass   we went off under a gentle breeze of wind.
Ordway   A clear and pleasant morning…we then Set off under a gentle Breeze which
            happened to be favourable.
Whitehouse   We then set out with a fair fine breese of Wind

SATURDAY, SEPTEMBER 29

	Sunrise			4 p.m.		Mississippi River		
Temp	Weather	Wind	Temp	Weather	Wind	Rise/Fall	Feet	Inches
45a	f	SE	67	f	SE			

Gass   had fair weather
Ordway   the weather fair
Whitehouse   We set off early this morning, fine clear weather

SUNDAY, SEPTEMBER 30

	Sunrise			4 p.m.			Mississippi River		
Temp	Weather	Wind	Temp	Weather	Wind	Rise/Fall	Feet	Inches
42a	c a r	SE	52	c a r	SE			

Clark   between 7 & 9 oClock we proceeded on under a Double reafed Sail, & Some rain. Some rain & hard wind at about 10 oClock. We proceeded on under a verry Stiff Breeze from the SE, the Stern of the boat got fast on a log and the boat turned & was verry near filling before we got her righted, the waves being verry high, The Chief on board was So fritined at the motion of the boat which in its rocking caused Several loose articles to fall…he rain off and hid himself   This day is cloudy & rainey— verry Cold & wind— a verrey Cold evening.

Gass   a cloudy morning. The wind was fair and we made 9 miles by 10 o'clock. We halted and spoke to them and then went on under a fine breeze of wind. A short time before night, the waves ran very high and the boat rocked a great deal, which so alarmed our old chief, that he would not go any further.

Ordway   we set off eairly under a fine Breeze of wind from the E. The weather being cool, cloudy a mist of rain our officers gave each man of the part a draghm. We hoisted our Sails & Salied…. The barge…she Swang round in the stream   the wind being So had from E that caused the waves to run high the Boat got in the trough & She rocked verry much before we could git hir Strait   we hoisted Sail and cam Strait. Sailed verry fast.

Whitehouse   a cloudy morning. We proceeded on with a favourable breeze of Wind, towards evening, the Waves rain very high and out boat Rocked exceedingly—

Note   *The expedition is in north central South Dakota, between present-day Chamberlain and Mobridge.*

MONDAY, OCTOBER 1

	Sunrise			4 p.m.			Mississippi River		
Temp	Weather	Wind	Temp	Weather	Wind	Rise/Fall	Feet	Inches
40	c	SE	46	c	SE			

Weather Diary   the leaves of the ash popular & most of the shrubs begin to turn yellow and decline

Clark   the wind blew hard from the SE all last night, verry Cold. Set out early, the wind Still hard. The Current jentle river still verry wide and falling a little. The wind So hard we Came to & Stayed 3 hours   after it Slackened a little we proceeded on round a bend, the wind in the after part of the day a head. Continued on with the wind imediately a head. Verry windy & Cold—

Gass   the morning was cloudy but the wind fair and we sailed rapidly. The sand bars are so numerous, that we had great difficulty to get along.

Ordway   we Set of as usal under a hard Breeze from E Sailed on verry well past an island the wind Blew so hard that it was difficult to find the channel   hoisted Sail   at 2 oC where the wind came a head.

Whitehouse   a cloudy morning. Fare wind. Hoisted all sail and made great headway. Sailed on rapidly

TUESDAY, OCTOBER 2

	Sunrise			4 p.m.		Mississippi River		
Temp	Weather	Wind	Temp	Weather	Wind	Rise/Fall	Feet	Inches
39	f	SE	75	c	NW*			

Clark   a Violent wind all night from the SE. Slackened a little we proceeded on. Some black clouds flying. The mid day verry worm. The after part of this day is pleasent then verry windy and Cold. The wind changed to the NW & rose verry high and hard and Cold from the NW Continud. Cold, the current of the river less rapid as below the Chien, & retains less Sediment than below.

Gass   we set sail before day light

Ordway   the wind Shifted to NW

WEDNESDAY, OCTOBER 3

	Sunrise			4 p.m.		Mississippi River		
Temp	Weather	Wind	Temp	Weather	Wind	Rise/Fall	Feet	Inches
40	c	NW	45	c a r & f	NW*			

Weather Diary   the earth and sand which form the bars of the river are so fully impregnated with salt that it shoots and adhers to the little sticks which appear on the serface   it is pleasent & seems niterous

---

*According to Moulton (1987, 3: 223), Clark's Journal Codex C lists this wind direction as "N."

Clark   the NW wind blew verry hard all night with Some rain. A Cold morning. Some rain this after noon– saw Brant & white fulls flying Southerly in large flocks—

Gass   The morning was cloudy, and some rain fell. About 12 o'clock the wind began to blow so hard down stream, that we were unable to proceed, and we halted under some high bluffs, At 3 we continued our voyage.

Ordway   the wind raised at 1 oClock last night & blew hard from NW & continues to blow this morning. So that it detained us untill ½ past 7 oClock. Cloudy. Some Thunder last night. A little rain this morning. The wind so hard a head that we halted about noon at a black Bluff SS. Delayed about 3 hours & proceeded on 3 miles found we had the rong channel the water shallow

Whitehouse   This morning it was cloudy, attended with some Rain. We sett of at half past 7 oClock and proceeded on, the Wind blowing hard down the River from the West; we came too at 9 oClock AM and lay by 'till 3 oClock PM.

THURSDAY, OCTOBER 4

	Sunrise			4 p.m.			Mississippi River	
Temp	Weather	Wind	Temp	Weather	Wind	Rise/Fall	Feet	Inches
38	c a r	NW	50	c	NW			

Clark   the Wind blew all night from the SW, Some rain. Set out, the wind hard a head. The day verry Cool. The evening verry Cold and wood Scerce, make use of the Drift wood.

Gass   the water was so shallow and sand bars so numerous.

FRIDAY, OCTOBER 5

	Sunrise			4 p.m.			Mississippi River	
Temp	Weather	Wind	Temp	Weather	Wind	Rise/Fall	Feet	Inches
36	f	NW	54	f	NW			

Weather Diary   slight white frost last night— brant & geese passing to South

Clark   frost this morning. the evening is calm and pleasant. River about the Same width, the Sand bars as noumerous, the earth Black and many of the Bluffs have the appearance of being on fire.

Gass   This morning there was a white frost; the day clear and pleasant.

Ordway   a white frost this morning. Clear & Cool.

Whitehouse   Some whight frost last night. The day clear and pleasant.

SATURDAY, OCTOBER 6

	Sunrise			4 p.m.		Mississippi River		
Temp	Weather	Wind	Temp	Weather	Wind	Rise/Fall	Feet	Inches
43	f	NW	60	f	NW			

Weather Diary   frost as last night— saw teal, mallard, & Gulls large.

Clark   a cool morning, Cold Wind from the North.

Gass   had a clear day.

Whitehouse   clear pleasant weather this morning

SUNDAY, OCTOBER 7

	Sunrise			4 p.m.		Mississippi River		
Temp	Weather	Wind	Temp	Weather	Wind	Rise/Fall	Feet	Inches
45	c	SE	58	f	SE			

Clark   a Cloudy morning. Some little rain frost last night. We proceeded on under a gentle Breeze from the SW before 10 oClock AM. Wind hard from the South, a fine evening.

Gass   had a clear day. Came to the Cerwercerna [Moreau] River about 90 yards wide. It is not so sandy as the Missouri, and the water is clear with a deep channel.

Ordway   a clear and pleasant morning   a Small Shower of rain   The wind more from the S   Sailed on

Whitehouse   clear weather this day

Note   *The expedition passes present-day Mobridge, South Dakota.*

MONDAY, OCTOBER 8

	Sunrise			4 p.m.		Mississippi River		
Temp	Weather	Wind	Temp	Weather	Wind	Rise/Fall	Feet	Inches
48	f	NW	62	f	NW			

Weather Diary   arrived at Recare vilage,

Clark   a cool Morning, wind from the NW.
Gass   The morning was pleasant   The river here is very shallow and full of sand bars.
Ordway   a pleasant morning   the wind from the North.
Whitehouse   pleasant weather this morning

TUESDAY, OCTOBER 9

Sunrise			4 p.m.			Mississippi River		
Temp	Weather	Wind	Temp	Weather	Wind	Rise/Fall	Feet	Inches
45	c	NE	50	c a r	N			

Weather Diary   wind blew hard this morning drove the boat from her anker, came
to Shore, some brant & geese passing to the south, (spoke to them
recares)

Clark   A windey rainey night, and cold, Some rain, and the [wind] Continued So high &
cold So much So we Could not Speek with the Indians to day. The day continued
Cold and windey   Some rain. The wind verry high. I observed Canoos...cross the
river at the time the waves were as high as I ever Saw them in the Missouri—   This
evening very cold &c &c
Gass   The day was stormy, and we remained here
Ordway   blustering cold wind this morning. Some Showers of rain. Some chiefs & other
Indians came to See us; but it being So cold & windy that they did not assemble
for counsel.
Whitehouse   This day we had Stormy weather, we lay by. A Stormy day.

WEDNESDAY, OCTOBER 10

Sunrise			4 p.m.			Mississippi River		
Temp	Weather	Wind	Temp	Weather	Wind	Rise/Fall	Feet	Inches
42	f a r	NW	67	f	NW			

Weather Diary   had the mill erected shewed the savages its operation, spoke to them
shot my airgun. The men traded some articles for robes, the savages
much pleased, the French chief lost his presents by his canoe
overseting

Clark   a fine forming wind from the SE. at 11 oClock the wind Shifted to NW
Ordway   a pleasant morning

THURSDAY, OCTOBER 11

	Sunrise			4 p.m.		Mississippi River		
Temp	Weather	Wind	Temp	Weather	Wind	Rise/Fall	Feet	Inches
43	f	NW	59	f	NW			

Weather Diary no fogg or dew this morning nor have we seen either for many days since the 21st of Septr.— received the answer at the 1st Chief, set out

Clark a fine morning the wind from the SE. all Tranquillity—

Gass A clear day.

Ordway a clear and pleasant morning the wind from the NW

FRIDAY, OCTOBER 12

	Sunrise			4 p.m.		Mississippi River		
Temp	Weather	Wind	Temp	Weather	Wind	Rise/Fall	Feet	Inches
42	f	S	65	f	SE			

Weather Diary set out at 2 in the evening

Clark the evening Clear & pleasant Cooler

Gass We had a pleasant morning

Ordway a clear and pleasant morning

Whitehouse this morning we had pleasant weather

SATURDAY, OCTOBER 13

	Sunrise			4 p.m.		Mississippi River		
Temp	Weather	Wind	Temp	Weather	Wind	Rise/Fall	Feet	Inches
43	f	SW	49	c a r	NE			

Weather Diary cottonwood all yellow and the leaves begin to fall, abundance of grapes and red burries—

Clark a fine Breez from SE. Camped. Cold & Some rain this evening. The river narrow current jentle

Gass had a cloudy day

Ordway Cloudy about 12 oClock it rained Some.

Whitehouse we Set off eairly clouday, about 12 oClock it rained some.

Note *The expedition crosses the present-day South Dakota–North Dakota border.*

SUNDAY, OCTOBER 14

Sunrise			4 p.m.			Mississippi River		
Temp	Weather	Wind	Temp	Weather	Wind	Rise/Fall	Feet	Inches
42	r	SE	40	r	SE			

Weather Diary   the leaves of all the trees as ash, elm &c except the cottonwood is now fallen—

Clark   Some rain last night all wet & Cold, we Set our early the rain contind all day. The wind a head from NW. The evening wet and disagreeable. The river Something wider more timber on the banks

Gass   We had a cloudy morning and some rain. It rained slowly during the whole of the day.

Ordway   Cloudy & rain. It rained slowly the greater part of the Day.

Whitehouse   Cloudy. Some rain. It continued raining all this day

MONDAY, OCTOBER 15

Sunrise			4 p.m.			Mississippi River		
Temp	Weather	Wind	Temp	Weather	Wind	Rise/Fall	Feet	Inches
46	r	N	57	f a r	NW			

Clark   Rained all last night. The evening was pleasent, wind from the NE.

Gass   It rained all last night, and we set out early in a cloudy morning.

Ordway   Some rain last night. Cloudy morning.

Whitehouse   rained all last night.

TUESDAY, OCTOBER 16

Sunrise			4 p.m.			Mississippi River		
Temp	Weather	Wind	Temp	Weather	Wind	Rise/Fall	Feet	Inches
45	c	NE	50	f	NE			

Clark   Some rain this morning. Wind hard a ahead from the NW.

Gass   had a clear morning.

Ordway   a clear and pleasant morning   we proceeded on under a gentle breeze from the SW

WEDNESDAY, OCTOBER 17

	Sunrise			4 p.m.		Mississippi River		
Temp	Weather	Wind	Temp	Weather	Wind	Rise/Fall	Feet	Inches
47	f	NW	54	f	NW			

Weather Diary   saw a large flock of White geese with Black wings, Antilopes are passing to the black hills to winter, as is their custom

Clark   a fine morning the wind from the NW. The wind rose So high that the Boat lay too all Day. The leaves are falling fast—   the river wide and full of Sand bars—

Gass   had a clear morning. At half past ten the wind blew so hard down the river that we were obliged to halt.

Ordway   the weather clear. The wind from NW   towards evening the wind abated So that we proceeded on untill some time after dark before we found a good place to camp.

Whitehouse   the Wind blowing hard against us at West, so that it occasion'd our getting on slowly, part of the day. The River running strong against us

THURSDAY, OCTOBER 18

	Sunrise			4 p.m.		Mississippi River		
Temp	Weather	Wind	Temp	Weather	Wind	Rise/Fall	Feet	Inches
30	f	NW	68	f	NW			

Weather Diary   hard frost last night, the clay near the water edge was frozen, as was the water in the vessels exposed to the air.

Clark   a fine Day

Gass   We had a clear pleasant morning with some frost.

Ordway   a clear and pleasant morning   white frost & forst Some last night.

FRIDAY, OCTOBER 19

	Sunrise			4 p.m.		Mississippi River		
Temp	Weather	Wind	Temp	Weather	Wind	Rise/Fall	Feet	Inches
43	f	SE	62	f	S			

Weather Diary   no Mule deer seen above the dog river none at the recares

Clark   a fine morning. Set out early under a gentle Breeze from the SE. All the Streems falling from the Hills or high lands So brackish that the water Can't be Drank. This day is pleasant.

Gass   Early this morning we renewed our voyage, having a clear day and a fair wind

Ordway   a clear and pleasant morning   a gentle breeze from the South

Whitehouse   This morning being clear, we sent our Hunters out, and proceeded on our Voyage with a fair wind;

Note   *On October 20–21, the expedition passes present-day capital of North Dakota, Bismarck.*

SATURDAY, OCTOBER 20

Sunrise			4 p.m.			Mississippi River		
Temp	Weather	Wind	Temp	Weather	Wind	Rise/Fall	Feet	Inches
44	f	NW	48	f	N			

Weather Diary   much more timber than usual—   Saw the first black haws that we have seen for a long time—

Clark   Set our early this morning and proceeded on, the wind from the SE. The wind hard all Day from the N.E. & East

Gass   We were early under way this morning, which was very pleasant.

Ordway   a pleasant morning

Whitehouse   this morning, having pleasant weather

SUNDAY, OCTOBER 21

Sunrise			4 p.m.			Mississippi River		
Temp	Weather	Wind	Temp	Weather	Wind	Rise/Fall	Feet	Inches
31	s	NW	34	s	NW			

Weather Diary   the snow $1/2$ inch deep (Clark)

Clark   a verry Cold night. Wind hard from the NE. Some rain in the night which freesed as it fell, at Day light it began to Snow and Continud all the fore part of the day. A Cloudy afternoon. Camped. verry Cold ground Covered with Snow.

Gass   We had a disagreeable night of sleet and hail. It snowed during the forenoon, but we proceeded early on our voyage

Ordway   Some frozen rain last night   Snow this morning. The wind from NE   the current swift.   Snowed Slowly untill 12 oClock. A cool and chilley day.

Whitehouse   Last night we had rainy disagreeable Weather, We set out early this morning, Shortly after we had some Snow,

MONDAY, OCTOBER 22

	Sunrise			4 p.m.		Mississippi River		
Temp	Weather	Wind	Temp	Weather	Wind	Rise/Fall	Feet	Inches
35	c a s	NE	42	c	NE			

Weather Diary   the snow $1/2$ inch deep. (Lewis)

Clark   we Set out early, the morning Cold

Gass   Some snow fell last night, and the morning was cloudy and cold. At 9 we saw 11 Mandans, who, notwithstanding the coldness of the weather, had not an article of clothing except their breechclouts. At 10 o'clock the day became clear and pleasant.

Ordway   Some Snow last night. We set off eairly Cloudy & cool this morning.

Whitehouse   This morning was Cold & Cloudy,   at One oClock PM the weather cleared off and became pleasant.

TUESDAY, OCTOBER 23

	Sunrise			4 p.m.		Mississippi River		
Temp	Weather	Wind	Temp	Weather	Wind	Rise/Fall	Feet	Inches
32	s	NW	45	c	NE			

Clark   a cloudy morning Some Snow. Camped. Cold & Cloudy.

Gass   Some snow again fell last night, and the morning was cloudy. At 8 it began to snow, and continued snowing to 11, when it ceased.

Ordway   a little Snow last night   A cloudy morning

Whitehouse   about 8 oClock [AM] it began to snow

WEDNESDAY, OCTOBER 24

	Sunrise			4 p.m.		Mississippi River		
Temp	Weather	Wind	Temp	Weather	Wind	Rise/Fall	Feet	Inches
33	s a f	NW	51	c a s	NW			

Weather Diary   arrived at a mandane hunting camp visited the lodge of the chief

Clark   a Cloudy day. Some little Snow in the morning.

Gass   We set out early in a cloudy morning. At 9 it began to rain and continued to rain for an hour.

Whitehouse   This morning was Cloudy, we set off early as usual. At 9 oClock AM it began to rain

Note   The expedition arrives in the vicinity of their intended winter camp, west of the present-day town of Washburn, North Dakota. After traveling upriver through October 30, looking for a suitable location for winter quarters, they will return downriver a few miles, where they will commence building Fort Mandan.

THURSDAY, OCTOBER 25

Sunrise			4 p.m.			Mississippi River		
Temp	Weather	Wind	Temp	Weather	Wind	Rise/Fall	Feet	Inches
31	c	SE	50	c	SE			

Weather Diary   this evening passed a rapid and sholde place in the river were obliged to get out and dragthe boat— all the leaves of the trees have now fallen— the snow did not lye.

Clark   a Cold morning. Set our early under a gentle Breeze from the SE by E. Wind Shifted to the SW at about 11 oClock and blew hard untill 3 OCk. Clouded up. The wind blew verry hard this evening from the SW. Verry Cold.

Gass   The morning was pleasant, and we set sail early with a fair wind.

Ordway   a clear morning. We Set off eairly under a fine breeze from the S. Sailed on

Whitehouse   a fair wind & pleasant weather this morning

FRIDAY, OCTOBER 26

Sunrise			4 p.m.			Mississippi River		
Temp	Weather	Wind	Temp	Weather	Wind	Rise/Fall	Feet	Inches
42	f	SE	57	f	SE			

Clark   wind from the SE.

Gass   had a clear morning.

Ordway   a clear morning

Whitehouse   This morning we had clear & pleasant Weather

SATURDAY, OCTOBER 27

Sunrise			4 p.m.			Mississippi River		
Temp	Weather	Wind	Temp	Weather	Wind	Rise/Fall	Feet	Inches
39	f	SW	58	f	SW			

Weather Diary   camp for the purpose of speaking to the five villages....at the place we intended to fix our camp

Clark   a fine w[a]rm Day

Gass   The morning was clear and pleasant

Ordway   a clear and pleasant morning

Whitehouse   We had pleasant weather, and we set out early, and proceeded on our Voyage.

SUNDAY, OCTOBER 28

Sunrise			4 p.m.			Mississippi River		
Temp	Weather	Wind	Temp	Weather	Wind	Rise/Fall	Feet	Inches
34	f	SW	54	f	SW			

Weather Diary   wind so [hard] that we could not go into council

Clark   a windey Day, fair and Clear. The wind So violently hard from the SW we could not meet the Indians in Council

Gass   The day was clear, and we remained here; but could not sit in council, the wind blew so violent.

Ordway   a clear morning. The blew verry high from the NW

Whitehouse   This morning, we had fine clear weather, which continued the whole day, the wind commenced blowing and blew so hard, that we could not sit in Council with the Savages

MONDAY, OCTOBER 29

Sunrise			4 p.m.			Mississippi River		
Temp	Weather	Wind	Temp	Weather	Wind	Rise/Fall	Feet	Inches
32	f	SW	59	f	SW			

Weather Diary   we Spoke to the Indians in council— tho' the wind was so hard it was extremely disagreeable. The sand was blown on us in clouds—

Clark   A fair fine morning. At 10 oClock the SW wind rose verry high— this evening…The Prarie was Set on fire (or Cought by accident) by a young man of the Mandins, the fire went with Such velocity that it burnt to death a man and woman, who Could not Get to an place of Safty, one man a woman & Child much burnt and Several narrowly escaped the flame— This fire passed our Camp last about 8 oClock PM it went with great rapidity and looked Tremendious

Gass   We had again a clear day

Ordway   a clear and pleasant morning

Whitehouse   This morning we had fine clear weather.

TUESDAY, OCTOBER 30

	Sunrise			4 p.m.			Mississippi River	
Temp	Weather	Wind	Temp	Weather	Wind	Rise/Fall	Feet	Inches
32	f	SW	52	f	SW			

Weather Diary   Capt. Clark visited the island above to look out a place for winter encampment, but did not succeed

Clark   in the evening…the Wind SE.

Gass   The day was clear and pleasant.

Ordway   a clear and pleasant morning

Whitehouse   This day we had clear & pleasant weather,

WEDNESDAY, OCTOBER 31

	Sunrise			4 p.m.			Mississippi River	
Temp	Weather	Wind	Temp	Weather	Wind	Rise/Fall	Feet	Inches
33	f	W	48	f	W			

Lewis   The river being very low and the season so far advanced that it frequently shuts up with ice in this climate we determined to spend the Winter in this neighbourhood.

Clark   a fine morning. The wind blew hard all the after part of the day from the NE and Continued all night to blow hard from that point, in the morning it Shifted NW.

Gass   A pleasant morning.

Ordway   a clear and pleasant morning. The wind Blew high from the South

Whitehouse   This morning we had fine pleasant Weather

Note   *The expedition is in the vicinity of the Mandan villages, west of present-day Washburn, North Dakota, in search of a location to build their winter camp.*

THURSDAY, NOVEMBER 1

Sunrise			4 p.m.			Mississippi River		
Temp	Weather	Wind	Temp	Weather	Wind	Rise/Fall	Feet	Inches
31	f	NW	47	f	NW			

Weather Diary   the winds blue so [hard] this day that we could not decend the river untill after 5 PM when we left our

Lewis   The wind blew so violently during the greater part of this day that we were unable to quit our encampment; in the evening it abated.

Clark   the wind hard from the NW.

Ordway   a clear & [illegible] morning   the wind high from the NW, cool   the wind abated. But the River So Shallow that we Struck the Sand bars.

Whitehouse   This morning the wind blew So fresh and hard from the South that we Could not set off at the time appointed. At 3 oClock we set of,

SECTION 4 Fort Mandan

November 2, 1804 to April 6, 1805

The expedition established Fort Mandan on November 2, 1804, near the Mandan and Hidatsa villages, the last mapped outposts in a largely unexplored territory. They remained at the fort, their winter quarters, until April 7, 1805. At Fort Mandan, located near today's Washburn, North Dakaota, they would endure a harsh winter with the rigors of prairie winter, including frequent snows, blizzards, and extreme temperatures—indeed, the coldest temperature of the year. They saw a variety of natural phenomena, including sun dogs, mirages, northern lights, and even an eclipse of the moon. They interacted with the villagers—even partook in hunting ceremonies—and met the Shoshone woman who would become a tremendous asset during the journey: Sakagawea. As spring neared, the party moved quickly to free the keelboat and pirogues from the icy barrier of the Missouri River before breakup. By late March, the ice was flowing and temperatures were warming.

As in the previous section, each day's entry begins with the data from the Weather Diary's Observation Tables (when available), followed by the corresponding remarks (when available), and finally excerpts that pertain to weather and climate from the expedition members' daily narrative journals.

White at Fort Mandan, river observations on the Missouri River were taken at sunrise for a 24-hour period.

Weather data included in this section are from Coues (1893, 3: 1268–1288), Moulton (1987, vols. 3 and 4), and Thwaites (1904, 6, part 2: 178–188).

Different journals and notebooks were used during the expedition. For a more detailed explanation on the journals and entry practices consult Cutright (1976) and Moulton (1986, 2: 8–48, 530–567).

Note   *On November 2, 1804, the expedition establishes its winter quarters, Fort Mandan.*

FRIDAY, NOVEMBER 2

	Sunrise			4 p.m.			Missouri River	
Temp	Weather	Wind	Temp	Weather	Wind	Rise/Fall	Feet	Inches
32	*f*	SE	63	*f*	SE			

Weather Diary   the boat dropped down to our winter station & formed a camp

Clark   the wind from the SE, a fine day

Ordway   a cloudy morning

Whitehouse   This morning we began to build the Fort, having pleasant weather for 14 days

123

SATURDAY, NOVEMBER 3

	Sunrise			4 p.m.		Missouri River		
Temp	Weather	Wind	Temp	Weather	Wind	Rise/Fall	Feet	Inches
32	f	NW	53	f	NW			

Weather Diary   wind blew hard all day (Lewis)   wind hard this evening
(Clark)

Clark   a fine morning wind hard from the west.

Gass   A clear day

Ordway   a clear and pleasant morning

Whitehouse   Pleasant

SUNDAY, NOVEMBER 4

	Sunrise			4 p.m.		Missouri River		
Temp	Weather	Wind	Temp	Weather	Wind	Rise/Fall	Feet	Inches
31	f	NW	43	c	W			

Weather Diary   wind hard this evening.

Clark   a fine morning. The wind rose this evening from the East & Clouded
up—

Ordway   cold last night & white frost this morning. Clear and pleasant.

Whitehouse   Pleasant

MONDAY, NOVEMBER 5

	Sunrise			4 p.m.		Missouri River		
Temp	Weather	Wind	Temp	Weather	Wind	Rise/Fall	Feet	Inches
30	c	NW	58	c	NW			

Clark   the Greater part of this day Cloudy, wind moderate from the NW.

Ordway   a clear and pleasant morning

Whitehouse   Pleasant

TUESDAY, NOVEMBER 6

	Sunrise			4 p.m.			Missouri River	
Temp	Weather	Wind	Temp	Weather	Wind	Rise/Fall	Feet	Inches
31	c	SW	43	c	W			

Weather Diary   some little hail about noon—

Clark   last night late we wer awoke by Sergeant of the Guard to See a northern light, which was light, (but) not red, and appeared to Darken and Some times nearly obscured (about 20 degrees above horizon — various Shapes), and open, many times appeared in light Streeks, and at other times a great Space light & containing floating Collomns which appeared to approach each other & retreat leaving the lighter space at no time of the Same appearence. This morning I rose at Day light the Clouds to the North appeared black. At 8 oClock the [wind] begun to blow hard from the NW and Cold, and Continud all Day.

Ordway   it was uncommon light in the north the Greater part of last night. A clear morning. About 9 oC it clouded up   cold   look likely for Snow

Whitehouse   Pleasant

WEDNESDAY, NOVEMBER 7

	Sunrise			4 p.m.			Missouri River	
Temp	Weather	Wind	Temp	Weather	Wind	Rise/Fall	Feet	Inches
43	c	s	62	c	S			

Weather Diary   a few drops of rain this evening—   saw the arrora. borialis at 10 PM it was very briliant in perpendiculer collums frequently changing position—

Clark   a temperate day. Cloudy and fogging all day

Ordway   a Cloudy morning.

Whitehouse   Pleasant

THURSDAY, NOVEMBER 8

	Sunrise			4 p.m.			Missouri River	
Temp	Weather	Wind	Temp	Weather	Wind	Rise/Fall	Feet	Inches
38	c	S	39	c	W			

Weather Diary   Since we have been at our present station the River has fallen about nine inches

Clark   a Cloudy morning
Ordway   Cloudy
Whitehouse   Pleasant

FRIDAY, NOVEMBER 9

	Sunrise			4 p.m.			Missouri River		
Temp	Weather	Wind	Temp	Weather	Wind	Rise/Fall	Feet	Inches
27	f	NW	43	f	NW			

Weather Diary   very [hard] frost this morning—

Clark   a verry hard frost this morning. Day Cloudy wind from the NW. Great number of wild gees pass to the South, flew verry high
Ordway   a hard white frost last night. a clear and pleasant morning
Whitehouse   Pleasant

SATURDAY, NOVEMBER 10

	Sunrise			4 p.m.			Missouri River		
Temp	Weather	Wind	Temp	Weather	Wind	Rise/Fall	Feet	Inches
34	f	NW	36	c	NW			

Weather Diary   many gees passing to the South—   saw a flock of the crested cherry birds passing to the south

Clark   the Day raw and Cold wind from the NW. the Gees Continue to pass in gangues as also brant to the South. Some Ducks also pass.
Ordway   Cloudy & cold.
Whitehouse   Pleasant

SUNDAY, NOVEMBER 11

	Sunrise			4 p.m.			Missouri River		
Temp	Weather	Wind	Temp	Weather	Wind	Rise/Fall	Feet	Inches
28	f	NW	60	f	NW			

Clark   a Cold Day. The large Ducks pass to the South.
Ordway   a clear and pleasant morning
Whitehouse   Pleasant

MONDAY, NOVEMBER 12

	Sunrise			4 p.m.			Missouri River	
Temp	Weather	Wind	Temp	Weather	Wind	Rise/Fall	Feet	Inches
18	f	N	31	f	NE			

Clark   a verry Cold night. Wind Changeable verry cold evening, freesing all day. Some ice on the edges of the river. Swans passing to the South.
Ordway   Clear & cold this morning. A verry hard frost. froze some last night.
Whitehouse   Pleasant

Note   **On November 13, Lewis "resumes noting the fall and rise of the river, which was only possible while they remained in one place during the day." (Moulton 1987, 3: 250)**

TUESDAY, NOVEMBER 13

	Sunrise			4 p.m.			Missouri River	
Temp	Weather	Wind	Temp	Weather	Wind	Rise/Fall	Feet	Inches
18	s	SE	28	c a s	SE	F	1	$^1/_2$

Weather Diary   large quantity of drift ice running this morning   the river has every appearance of closing for the winter
Clark   The Ice began to run in the river $^1/_2$ past 10 oClock PM. Snowed all day, the Ice ran thick and air Cold.
Ordway   Snowey morning   the Ice run considerable fast in the river. The Ice running against their legs. Their close frooze on them. One of them got 1 of his feet frost bit. It hapned that they had some wiskey with them to revive their Spirits.
Whitehouse   Pleasant

WEDNESDAY, NOVEMBER 14

	Sunrise			4 p.m.			Missouri River	
Temp	Weather	Wind	Temp	Weather	Wind	Rise/Fall	Feet	Inches
24	s	SE	32	c a s	SE	R	1	0

Clark   a Cloudy morning. Ice running verry thick, river rose $^1/_2$ Inch last night. Some Snow falling.
Ordway   a Snowey morning.
Whitehouse   Pleasant

THURSDAY, NOVEMBER 15

	Sunrise			4 p.m.		Missouri River		
Temp	Weather	Wind	Temp	Weather	Wind	Rise/Fall	Feet	Inches
22	c	NW	31	c a s	NW	R		1/2

Clark   a Cloudy morning. The ice run much thicker than yesterday. Dispatched a man with order to the hunters to proceed without Delay thro the floating ice, The wind Changeable— Swans passing to the South— but fiew fowls water to be Seen—

Ordway   Cloudy   The Bow of the pearogue was cut with the Ice &.C

Whitehouse   Pleasant

FRIDAY, NOVEMBER 16

	Sunrise			4 p.m.		Missouri River		
Temp	Weather	Wind	Temp	Weather	Wind	Rise/Fall	Feet	Inches
25	c	NW	30	f	SE	R		1/4

Weather Diary   very hard frost this morning attatched to the limbs and boughs of the trees—

Clark   a verry white frost, all the trees all Covered with ice, Cloudy

Gass   about now, the weather became very cold, and the ice began to run in the river.

Ordway   a cold frosty night. the Trees were covered with frost which was verry course white & thick even on the Bows of the trees all this day. Such a frost I never Saw in the States. The air verry thick with fogg from the R.

SATURDAY, NOVEMBER 17

	Sunrise			4 p.m.		Missouri River		
Temp	Weather	Wind	Temp	Weather	Wind	Rise/Fall	Feet	Inches
28	f	SE	34	f	SE	R		1/4

Weather Diary   the frost of yesterday remained on the trees untill 2 PM when it descended like a shower of snow— swans passing from the N

Clark   a fine morning, last night was Cold. The ice thicker than yesterday.
Ordway   a cold clear morning. The frost fell from the trees by the Sun Shineing upon
them.

SUNDAY, NOVEMBER 18

Sunrise			4 p.m.			Missouri River		
Temp	Weather	Wind	Temp	Weather	Wind	Rise/Fall	Feet	Inches
30	*f*	SE	38	*f*	W	R		¼

Clark   a Cold morning. Some wind.
Ordway   clear & cold.

MONDAY, NOVEMBER 19

Sunrise			4 p.m.			Missouri River		
Temp	Weather	Wind	Temp	Weather	Wind	Rise/Fall	Feet	Inches
32	*f*	NW	48	*f*	NW	R	1	0

Weather Diary   the runing ice had declined

Clark   a Cold day. The ice continue to run. The wind bley hard from the NW by W
Ordway   the River Riscing   the wind from SW the weather moderates as the day is
pleasant.

TUESDAY, NOVEMBER 20

Sunrise			4 p.m.			Missouri River		
Temp	Weather	Wind	Temp	Weather	Wind	Rise/Fall	Feet	Inches
35	*f*	NW	50	*f*	W	R	1	¼

Weather Diary   little soft ice this morning; that from the boarder of the river came down
in such manner as to endanger the boat

Clark   a verry hard wind from the W. all the after part of the day. A temperate day.
Gass   had fine pleasant weather
Ordway   clear & pleasant   the day warm. The work go on as usal.

WEDNESDAY, NOVEMBER 21

Sunrise			4 p.m.			Missouri River		
Temp	Weather	Wind	Temp	Weather	Wind	Rise/Fall	Feet	Inches
33	c	S	49	f	SE	R		

Weather Diary   we got into our hut yesterday evening.—
Clark   a fine Day. Some wind from the SW.
Gass   had fine pleasant weather
Ordway   cloudy & warm

THURSDAY, NOVEMBER 22

Sunrise			4 p.m.			Missouri River		
Temp	Weather	Wind	Temp	Weather	Wind	Rise/Fall	Feet	Inches
37	f	W	45	f	NW	R		1/2

Clark   a fine morning. A warm Day fair afternoon.
Gass   had fine pleasant weather
Ordway   pleasant & warm

FRIDAY, NOVEMBER 23

Sunrise			4 p.m.			Missouri River		
Temp	Weather	Wind	Temp	Weather	Wind	Rise/Fall	Feet	Inches
38	f	W	48	f	NW			

Clark   a fair warm Day, wind from the SE. The river on a Stand having rose 4 Inches
        in all
Gass   had fine pleasant weather
Ordway   pleasant & warm

SATURDAY, NOVEMBER 24

Sunrise			4 p.m.			Missouri River		
Temp	Weather	Wind	Temp	Weather	Wind	Rise/Fall	Feet	Inches
36	f	NW	34	f	NW			

Clark   a warm Day. The wind from the SE—
Gass   had fine pleasant weather
Ordway   warm & pleasant. The work continued on as usal.

SUNDAY, NOVEMBER 25

Sunrise			4 p.m.			Missouri River		
Temp	Weather	Wind	Temp	Weather	Wind	Rise/Fall	Feet	Inches
34	f	W	32	f	SW			

Clark   a fine day warm & pleasant. River fall 1 $1/2$ inch.
Gass   had fine pleasant weather
Ordway   a pleasant morning.

MONDAY, NOVEMBER 26

Sunrise			4 p.m.			Missouri River		
Temp	Weather	Wind	Temp	Weather	Wind	Rise/Fall	Feet	Inches
15	f	SW	21	f	W			

Weather Diary   wind bleue verry hard,

Clark   a little before day light the wind shifted to the NW and blew hard and the air Keen & Cold all day, Cloudy and much the appearance of Snow; but little work done to day it being Cold &c.
Gass   had fine pleasant weather
Ordway   cold & windy.

TUESDAY, NOVEMBER 27

Sunrise			4 p.m.			Missouri River		
Temp	Weather	Wind	Temp	Weather	Wind	Rise/Fall	Feet	Inches
10	f	SE	19	c	SE	F	3	0

Weather Diary   much drift ice running in the river—

Clark   a cloudy morning after a verry Cold night, the River Crouded with floating ice. Wind from the NW. The river fall 2 inches. Very Cold and began to Snow at 8 oClock PM and Continued all night— at day the Snow seased.

Gass   on this night the snow fell seven inches deep

Ordway   cold & chilly, the Ice Ran in the River thick.

WEDNESDAY, NOVEMBER 28

Sunrise			4 p.m.			Missouri River		
Temp	Weather	Wind	Temp	Weather	Wind	Rise/Fall	Feet	Inches
12	s	SE	15	s	E	F	4	0

Clark   a cold morning wind from the NW. river full of floating ice, began to Snow at 7 oClock AM and continued all day. A verry disagreeable day— no work done to day. River fall 1 Inch to day.

Gass   stormy

Ordway   Snowed hard the Greater part of last night. Snow this morning. the wind from N.E. the River falling.

THURSDAY, NOVEMBER 29

Sunrise			4 p.m.			Missouri River		
Temp	Weather	Wind	Temp	Weather	Wind	Rise/Fall	Feet	Inches
14	c a s	NE	18	f	W	F	2	1/2

Weather Diary   the snow fell 8 inches deep— it drifted in heeps in the open growns—

Clark a verry Cold windey day wind from the NW by W. Some Snow last night. The depth of the Snow is various in the wood about 13 inches. The river Closed [frozen over] at the village above and fell last night two feet. A Cold after noon wind as usial NW. River begin to rise a little.

Gass This day was clear, but cold.

Ordway the Snow fell yesterday and last night about 12 Inches on a level. a cold Frosty clear morning....the river fell abt. 2 feet last night So that our Boat lay dry on Shore.

Friday, November 30

	Sunrise			4 p.m.		Missouri River		
Temp	Weather	Wind	Temp	Weather	Wind	Rise/Fall	Feet	Inches
17	f	W	23	f	W	F		

Weather Diary the indians pass over the river on the ice— (Lewis) returned in the evening on the ice. (Clark)

Clark Chief said...The Snow is deep and it id Cold, our horses Cannot Travel thro the plains in pursute. I then Paraded & Crossed the river on the ice and Came down on the N. Side, the Snow so deep, it was verry fatigueing arived at the fort after night. A Cold night. The river rise to its former hite

Gass This day was clear, but cold.

Ordway a clear Sharp frosty morning. Froze hard last night.

Whitehouse On our arrival at the Village, the Chiefs of both nations, concluded, not to go fight those Indians with us, they saying the Weather was cold, and the Snow was deep, (being upwards of 18 Inches on the Ground)

Saturday, December 1

	Sunrise			4 p.m.		Missouri River		
Temp	Weather	Wind	Temp	Weather	Wind	Rise/Fall	Feet	Inches
1b0	f	E	6	f	SE	R	1	0

Weather Diary Ice thick   wind hard

Clark wind from the NW.

Gass The day was pleasant

Ordway the morning fair.

SUNDAY, DECEMBER 2

	Sunrise			4 p.m.		Missouri River		
Temp	Weather	Wind	Temp	Weather	Wind	Rise/Fall	Feet	Inches
38a	f	NW	36	f	NW	R		1

Clark   The latter part of last night was verry warm and Continued to Thaw when the wind shifted to the North…before 11 oClock [a.m.]. River rise one inch.
Gass   The day was pleasant, and the Snow melted fast.
Ordway   a pleasant thoughy [thawing?] morning

MONDAY, DECEMBER 3

	Sunrise			4 p.m.		Missouri River		
Temp	Weather	Wind	Temp	Weather	Wind	Rise/Fall	Feet	Inches
26a	f	NW	30	f	NW	R		1

Clark   a fine morning, the after part of the day Cold & windey, the wind from the NW.
Gass   moderate weather
Ordway   cold & windy.

TUESDAY, DECEMBER 4

	Sunrise			4 p.m.		Missouri River		
Temp	Weather	Wind	Temp	Weather	Wind	Rise/Fall	Feet	Inches
18	f	N	29	f	N	R		1

Clark   a Cloudy raw Day. Wind from the NW. The river rise one inch.
Gass   moderate weather
Ordway   clear & cold

WEDNESDAY, DECEMBER 5

	Sunrise			4 p.m.			Missouri River		
Temp	Weather	Wind	Temp	Weather	Wind	Rise/Fall	Feet	Inches
14	c	NE	27	s	NE			

Weather Diary   Wind blew excessively hard this (morning) night from NW

Clark   a Cold raw morning, wind from the SE. Some Snow. A little Snow fell in the evening at which time the wind Shifted round to NE.

Gass   moderate weather

Ordway   cloudy & cold   look likely for Snow.

THURSDAY, DECEMBER 6

	Sunrise			4 p.m.			Missouri River		
Temp	Weather	Wind	Temp	Weather	Wind	Rise/Fall	Feet	Inches
10a	s	NW	11	c a s	NW			

Weather Diary   Capt. Clark was hunting the Buffaloe this day with 16 Men— severall of the men frosted killed 3 buffaloe himself and the party killed 4 others

Clark   The wind blew violently hard from the NNW with Some Snow. The air Keen and Cold. The Thermometer at 8 oClock AM Stood at 10 dgs. above 0—   Cold after noon. River rise 1 $1/2$ inch to day.

Gass   was so cold and stormy, we could do nothing. In the night the river froze over, and in the morning was covered with solid ice an inch and a half thick.

Ordway   a cold Blustry morning. Some Squalls of Snow & wind high it being So disagreeable weather that we delayed on the work.

FRIDAY, DECEMBER 7

	Sunrise			4 p.m.			Missouri River		
Temp	Weather	Wind	Temp	Weather	Wind	Rise/Fall	Feet	Inches
a0*	f	NW	1	c	NW	R	2	$1/2$

*According to Moulton (1987a, 3: 266), Clark's Journal Codex C lists this temperature as "0 a."

Weather Diary   last night the river blocked up with ice which was 1 $1/2$ inches thick in the part that had not previously frosen—

Clark   The weather so excesive Cold. A verry Cold day. Wind from the NW. The Thermormeter Stood this morning at 44 d. below [freezing]. The Thermometer Stood this Morning at 1 d. below 0. Three men badly frost bit to day—

Gass   A clear cold morning.

Ordway   a clear cold frosty morning. as the River Shut up last night the Ice had not Got Strong enofe to bear the Buffalow out in the middle of the R.

SATURDAY, DECEMBER 8

Sunrise			4 p.m.			Missouri River		
Temp	Weather	Wind	Temp	Weather	Wind	Rise/Fall	Feet	Inches
12b	s	NW	5	f a s	NW			

Weather Diary   The ice 1 $1/2$ inch thick on the part that had not previously frosen. (Lewis)   I hunt   3 men frosted. (Clark)

Clark   a verry Cold morning, The Thermometer Stood at 12 d. below zero, which is 42 d. [44 is correct reading] degrees below the freezing point, wind from the NW. This day being Cold Several men returned a little frost bit, one of [the] men with his feet badly frost bit, my Servents feet also frosted & his P— s a little, I feel a little fatigued haveing run after the Buffalow all day in Snow many Places 10 inches Deep, generally 6 or 8. 2 reflecting Suns to Day [parhelion].

Gass   In our hunt of yesterday, two men had their feet frost bitten. Captain Clark and another party went out though the cold was extreme, to hunt   One man got his hand frozen; another his foot; and some more got a little touched.

Ordway   the weather is 12 degrees colder this morning than I ever new it to be in the States. Clear   the wind NW   the air thick with Ice all this day, like a fog—

SUNDAY, DECEMBER 9

Sunrise			4 p.m.			Missouri River		
Temp	Weather	Wind	Temp	Weather	Wind	Rise/Fall	Feet	Inches
7a	f	E	10	f	NW			

Weather Diary   went hunting with a part [y] of fifteen men killed 10 Buffaloe and 1 deer staid out all night (Lewis)   "no blanket" (after referring to Lewis's remaining out all night) (Clark)

Clark   The thermometer Stood this morning at 7° above 0, wind from the E. The Sun Shown to day Clear.

Ordway   the morning pleasant but not So cold as it was yesterday

MONDAY, DECEMBER 10

	Sunrise			4 p.m.			Missouri River		
Temp	Weather	Wind	Temp	Weather	Wind	Rise/Fall	Feet	Inches
10b	*c*	*N*	*11*	*c*	*N*	*R*		*1 ¹/₂*

Clark   a verry Cold Day, The Thermometer to day at 10 & 11 Degrees below 0. Capt. Lewis had a Cold Disagreeable night last in the Snow on a Cold point with one Small Blankett. The Buffaloe Crossed the river below in emence herds without brakeing in. The men which was frost bit is gitting better. The rise 1 ¹/₂ inch. Wind North.

Gass   This day was very cold; an experiment was made with proof spirits, which in fifteen minutes froze into hard ice.

Ordway   a Cloudy cold morning. The Weather Gits colder verry fast So that the Sentinel had to be relieved every hour. The weather is [blank] degrees colder this evening than it was this morning. Blanket cappoes provided for each man who Stood in need of them &.C.

TUESDAY, DECEMBER 11

	Sunrise			4 p.m.			Missouri River		
Temp	Weather	Wind	Temp	Weather	Wind	Rise/Fall	Feet	Inches
21b	*f*	*N*	*18*	*f*	*N*	*F*		*¹/₂*

Clark   a verry Cold morning. Wind from the North. The Thermomettr at (4 oClock AM at 21°) Sunrise at 21°. Below 0 which is 53° below the freezing point and getting colder, the Sun Shows and reflects two imigies, the ice floating in the atmespear being So thick that the appearance is like a fog Despurceing [mirage?]. At night all the hunters returned, Several a little frosted. Continued Cold all day. River at a Stand.

Gass   Captain Lewis and Captain Clarke thinking the weather too cold to hunt, set men down to the camp to bring up the remainder of the meat   The cold was so severe they could do nothing with the other two buffaloe

Ordway   a clear cold morning.

WEDNESDAY, DECEMBER 12

Sunrise			4 p.m.			Missouri River		
Temp	Weather	Wind	Temp	Weather	Wind	Rise/Fall	Feet	Inches
38b	f	N	16	f	N			

Clark   a Clear Cold morning, wind from the North. The Thormometer at Sun rise Stood at 38° below 0, moderated untill 6 oClock at which time it began to get Colder. I line my Gloves and have a cap made of the Skin of the Lynx. the weather is So Cold that we do not think it prudent to turn out to hunt in Such Cold weather, or at least untill our Consts. are prepared to under go this Climate. I measure the river from bank to bank on the ice and make it 500 yards

Gass   We all remained at the garrison, the weather being intensely cold.

Ordway   Clear and cold. The frost was white in the Guard chimney where their was a fire kept all last night. It is Several degrees colder this morning than it has been before, so that we did nothing but git wood for our fires. our Rooms are verry close and warm, So we can keep ourselves warm and comfortable, but the Sentinel who Stood out in the open weather had to be relieved every hour all this day.

THURSDAY, DECEMBER 13

Sunrise			4 p.m.			Missouri River		
Temp	Weather	Wind	Temp	Weather	Wind	Rise/Fall	Feet	Inches
20b	f	SE	4	c	SE			

Clark   The last night was verry cold & the frost which fell Covered the ice old Snow & those parts which was naked ¹/₆ of an inch, The Thermometer Stands this morning at 20° below 0, a fine day. Find it impossible to make an Observation with an artifical Horsison. River falls

Gass   The weather this day, began to be more moderate.

Ordway   clear frosty morning but not So cold as it was yesterday

Whitehouse   Extremely cold

FRIDAY, DECEMBER 14

Sunrise			4 p.m.			Missouri River		
Temp	Weather	Wind	Temp	Weather	Wind	Rise/Fall	Feet	Inches
2b	c	SE	2	s	SE	F		1

Weather Diary   Capt. Clark sets out with a hunting party on the ice with three small sleds—

Lewis   out hunting…much Snow, verry cold. 52° below freesinge.

Clark   a fine morning. Wind from the SE. The murckerey Stood at 0 this morning. A verry Cold night, Snowed.

Gass   This day was more moderate, and light snow showers fell. The snow fell about three inches deep.

Ordway   Cloudy & moderate this morning

Whitehouse   Extremely cold

SATURDAY, DECEMBER 15

	Sunrise			4 p.m.		Missouri River		
Temp	Weather	Wind	Temp	Weather	Wind	Rise/Fall	Feet	Inches
8b	c a s	W	4	c a s	W			

Weather Diary   snow fell one $1/2$ inch— ….inform me that many buffaloe have visited ….they came from the west.

Lewis   out hunting…much Snow, verry cold.

Clark   a Cold Clear morning. The Snow fell 1 $1/2$ inches deep last night. Wind North—

Gass   A cloudy day. Some slight showers of snow fell during the day.

Ordway   Cloudy cold and Snowey. Although the day was cold & Stormy we Saw Several of the chiefs and warries were out at a play which they call [a Mandan hoop and pole game]

Whitehouse   Extremely cold. Some Snow fell that made the Air warmir to night

SUNDAY, DECEMBER 16

	Sunrise			4 p.m.		Missouri River		
Temp	Weather	Wind	Temp	Weather	Wind	Rise/Fall	Feet	Inches
22b	f	NW	4	f	NW	F		1

Clark   a clear Cold morning, the Thermtr. at Sun rise Stood at 22° below 0, a verry Singaler appearance of the Moon last night, as She appeared thro: the Frosty atmispear—

Gass   A clear cold day.

Ordway   Clear & cold.

Whitehouse   Captain Clark and his party…had remain'd all night in the woods, & some Snow falling last night, the Weather became more moderate, it being extremely Cold the three preceeding days—

MONDAY, DECEMBER 17

Sunrise			4 p.m.			Missouri River		
Temp	Weather	Wind	Temp	Weather	Wind	Rise/Fall	Feet	Inches
43b*	f	N	28	f	N	R		3

Weather Diary   at 8 PM this evening the Thertr. Stood at 42 b.o.

Clark   A very cold morning. The Thrmt. Stood at 43° (45) below 0. about
8 oClock P M. the thermometer fell to 74° below the freesing
pointe—

Gass   This was a cold clear day and we all remained in the garrison.

Ordway   a clear & cold morning. The Thurmometer Stood at about 35 fat. It
has been Several degrees lower Some days past.

Whitehouse   This day was clear and cold weather.

TUESDAY, DECEMBER 18

Sunrise			4 p.m.			Missouri River		
Temp	Weather	Wind	Temp	Weather	Wind	Rise/Fall	Feet	Inches
32b	f	W	16	f	SW	R		1

Weather Diary

Clark   The Thermometer the Same as last night. Mr. Haney & La Rocke left us
for the Grosventre Camp, Sent out 7 men to hunt for the Buffalow, they
found the weather too cold & returned. The River rise a little.

Gass   A very cold day. At 9 we returned and found the men from the NW
Company had set out on their return, notwithstanding the severity of
the weather.

Ordway   verry cold last night So that the Sentinel had to be relieved everry half hour
dureing all last night. A clear Sharp morning. The Thurm. S,. at 42 ds.
Hunters.…some of them went out on the hills but found it So cold that
they would not follow the Buffo in the praries so they returned to the
Fort.

Whitehouse   This day we had very Cold weather. The North West Traders left us this
morning, having come to take their leave of our Officers and Men, and
proceeded on their Journey notwithstanding the coldness of the
weather—

---

*According to Moulton (1987a, 3: 266), Clark's Journal Codex C lists this temperature as "45 b" and "43 b" in
Voorhis No. 4.

WEDNESDAY, DECEMBER 19

	Sunrise			4 p.m.		Missouri River		
Temp	Weather	Wind	Temp	Weather	Wind	Rise/Fall	Feet	Inches
2b	c	Sw	16	f	S	R		1

Weather Diary   began to Piquet the Fort on the river side—

Clark   The wind from SW the weather moderated a little. River rise a little.

Gass   This was a more pleasant day

Ordway   the weather has moderated Some Since yesterday morning. Half the men out at a time & relieved every hour, it being too cold to be out all the time.

Whitehouse   We had clear weather & a pleasant day

THURSDAY, DECEMBER 20

	Sunrise			4 p.m.		Missouri River		
Temp	Weather	Wind	Temp	Weather	Wind	Rise/Fall	Feet	Inches
24a	c 5	NW	37 5	f 5	NW 5	R 5		3 1/2*

Clark   The wind from the NW, a moderate day. The Thermometr 37° (24°) above 0, which givs an oppertunity of putting up our pickets next to the river, nothing remarkable took place to Day. River fall a little.

Gass   quite warm and pleasant

Ordway   Some cloudy & Warm this morning, but a pleasant day

Whitehouse   a quite warm day. Moderate weather. The Snow melted fast.

FRIDAY, DECEMBER 21

	Sunrise			4 p.m.		Missouri River		
Temp	Weather	Wind	Temp	Weather	Wind	Rise/Fall	Feet	Inches
22a	f	NW	22	c	W	R		2

---

*According to Moulton (1987, 3: 266), "There are several discrepancies between Lewis and Clark on this date; in Codex C Clark's sunrise weather is 'f'; his 4 p.m. temperature is '22a'; his 4 p.m. weather is 'c'; his 4 p.m. wind is 'W'; his river rise is 2 inches. His table in Voorhis No. 4 agrees with Lewis."

Clark   a fine Day w[a]rm and wind from the NW by W.
Gass   quite warm and pleasant
Ordway   the morning clear & warm
Whitehouse   We had still pleasant and warm Weather, and the Snow continued melting

SATURDAY, DECEMBER 22

	Sunrise			4 p.m.		Missouri River		
Temp	Weather	Wind	Temp	Weather	Wind	Rise/Fall	Feet	Inches
10a	f	NW	23	f	NW	R		2 ½

Gass   The weather continued clear, pleasant and warm.
Ordway   pleasant moderate weather.
Whitehouse   clear fine pleasant warm weather to day

SUNDAY, DECEMBER 23

	Sunrise			4 p.m.		Missouri River		
Temp	Weather	Wind	Temp	Weather	Wind	Rise/Fall	Feet	Inches
18a	c	SW	27	c	W	F		1

Clark   a fine Day.
Gass   The weather continued pleasant
Ordway   a clear and pleasant morning.
Whitehouse   The weather still continued clear and pleasant

MONDAY, DECEMBER 24

	Sunrise			4 p.m.		Missouri River		
Temp	Weather	Wind	Temp	Weather	Wind	Rise/Fall	Feet	Inches
22a	s	SW	31	c a s	W	F		2 ½

Weather Diary   Snow verry inconsiderable   complete the fort (Lewis) Snow very much
(Clark)

Clark   A fine Day

Gass   Some snow fell this morning; about 10 it cleared up, and the weather became
pleasant.

Ordway   Some Snow this morning   the afternoon pleasant.

Whitehouse   Some Snow fell this morning, at about 10 oClock AM it cleared away, and
we had fair weather the remainder of the day

TUESDAY, DECEMBER 25

Sunrise			4 p.m.			Missouri River		
Temp	Weather	Wind	Temp	Weather	Wind	Rise/Fall	Feet	Inches
15	s	NW	20	c a s	NW	F		1

Weather Diary   Snow verry inconsiderable

Gass   we hoisted the American flag in the garrison, and its first waving in fort Mandan was
celebrated with another glass—

Ordway   cloudy

WEDNESDAY, DECEMBER 26

Sunrise			4 p.m.			Missouri River		
Temp	Weather	Wind	Temp	Weather	Wind	Rise/Fall	Feet	Inches
18	c	NW	21	f	NW			

Clark   a temperate day

Ordway   pleasant

THURSDAY, DECEMBER 27

Sunrise			4 p.m.			Missouri River		
Temp	Weather	Wind	Temp	Weather	Wind	Rise/Fall	Feet	Inches
4b	c	NW	14	c	NW			

Weather Diary   The trees all this day with the white frost which attached itself to their boughs

Clark   a little fin Snow, weather something Colder than yesterday. Wind hard from the NW.

Ordway   cloudy Some Snow

FRIDAY, DECEMBER 28

	Sunrise			4 p.m.		Missouri River		
Temp	Weather	Wind	Temp	Weather	Wind	Rise/Fall	Feet	Inches
12a	f	N	13	f	NW	R		2 1/2

Clark   blew verry hard last night, the frost fell like a Shower of Snow, nothing remarkable to day, the Snow Drifting from one bottom to another and from the leavel plains into the hollows &c.

Ordway   clear & cold. High wind.

SATURDAY, DECEMBER 29

	Sunrise			4 p.m.		Missouri River		
Temp	Weather	Wind	Temp	Weather	Wind	Rise/Fall	Feet	Inches
9b	f	N	3	f	N	R		1

Weather Diary   the wind blue verry hard last night. the frost fell like a shower of snoe

Clark   The frost fell last night nearly a 1/4 of an inch Deep and Continud to fall untill the Sun was of Some hite, the Murcurey Stood this morning at 9 d below 0 which is not considered Cold, as the Changes take place gradually without long intermitions

Ordway   clear & cold.

SUNDAY, DECEMBER 30

	Sunrise			4 p.m.		Missouri River		
Temp	Weather	Wind	Temp	Weather	Wind	Rise/Fall	Feet	Inches
20b	f	N	11	f	N	R		1/2

Clark   Cold the Termtr. At 20 d below 0

Ordway   clear & cold this morning.

MONDAY, DECEMBER 31

	Sunrise			4 p.m.		Missouri River		
Temp	Weather	Wind	Temp	Weather	Wind	Rise/Fall	Feet	Inches
10b	f	SE	12	c	SW	R		1 ¹/₂

Clark   a fine Day, some wind last night which mixed the Snow and Sand in the bend of the river, which has the appearance of hillocks of Sand on the ice, which is also Covered with Sand & Snow, the frost which falls in the night continues on the earth & old Snow &c &c.

Ordway   a clear & cold morning.

Whitehouse   nothing particular occured Since christmas but we live in peace and tranquillity in our fort. The weather continued pleasant & the Air Serene—

Note   *Expedition members celebrate the New Year by firing two cannonballs. Some dance with the Mandans.*

TUESDAY, JANUARY 1

	Sunrise			4 p.m.		Missouri River		
Temp	Weather	Wind	Temp	Weather	Wind	Rise/Fall	Feet	Inches
18a	s	SE	34a	f	NW	R	1	0

Clark   The Day was w[a]rm. Themtr 34° above 0. Some fiew Drops of rain about Sunset, at Dark it began to Snow, and Snowed the greater part of the night. (The temptr for Snow is about 0)

Gass   The day was warm and pleasant.

Ordway   cloudy but moderate. Rained a little in the eve.

Whitehouse   the day was warm and pleasant.

WEDNESDAY, JANUARY 2

	Sunrise			4 p.m.		Missouri River		
Temp	Weather	Wind	Temp	Weather	Wind	Rise/Fall	Feet	Inches
4b	s	NW	8b	f a s	N			

Clark   a Snowey morning. Some Snow to Day. Verry Cold in the evening.

Gass   Some snow fell this morning. This day I discovered how the Indians keep their horses during the winter. In the day time they are permitted to run out and gather what they can; and at night are brought into the lodges, with the natives themselves

Ordway   Snowed fast this morning.

Whitehouse   This morning some Snow fell. The Mandan Indians in this Second Village had a number of horses, which they keep in their lodges with them, every Cold night during the Winter.

THURSDAY, JANUARY 3

Sunrise			4 p.m.			Missouri River		
Temp	Weather	Wind	Temp	Weather	Wind	Rise/Fall	Feet	Inches
14b	c	N	4b	s	SE			

Weather Diary   the Snow was not considerable   the ground is now covered 9 inches deep—

Clark   some Snow to day

Gass   The weather was generally very cold.

Ordway   Snowed this morning

FRIDAY, JANUARY 4

Sunrise			4 p.m.			Missouri River		
Temp	Weather	Wind	Temp	Weather	Wind	Rise/Fall	Feet	Inches
28a	c a s	W	4b	c	NW	R		2 1/2

Clark   a w[a]rm Snowey morning, the Themtr. at 28° abov 0, Cloudy. The evening the weather became cold and windey, wind from the NW.

Gass   The weather was generally very cold.

Ordway   Cloud, warm morning.   The afternoon blustry.

Whitehouse   This morning Clear, the weather is not as cold, the weather was moderate to what it had been some days past. In the Evening, the weather grew verry cold and the Wind blew hard from the NW all night—

Saturday, January 5

	Sunrise			4 p.m.		Missouri River		
Temp	Weather	Wind	Temp	Weather	Wind	Rise/Fall	Feet	Inches
20b	c	NW	18b	s	NE	R		2

Clark a cold day. Some Snow.
Gass The weather was generally very cold.
Ordway high blustry winds all last night & verry cold three of our hunters Stayed out all
         night.   A cold morning.
Whitehouse a cloudy cold day. The weather continued verry Cold. The Weather
         continuing very Cold—

Sunday, January 6

	Sunrise			4 p.m.		Missouri River		
Temp	Weather	Wind	Temp	Weather	Wind	Rise/Fall	Feet	Inches
11b	c a s	NW	16b	f	NW	R		3

Weather Diary at 12 oC. Today two Luminous spots appeared on either side of the sun
         extreemly bright

Clark a Cold day
Gass The weather was generally very cold.
Ordway a clear cold morning.   The wind high & blustry.
Whitehouse Cloudy, Cold weather

Monday, January 7

	Sunrise			4 p.m.		Missouri River		
Temp	Weather	Wind	Temp	Weather	Wind	Rise/Fall	Feet	Inches
22b	f	NW	14b	f	W	F		1

Clark   a verry Cold clear Day, the Thermtr Stood at 22 d below 0, wind NW. The river fell
        1 inch.
Gass   The weather was generally very cold.
Ordway   a clear cold morning.   The wind high from NW
Whitehouse   Cloudy, Cold weather

Tuesday, January 8

	Sunrise			4 p.m.		Missouri River		
Temp	Weather	Wind	Temp	Weather	Wind	Rise/Fall	Feet	Inches
20b	f	NW	10b	f	NW	R		1

Weather Diary   the snow is now ten inches deep. (Lewis) accumolateing by frosts
        (Clark)
Clark   a Cold Day. Wind from the NW.
Gass   The weather was generally very cold.
Ordway   the wind blew cold from NW
Whitehouse   Cloudy, Cold weather

Wednesday, January 9

	Sunrise			4 p.m.		Missouri River		
Temp	Weather	Wind	Temp	Weather	Wind	Rise/Fall	Feet	Inches
21b	f	W	18b	f a c	NW	R		1

Clark   A Cold Day, Thermometer at 21° below 0. The after part of this day verry
        Cold, and wind Keen.
Gass   The weather was generally very cold.
Ordway   Some Snow this morning   Squally the after part of the day   blustry and
        exceeding cold. A nomber of the Savages out hunting the Buffalo [again] &
        came in towards evening with their horses loaded with meat and told us that
        two of their young men was froze to death in the prarie.   Had Suffered
        considerable with the cold. We expected nothing else but the other man had
        froze or would freeze this night.   A young Indian came in the Garrison with
        his feet frost bit.
Whitehouse   the day proved to be very cold & Stormey, one of the them (hunters)
        returned to the fort about 8 oClock in the evening with one of his feet
        frost bit. The other Stayed out all night.

THURSDAY, JANUARY 10

Sunrise			4 p.m.			Missouri River		
Temp	Weather	Wind	Temp	Weather	Wind	Rise/Fall	Feet	Inches
40b	f	NW	28*	f	NW	R		1

Clark   last night was excessively Cold. The murkery this morning Stood at 40° below 0 which is 72 below the freezing point. The Indians...turned out to hunt for a man & boy who had not returned from the hunt of yesterday, and borrowed a Slay to bring them in, expecting to find them frozed to death. about 10 oClock the boy about 13 years of age Came to the fort with his feet frosed and had layed out last night without fire with only a Buffalow Robe to cover him. We had his feet put in Cold water and they are Comeing too— (this boy lost his Toes only) ...a man Came in who had also Stayed ou[t] without fire, and verry thinly Clothed, this man was not the least injured— Customs & the habits of those people has ancered to bare more Cold than I thought it possible for man to indure—

Gass   The weather was generally very cold. A number of the natives being out hunting in a very cold day, one of them gave out on his return in the evening; and was left in the plain or prairie covered with a buffalo robe. After some time he began to recover and removed to the woods, where he broke a number of branches to lie on, and to keep his body off the snow. In the morning he came to the fort, with his feet badly frozen, and the officers undertook his cure.

Ordway   a clear cold morning. It is the Same Boy that the Indians had left last night & expected that he was froze to death in the praries.

Whitehouse   This day we had severe cold weather. The weather still continued to be extremely Cold and Stormy, the Officers had some of our party preparing to go in search of the Man, who Staid out all night, believing from the severty of the weather that he had been froze to death— but fortunately he returned to the Fort, before they had started in good health— Some of the Natives came to our Fort, bringing with them one of their Nation, that was frost bitten— His feet very much bit by the frost.

FRIDAY, JANUARY 11

Sunrise			4 p.m.			Missouri River		
Temp	Weather	Wind	Temp	Weather	Wind	Rise/Fall	Feet	Inches
38b	f	NW	14b	f	NW	F		1/2

*According to Moulton (1987a, 3: 283), Clark's Journal Codex C lists this temperature as "28 b."

Clark   verry Cold
Gass   The weather was generally very cold.
Ordway   clear cold morning. Nothing extroordnary accured.—
Whitehouse   This day the weather still continued Cold & the Air very thin;

Note   *On January 12, the captains note in the Weather Diary the "singular appearance of three distinct Halo or luminus rings about the moon." This is caused by ice crystals in the atmosphere, which refract light (Moulton 1986, 2: 284).*

Saturday, January 12

	Sunrise			4 p.m.			Missouri River		
Temp	Weather	Wind	Temp	Weather	Wind	Rise/Fall	Feet	Inches
20b	f	NW	16*	f	NW	R		1

Weather Diary   singular appearance of three distinct Halo or luminus rings about the moon appeared this evening at half after 9 PM and continued one hour.   the moon formed the center of the middle ring, the other two which lay N & S of the moon & had each of them a limb passing through the Moons Center and projecting N & S a simidiameter beyond the middle ring to which last they were equal in dimentions, each ring appearing to subtend an angle of 15 degrees of a great circle

Clark   a verry Cold Day
Gass   The weather was generally very cold.
Ordway   cloudy
Whitehouse   The weather still continued clear and cold;

Sunday, January 13

	Sunrise			4 p.m.			Missouri River		
Temp	Weather	Wind	Temp	Weather	Wind	Rise/Fall	Feet	Inches
34b	f	NW	20†	f	NW	R		2

Clark   a Cold Clear Day

---

*According to Moulton (1987a, 3: 283), Clark's Journal Codex C lists this temperature as "16 b."
†According to Moulton (1987a, 3: 283), Clark's Journal Codex C lists this temperature as "20 b."

Gass   A clear cold day. Two frenchmen came by, they had their faces so badly frost bitten that the skin came off, and their guide was so badly froze that they were obliged to leave him with the Assiniboins

Ordway   a clear cold morning.

Whitehouse   The weather still continues clear & Cold. The Indians infrom'd us that the Guide who went with them, had got so bad frost bitten on their faces, that the whole of the skin came off—

MONDAY, JANUARY 14

	Sunrise			4 p.m.		Missouri River		
Temp	Weather	Wind	Temp	Weather	Wind	Rise/Fall	Feet	Inches
16b	s	SE	8b	c a s	SE			

Lewis   Observed an Eclips of the Moon. I had no other glass to assist me in this observation but a small refracting telescope belonging to my sextant, which however was of considerable service, as it enabled me to define the edge of the moon's immage with much more precision that I could have done with the natural eye. The commencement of the eclips was obscured by clouds, which continued to interrupt me throughout the whole observation; to this cause is also attributable the inaccuracy of the observation of the commencement of total darkness. I do not put much confidence in the observation of the middle of the Eclips, as it is the worst point of the eclips to distinguish with accuracy. The two last observations (i.e.) The end of total darkness, and the end of the eclips, were more satisfactory; they are as accurate as the circumstance under which I laboured would permit me to make them—

Clark   this morning early a number of indians...passed down on the ice. Our hunters...informs that one Man (Whitehouse) is frost bit and Can't walk home—

Gass   Some snow fell this morning. Hunters, one of men had got his feet so badly frozen that he was unable to come to the fort.

Ordway   Whitehouse had his feet frost bit & could not come in without a horse

Whitehouse   Some Snow fell this morning. I got my feet so Froze that I could not walk to the fort.

TUESDAY, JANUARY 15

	Sunrise			4 p.m.		Missouri River		
Temp	Weather	Wind	Temp	Weather	Wind	Rise/Fall	Feet	Inches
10b	f	E	3a	c	SW	R		1

Weather Diary   an eclips of the moon total last night, visible here but partially obscured by the clouds.

Lewis   I do not place much confidence in this observation in consequence of loosing the observation of the Altitude of the Sun…was somewhat obscured by a cloud. The weather was so [cold] that I could not use water as the reflecting surface, and I was obliged to remove my glass horizon from it's first adjustment lest the savages should pilfer it.

Clark   between 12 & 3 oClock this morning we had total eclips of the moon. This morning not So Cold as yesterday, wind from the SE. Wind choped around to the NW. Still temperate.

Gass   the weather was warm, and the snow melted fast

Ordway   a warm pleasant day   the weather is thoughy [thawing] so that the Snow melts off the huts &.C.

Whitehouse   This day the weather had moderated considerable, warm to what it has been. The day kept warm & pleasant—

WEDNESDAY, JANUARY 16

	Sunrise			4 p.m.		Missouri River		
Temp	Weather	Wind	Temp	Weather	Wind	Rise/Fall	Feet	Inches
36a	c	W	16a	f	SW	R		2 ½

Clark   4 men…hunting returned one frost'd (but not bad) [Whitehouse]

Gass   the weather was warm, and the snow melted fast

Ordway   cloudy & warm.

Whitehouse   quite warm weather for the time a year & pleasant and the snow melted fast— The Man that was frost bitten informed us that he felt much easier than he had done, since he was frost bitten

THURSDAY, JANUARY 17

	Sunrise			4 p.m.		Missouri River		
Temp	Weather	Wind	Temp	Weather	Wind	Rise/Fall	Feet	Inches
2b	c	W	12b	f	NW			

Clark   a verry windey morning hard from the North. Thermometer at 0

Gass   it became cold; the wind blew hard from the north, and it began to freeze.

Ordway   a clear cold morning. The wind high from the NW

Whitehouse   This morning about 3 oClock the Wind began to blow from the North, & began to freeze. This wind continued all this day, the Weather being very Cold—

FRIDAY, JANUARY 18

	Sunrise			4 p.m.			Missouri River	
Temp	Weather	Wind	Temp	Weather	Wind	Rise/Fall	Feet	Inches
1b*	f	NW	7a	f a c	NW	F		1

Weather Diary   at Sun rise 12° below 0

Clark   a fine w[a]rm morning
Gass   Clear cold weather.
Ordway   moderate weather
Whitehouse   This day we had clear cold Weather

SATURDAY, JANUARY 19

	Sunrise			4 p.m.			Missouri River	
Temp	Weather	Wind	Temp	Weather	Wind	Rise/Fall	Feet	Inches
12a	c	NE	6b	f	NW	R		1

Weather Diary   Ice now 3 feet thick on the most rapid part of the river—

Clark   a fine Day
Ordway   cloudy
Whitehouse   The weather continued Cold and Clear. Two hunters proceeded on the Ice the River being fast froze over for some time past—

SUNDAY, JANUARY 20

	Sunrise			4 p.m.			Missouri River	
Temp	Weather	Wind	Temp	Weather	Wind	Rise/Fall	Feet	Inches
28a†	f	NE	9b	c	SE	R		3/4

Clark   a Cold fair day
Ordway   a pleasant morning
Whitehouse   We still continued to have clear cold weather.

---

* According to Moulton (1987a, 3: 283), In Voorhis No. 4 Clark lists this temperature as "20 b."
† According to Moulton (1987a, 3: 283), Clark's Journal Codex C lists this temperature as "28 a."

MONDAY, JANUARY 21

Sunrise			4 p.m.			Missouri River		
Temp	Weather	Wind	Temp	Weather	Wind	Rise/Fall	Feet	Inches
2b	c	NE	8a	f	SE	R		

Clark   a fine day
Gass   A clear cold day.
Ordway   moderate weather.
Whitehouse   The weather still continued Clear and Cold,

TUESDAY, JANUARY 22

Sunrise			4 p.m.			Missouri River		
Temp	Weather	Wind	Temp	Weather	Wind	Rise/Fall	Feet	Inches
10a	f a h	NW	19a	c	NW	R		1 3/4

Weather Diary   [missed] the afternon observation.

Clark   a fin[e] warm Day. Attempted to Cut the Boat & perogues out of the Ice, found water at about 8 inches under the 1st Ice, the next thickness about 3 feet

Gass   The weather was warm.   We commenced cutting ice from about our craft, in order to get them out of the river.

Ordway   a pleasant morning.   All hands Employed at cutting away the Ice from round the Barge & pearogues. They soon cut through the Ice in places.   The water Gushed over where they had cut so they had to quit cutting with axes—

Whitehouse   This day all our Men who were at the fort was employed to cut the Ice in order to get the boat & Pettyaugers out of the River, in the night we had a heavy fall of Snow, which made it difficult to work in the Ice for some days—

WEDNESDAY, JANUARY 23

Sunrise			4 p.m.			Missouri River		
Temp	Weather	Wind	Temp	Weather	Wind	Rise/Fall	Feet	Inches
2b	s	E	2b	c a s	N	F		2 1/2

Weather Diary   the snow feel about 4 inches deep last night and continues to snow

Clark   a Cold Day, Snow fell 4 Inches deep, the occurrences of this day is as is common—

Gass   The weather was warm.   The snow fell about 3 inches deep.

Ordway   Snowey this morning

Whitehouse   We had a continuation of Snow the greater part of this day, on its leaving off, it continued Cold to the 30th, all hands during this time were employed at work on the boats & Pettyaugers to get them free from the Ice, and hawled Stones on a Sled which they made warm in a fire, in order to thaw the Ice from about the said Crafts, when the Stones were put into the fire, they would not stand the heat of the fire but all of them broke, so that their labour was lost.

THURSDAY, JANUARY 24

	Sunrise			4 p.m.			Missouri River	
Temp	Weather	Wind	Temp	Weather	Wind	Rise/Fall	Feet	Inches
12b	c	NW	2b	f	NW	R		1/4

Clark   a fine day

Gass   A cold day.

Ordway   colder this morning than it has ben for Several days past.

Whitehouse   Cold

FRIDAY, JANUARY 25

	Sunrise			4 p.m.			Missouri River	
Temp	Weather	Wind	Temp	Weather	Wind	Rise/Fall	Feet	Inches
26b	f	NW	4b	f a c	W			

Weather Diary   it frequently happens that the sun rises fair and in about 15 or 20 minutes it becomes suddonly [cloudy] turbid, as if the had some chimical effect on the atmosphere.—

Clark   men employ'd in Cutting the Boat out of the ice

Gass   All hands employed in cutting away the ice, which we find a tedious business.

Ordway   clear & cold this morning.

Whitehouse   Cold

SATURDAY, JANUARY 26

	Sunrise			4 p.m.			Missouri River	
Temp	Weather	Wind	Temp	Weather	Wind	Rise/Fall	Feet	Inches
12a	c	NE	20a	f a c	SE			

Clark   a verry fine w[a]rm Day
Gass   A pleasant day
Ordway   Cloudy & warm   the wind from the South.
Whitehouse   Cold

SUNDAY, JANUARY 27

	Sunrise			4 p.m.			Missouri River	
Temp	Weather	Wind	Temp	Weather	Wind	Rise/Fall	Feet	Inches
20a	c	SE	16a	c	NW	R		2

Clark   a fine day, attempt to Cut our Boat and Conoos out of the Ice, a deficuelt Task I fear
       as we find water between the Ice
Gass   The weather has become much more settled, warm and pleasant than it had been for
       some time.
Ordway   Cloudy
Whitehouse   Cold

MONDAY, JANUARY 28

	Sunrise			4 p.m.			Missouri River	
Temp	Weather	Wind	Temp	Weather	Wind	Rise/Fall	Feet	Inches
2b	f	NW	15a	f	SW			

Lewis   Observed Equal altitudes...in which sp[i]rits were substituted for water, it being
       to[o] [cold] to use the later.

Clark   warm day
Gass   The weather warm and pleasant
Ordway   Clear and cold.
Whitehouse   Cold

TUESDAY, JANUARY 29

Sunrise			4 p.m.			Missouri River		
Temp	Weather	Wind	Temp	Weather	Wind	Rise/Fall	Feet	Inches
4a	f	SW	16a	f	W	R		1/2

Clark   we Sent & Collect Stones and put them on a large log heap to heet them with a View of warming water in the Boat and by that means, Sepperate her from the Ices, our attempt appears to be defeated by the Stones all breaking & flying to peaces in the fire, a fine warm Day.
Gass   We attempted another plan for getting our water craft disengaged from the ice: which as heated in the [boats], with hot stones; but in this project we failed, as the stones we found would not stand the fire, but broke to pieces.
Ordway   Clear & pleasant
Whitehouse   Cold

WEDNESDAY, JANUARY 30

Sunrise			4 p.m.			Missouri River		
Temp	Weather	Wind	Temp	Weather	Wind	Rise/Fall	Feet	Inches
6a	c	NW	14a	c	NW	R		1

Clark   a fine morning, Clouded up at 9 oClock
Gass   I went up river and found another kind of stones, which broke in the same manner; so our batteux and periogues remained fast in the ice.
Ordway   Some Cloudy. Sergt. Gass sent up the river to an other bluff in order to look for another kind of Stone that would not Split with heat   he brought one home & het it found it was the Same kind of the other   as soon as it was hot it bursted asunder So we Gave up that plan—
Whitehouse   Cold

THURSDAY, JANUARY 31

Sunrise			4 p.m.			Missouri River		
Temp	Weather	Wind	Temp	Weather	Wind	Rise/Fall	Feet	Inches
2b	c a s	NW	8a	f a c	NW	F		1

Weather Diary   the Snow fe[l]l 2 Inches last night.

Clark   Snowed last night, wind high from the NW. Cold disagreeable

Gass   Some snow fell last night.   In the morning the wind blew and was cold, toward the middle of the day the weather became moderate, and the afternoon was pleasant.

Ordway   Snowed the greater part of last night. The wind high from NW the Snow flew

Whitehouse   This morning we had a fresh wind from the NW, and the weather Cold,   In the afternoon it got warm & pleasant weather—

FRIDAY, FEBRUARY 1

Sunrise			4 p.m.			Missouri River		
Temp	Weather	Wind	Temp	Weather	Wind	Rise/Fall	Feet	Inches
6a	c	NW	16a	f	NW	R		2 1/2

Clark   a cold windey Day

Gass   A cold day.

Ordway   clear & cold.   The weather being bad they killed nothing—

Whitehouse   This morning we had pleasant weather.

SATURDAY, FEBRUARY 2

Sunrise			4 p.m.			Missouri River		
Temp	Weather	Wind	Temp	Weather	Wind	Rise/Fall	Feet	Inches
12b	f	NW	3a	f	S	F		1

Clark   a fine Day

Ordway   a clear morning.   My hat got burnt exedantly this morning. The river raiseing.

SUNDAY, FEBRUARY 3

	Sunrise			4 p.m.		Missouri River		
Temp	Weather	Wind	Temp	Weather	Wind	Rise/Fall	Feet	Inches
8b	f	SW	2a	f	W			

Lewis   a fine day. the situation of our boat and perogues is now allarming, they are firmly inclosed in the Ice and almost covered with snow – the ice which incloses them lyes in several stratas of unequal thickness which are seperated by streams of water. This [is] peculiarly unfortunate because so soon as we cut through the first strata of ice the water rushes up and rises as high as the upper surface of the ice and thus creates such a debth of water as renders it impracticable to cut away the lower strata which appears firmly attached to, and confining the bottom of the vessels. We then determined to attempt freeing them from the ice by means of boiling water which we purposed heating in the vessels by means of hot stones, [b]ut this expedient proved also fruitless, as every species of stone which we could procure in the neighbourhood partook so much of the calcarious genus that they burst into small particles on being exposed to the heat of the fire.

Gass   A cold day.

Whitehouse   This day we had Clear cold weather, nothing of consequence happened at the Fort worth mentioning—

MONDAY, FEBRUARY 4

	Sunrise			4 p.m.		Missouri River		
Temp	Weather	Wind	Temp	Weather	Wind	Rise/Fall	Feet	Inches
18b	f	NW	9b	f	W			

Lewis   This morning fair tho' could, the thermometer stood at 18° below Naught, wind from NW.

Clark   our provisions of meat being nearly exorsted I concluded to Decend the River on the Ice & hunt

Gass   A fine day.

Ordway   clear and pleasant

Whitehouse   This day we had Clear weather but cold

TUESDAY, FEBRUARY 5

Sunrise			4 p.m.			Missouri River		
Temp	Weather	Wind	Temp	Weather	Wind	Rise/Fall	Feet	Inches
10a	f	NW	20a	f	NW	R		1

Lewis   Pleasant morning wind from NW, fair
Clark   the morning verry Cold & windey [entered on the 13th]
Ordway   the morning clear. The River Riseing So that the water Spreads over the Ice in Sundry places near this.
Whitehouse   We had fair Weather

WEDNESDAY, FEBRUARY 6

Sunrise			4 p.m.			Missouri River		
Temp	Weather	Wind	Temp	Weather	Wind	Rise/Fall	Feet	Inches
4b	f	NW	12a	f	W	R		1/2

Lewis   Fair morning, wind from NW
Clark   Cold morning the after part of the Day w[a]rm [entered on the 13th]
Ordway   clear and pleasant
Whitehouse   This day was clear & pleasant Weather

THURSDAY, FEBRUARY 7

Sunrise			4 p.m.			Missouri River		
Temp	Weather	Wind	Temp	Weather	Wind	Rise/Fall	Feet	Inches
18b	f	SE	29*	c	S	R		1/2

Lewis   This morning was fair, Thermometer at 18° above naught much warmer than it has been for some days; Wind SE
Ordway   pleasant & warm

---

*According to Moulton (1987a, 3: 307), Clark's Journal Codex C lists this temperature as "29 a."

FRIDAY, FEBRUARY 8

	Sunrise			4 p.m.		Missouri River		
Temp	Weather	Wind	Temp	Weather	Wind	Rise/Fall	Feet	Inches
18a	f	NW	28	c	NE	F		1

Weather Diary   the Black & white & Speckled woodpecker has returned—

Lewis   This morning was fair, wind SE. The weather still warm and pleasant—

Ordway   moderate weather

SATURDAY, FEBRUARY 9

	Sunrise			4 p.m.		Missouri River		
Temp	Weather	Wind	Temp	Weather	Wind	Rise/Fall	Feet	Inches
10a	f	SE	33a	c	SE			

Weather Diary   very little snow

Lewis   The morning fair and pleasant, wind from SE—

Ordway   Some cloudy, the water which run over the Ice in the River has froze Smoth.

SUNDAY, FEBRUARY 10

	Sunrise			4 p.m.		Missouri River		
Temp	Weather	Wind	Temp	Weather	Wind	Rise/Fall	Feet	Inches
18a	c a s	NW	12a	c	NW			

Lewis   This Morning was Cloudy after a slight snow which fell in the course of the night, the wind blue very hard from W. altho' the thermometer stood at 18° Above naught the violence of the wind caused a degree of [cold] that was much more unpleasant than that of yesterday when thermometer stood at 10° only above the same point.

Clark   a cold Day, wind blew hard from the NW. J. Fields got one of his ears frosed [entered on the 13th]

Ordway   high wind from NW Squawly flights of Snow.   On the river...the Ice being Smoth the horses could not Go on the Ice with out Shoes.

MONDAY, FEBRUARY 11

	Sunrise			4 p.m.			Missouri River		
Temp	Weather	Wind	Temp	Weather	Wind	Rise/Fall	Feet	Inches
8b	f	NW	2b	f	NW			

Lewis   this morning the weather was fair and could wind NW.
Clark   air keen [entered on the 13th]
Ordway   the day clear but cold—

TUESDAY, FEBRUARY 12

	Sunrise			4 p.m.			Missouri River		
Temp	Weather	Wind	Temp	Weather	Wind	Rise/Fall	Feet	Inches
14b	f	SE	2a	f	W			

Lewis   The morning was fair tho' [cold], thermometer at 14° below naught wind SE
Clark   The ice on the parts of the River which was verry rough, as I went down, was Smothe on my return, this is owing to the rise and fall of the water, which takes place every day or two, and Caused by partial thaws, and obstructions in the passage of the water thro the Ice, which frequently attaches itself to the bottom— the water when riseing forses its way thro the cracks & air holes above the old ice, & in one night becoms a Smothe Surface of ice 4 to 6 Inches thick— the river falls & the ice Sink in places with the water and attches itself to the bottom, and when it again rises to its former hite, frequently leavs a valley of Several feet to Supply with water to bring it on a leavel Surface. The water of the Missouri at this time is Clear with little Tinges. [entered on the 13th]
Ordway   a clear morning

WEDNESDAY, FEBRUARY 13

	Sunrise			4 p.m.			Missouri River		
Temp	Weather	Wind	Temp	Weather	Wind	Rise/Fall	Feet	Inches
2b	c	SE	10a	c	NW	F		1

Lewis   The morning cloudy, thermometer 2° below naught, wind from SE.

Clark   I returned last Night from a hunting party much fatigued, haveing walked 30 miles on the ice and through Points of wood land in which the Snow was nearly Knee Deep.

Ordway   Snow the later part of the day.

Whitehouse   This day clear & pleasant weather—

THURSDAY, FEBRUARY 14

	Sunrise			4 p.m.		Missouri River		
Temp	Weather	Wind	Temp	Weather	Wind	Rise/Fall	Feet	Inches
2a	c a s	NW	2b	f	NW			

Weather Diary   The Snow fell 3 Inches deep last night

Clark   The Snow fell 3 inches Deep last night, a fine morning

Ordway   Snowed the Greater part of last night. The day pleasant.

Whitehouse   This morning we had clear weather but pleasant—

FRIDAY, FEBRUARY 15

	Sunrise			4 p.m.		Missouri River		
Temp	Weather	Wind	Temp	Weather	Wind	Rise/Fall	Feet	Inches
16b	f	SW	6b	f	W			

Clark   the morning fine, the Thermometer Stood at 16° below 0, Noght. one Chief of the Mandans returned from Capt. Lewises Partly nearly blind— this Complaint is as I am informd. Common at this Season of the year and caused by the reflection of the Sun on the ice & Snow, it is cured by "jentilley Swetting the part affected, by throweng Snow on a hot Stone." Verry Cold part of the night—

Ordway   Some of the mens feet were sore walking 300 odd mls. on the Ice to day.

Whitehouse   This morning we had fine Clear weather.

SATURDAY, FEBRUARY 16

	Sunrise			4 p.m.		Missouri River		
Temp	Weather	Wind	Temp	Weather	Wind	Rise/Fall	Feet	Inches
2a	f	SE	8a	f	W	F		1

Clark   a fine morning

Gass   had a cold morning. This morning the Indians, who had come down with us and of our men whose feet had been a little frozen, returned home.
Ordway   a clear cold morning.
Whitehouse   We had a fine Clear day, the weather being moderate—

Sunday, February 17

Sunrise			4 p.m.			Missouri River		
Temp	Weather	Wind	Temp	Weather	Wind	Rise/Fall	Feet	Inches
4a	c	SE	12a	f	NW	F*		¹/₂

Clark   this morning w[a]rm & a little Cloudy. The after part of the day fair
Whitehouse   The weather continued Clear & moderate.

Monday, February 18

Sunrise			4 p.m.			Missouri River		
Temp	Weather	Wind	Temp	Weather	Wind	Rise/Fall	Feet	Inches
4a	s	NE	10a	f	S			

Clark   a cloudy morning, Some Snow
Ordway   clear and pleasant
Whitehouse   The hunters...brought the Sleds loaded with the Meat up the River on the Ice, it still being froze over the Mesouri, & the Ice very thick & strong—

Tuesday, February 19

Sunrise			4 p.m.			Missouri River		
Temp	Weather	Wind	Temp	Weather	Wind	Rise/Fall	Feet	Inches
4a	f	SE	20a	f	S			

Clark   a fine Day

---

*According to Moulton (1987a, 3: 307), Lewis does not record rise or fall but does record "¹/₂" inch. Clark gives neither a rise nor fall nor depth.

Ordway a clear pleasant morning.
Whitehouse We had fine clear moderate weather

Wednesday, February 20

	Sunrise			4 p.m.			Missouri River	
Temp	Weather	Wind	Temp	Weather	Wind	Rise/Fall	Feet	Inches
2a	f	S	22a	f	S			

Clark a Butifull Day
Ordway a clear and pleasant morning.
Whitehouse We had fine clear moderate weather

Thursday, February 21

	Sunrise			4 p.m.			Missouri River	
Temp	Weather	Wind	Temp	Weather	Wind	Rise/Fall	Feet	Inches
6a	f	S	30a	f	S			

Clark a Delightfull Day put our Clothes to Sun—
Gass Some rain fell to day, the first that has fallen since November. In the evening, the weather became clear and pleasant.
Ordway clear and pleasant. The Snow and Ice thoughed on the River considerable So that it was wet & Slopy halling the Sled. The men generally fatigued halling a heavy load 21 miles on the hard Ice & Snow in places which made the Sleds run hard except where the Ice was Smoth under—
Whitehouse This day the weather still continued Clear & pleasant

Friday, February 22

	Sunrise			4 p.m.			Missouri River	
Temp	Weather	Wind	Temp	Weather	Wind	Rise/Fall	Feet	Inches
8a	c	N	32a	c a r & s	NW			

Clark a Cloudy morning, at about 12 oClock it began to rain and Continud for a fiew minits, and turned to Snow, and Continud Snowing for about one hour, and Cleared away fair

Gass   Was a fine day and we again began to cut away the ice, and succeeded in getting out one of the periogues.

Ordway   rained a Short time and turned to Snow. Snowed a Short time and cleared off, the men came home last night

Whitehouse   We had still Clear, pleasant weather

SATURDAY, FEBRUARY 23

Sunrise			4 p.m.			Missouri River		
Temp	Weather	Wind	Temp	Weather	Wind	Rise/Fall	Feet	Inches
18a	f	NW	32a	f	W	R		½

Weather Diary   got the poplar perogue out of the ice.

Clark   All hands employed in Cutting the Perogues Loose from the ice, which was nearly even with their top; we found great difficuelty in effecting this work owing to the Different devisions of Ice & water. After Cutting as much as we Could with axes, we had all the Iron we Could get & Some axes put on long poles and picked through the ice, under the first water, which was not more the 6 to 8 inches deep— we disengaged one Perqogue, and nearly disingaged the 2nd in Course of this day which has been warm & pleasent

Gass   We had fine pleasant weather, and all hands were engaged in cutting the ice away from the boat and the other perioque. At 4 o'clock in the afternoon we had the good fortune to get both free from the ice

Ordway   a pleasant morning.

Whitehouse   The weather still continued fine and Clear, all our party were employ'd in cutting the Ice from round the Boat & Pettyaugers. At one oClock we got one of the Pettyaugers out of the Ice on Shore—

SUNDAY, FEBRUARY 24

Sunrise			4 p.m.			Missouri River		
Temp	Weather	Wind	Temp	Weather	Wind	Rise/Fall	Feet	Inches
8a	f	NW	32a	f	W			

Weather Diary   loosed the boat & large perogue from the ice.

Clark   The Day fine, we Commenced very early to day the Cutting loose the boat

Ordway   a beautiful morning. All hands employed cutting away Ice from round the Barge. Found that the Ice was verry thick clear under hir.

Whitehouse   We had pleasant weather. All our party were still employed in cutting the Ice round the Boat & pettyauger   we succeeded, and got both the boat & pettyauger on the bank....clear of danger, when the Ice broke up in the River—

MONDAY, FEBRUARY 25

	Sunrise			4 p.m.			Missouri River	
Temp	Weather	Wind	Temp	Weather	Wind	Rise/Fall	Feet	Inches
16a	f	W	38a	f	NW			

Clark   The day has been exceedingly pleasent
Whitehouse   This day was Clear & pleasant.

TUESDAY, FEBRUARY 26

	Sunrise			4 p.m.			Missouri River	
Temp	Weather	Wind	Temp	Weather	Wind	Rise/Fall	Feet	Inches
20a	f	NE	31a	f	N			

Weather Diary   got the Boat and perogues on the bank

Clark   a fine day   Drew up the Boat & perogus, after Cutting them out of the ice with great Dificuelty— & trouble
Ordway   a pleasant morning.
Whitehouse   The weather continued still clear & fine;

WEDNESDAY, FEBRUARY 27

	Sunrise			4 p.m.			Missouri River	
Temp	Weather	Wind	Temp	Weather	Wind	Rise/Fall	Feet	Inches
26a	f	SE	36a	f	E	F		1/2

Weather Diary   got the Boat and Perogues on the bank.

Clark   a fine day. I commence a Map of the Countrey on the Missouries & its waters &c &c—
Ordway   a beautiful pleasant morning.
Whitehouse   We have still fine weather;

THURSDAY, FEBRUARY 28

Sunrise			4 p.m.			Missouri River		
Temp	Weather	Wind	Temp	Weather	Wind	Rise/Fall	Feet	Inches
24a	f	E	38a	c	SE			

Clark   a fine morning
Whitehouse   This morning the weather cold but toward Noon it moderated

FRIDAY, MARCH 1

Sunrise			4 p.m.			Missouri River		
Temp	Weather	Wind	Temp	Weather	Wind	Rise/Fall	Feet	Inches
28a	c	W	38a	f	NW			

Weather Diary   the snow has disappeared in many place   the river partially broken up—
Clark   a fine Day
Ordway   cloudy & arm this morning. The after part of the day clear and pleasant—
Whitehouse   the weather continued Clear & Cold—

SATURDAY, MARCH 2

Sunrise			4 p.m.			Missouri River		
Temp	Weather	Wind	Temp	Weather	Wind	Rise/Fall	Feet	Inches
28a	f	NE	36a	f	NE	R		1 1/2

Clark   a fine Day. The river br[o]ke up in places
Ordway   a beautiful pleasant morning
Whitehouse   This day we had fine Clear weather.

SUNDAY, MARCH 3

Sunrise			4 p.m.			Missouri River		
Temp	Weather	Wind	Temp	Weather	Wind	Rise/Fall	Feet	Inches
28a	c	E	39a	f	NW			

Weather Diary   a flock of ducks pased up the river this morning—

Clark   a fine Day, wind from the W. A large flock of Ducks pass up the River—

Ordway   clear & pleasant.

Whitehouse   This day we had Clear Cold weather

MONDAY, MARCH 4

	Sunrise			4 p.m.			Missouri River	
Temp	Weather	Wind	Temp	Weather	Wind	Rise/Fall	Feet	Inches
26a	f	NW	36a	f	NW			

Clark   a Cloudy morning, wind from the NW. The after part of the day Clear

Ordway   the wind high from the NW   They day pleasant.

Whitehouse   the weather continued Clear & pleasant

TUESDAY, MARCH 5

	Sunrise			4 p.m.			Missouri River	
Temp	Weather	Wind	Temp	Weather	Wind	Rise/Fall	Feet	Inches
22a	f	E	40a	f	NW			

Clark   a fine Day. Themometer at 40° abo. 0

Ordway   clear and pleasant.   A light Squawl of Snow fell about 4 oClock this morning. Nothing extraordinary.

Whitehouse   the weather continued Clear & pleasant

WEDNESDAY, MARCH 6

	Sunrise			4 p.m.			Missouri River	
Temp	Weather	Wind	Temp	Weather	Wind	Rise/Fall	Feet	Inches
26a	c	E	36a	f	E	R		2

Clark   a Cloudy morning & Smokey all Day from the burning of the plains, which was Set on fire by the Minetarries for an early crop of Grass as an endusement for the Buffalow to feed on—   the river rise a little to day—

Ordway   moderate weather. The wind from NE   the Water has run over the Ice So that it is difficult crossing the river.

Whitehouse   the weather continued Clear & pleasant

THURSDAY, MARCH 7

Sunrise			4 p.m.			Missouri River		
Temp	Weather	Wind	Temp	Weather	Wind	Rise/Fall	Feet	Inches
12a	f	E	26a	c	E	R		2

Clark   a little Cloudy and windey NE.

Ordway   clear but Some colder than it has been for Several days past.

Whitehouse   the weather continued Clear & pleasant

FRIDAY, MARCH 8

Sunrise			4 p.m.			Missouri River		
Temp	Weather	Wind	Temp	Weather	Wind	Rise/Fall	Feet	Inches
7a	c	E	12a	f	E	R		2 1/2

Clark   a fair morning, Cold and windey, wind from the East

Ordway   Some cloudy & cold.

Whitehouse   the weather continued Clear & pleasant

SATURDAY, MARCH 9

Sunrise			4 p.m.			Missouri River		
Temp	Weather	Wind	Temp	Weather	Wind	Rise/Fall	Feet	Inches
2a	c	N	18a	f	NW	R		2

Weather Diary   wind hard all day.

Clark   a Cloudy Cold and windey morning, wind from the North. I walked up to see the Party that is makeing Perogues, about 5 miles above this, the wind hard and Cold

Ordway   the wind high from the NW

Whitehouse   the weather continued Clear & pleasant

SUNDAY, MARCH 10

	Sunrise			4 p.m.		Missouri River		
Temp	Weather	Wind	Temp	Weather	Wind	Rise/Fall	Feet	Inches
2b	f	NW	12a	f	NW	R		3 1/2

Clark   a Cold winday Day
Ordway   the day clear and cold high winds—
Whitehouse   the weather continued Clear & pleasant

MONDAY, MARCH 11

	Sunrise			4 p.m.		Missouri River		
Temp	Weather	Wind	Temp	Weather	Wind	Rise/Fall	Feet	Inches
12a	c	SE	26*	f a c	NW	R		4 1/2

Clark   A Cloudy Cold windey day, Some Snow in the latter part of the day
Ordway   clear Moderate weather
Whitehouse   the weather continued Clear & pleasant

TUESDAY, MARCH 12

	Sunrise			4 p.m.		Missouri River		
Temp	Weather	Wind	Temp	Weather	Wind	Rise/Fall	Feet	Inches
2b	f a s	N	10a	f	NW	R		5

Weather Diary   snow but slight disappeared to day

Clark   a fine day   Some Snow last night Wind NW
Ordway   a little Snow last night. A clear cold morning   the river raiseing fast.
Whitehouse   the weather continued Clear & pleasant

---

*According to Moulton (1987a, 3: 326), Clark's Journal Codex C lists this temperature as "26 a."

Wednesday, March 13

	Sunrise			4 p.m.			Missouri River		
Temp	Weather	Wind	Temp	Weather	Wind	Rise/Fall	Feet	Inches
1b	f	SE	28a	f	SW	R		3 ½

Clark   a fine day   The river riseing a little— wind SW
Ordway   clear and cold   the wind high from the South
Whitehouse   the weather continued Clear & pleasant

Thursday, March 14

	Sunrise			4 p.m.			Missouri River		
Temp	Weather	Wind	Temp	Weather	Wind	Rise/Fall	Feet	Inches
18a	f	SE	40a	f	W			

Clark   a fine day   wind Wet. River Still riseing
Ordway   clear & warm
Whitehouse   the weather continued Clear & pleasant

Friday, March 15

	Sunrise			4 p.m.			Missouri River		
Temp	Weather	Wind	Temp	Weather	Wind	Rise/Fall	Feet	Inches
24a	f	SE	38a	f	W	F		1

Clark   a fine day
Ordway   clear pleasant and warm.
Whitehouse   the weather continued Clear & pleasant

SATURDAY, MARCH 16

	Sunrise			4 p.m.			Missouri River		
Temp	Weather	Wind	Temp	Weather	Wind	Rise/Fall	Feet	Inches
32a	c	E	42a	c	W	F		3

Clark   a Cloudy day, wind from the SE
Ordway   Cloudy & warm. The wind high from the East. Look likely for rain—
Whitehouse   the weather continued Clear & pleasant

SUNDAY, MARCH 17

	Sunrise			4 p.m.			Missouri River		
Temp	Weather	Wind	Temp	Weather	Wind	Rise/Fall	Feet	Inches
30a	f	SE	46a	f	SW	R		2

Clark   a windey Day. The river riseing a little and Severall places open.
Ordway   clear and pleasant.
Whitehouse   the weather continued Clear & pleasant

MONDAY, MARCH 18

	Sunrise			4 p.m.			Missouri River		
Temp	Weather	Wind	Temp	Weather	Wind	Rise/Fall	Feet	Inches
24a	c	N	34a	c	N	F		1

Weather Diary   collected Some herbs plants in order to send by the boat. paticularly the root said to cure bites of the mad dog and rattlesnake.—

Clark   a cold cloudy Day, wind from the N
Ordway   Cloudy
Whitehouse   the weather continued Clear & pleasant

Tuesday, March 19

Sunrise			4 p.m.			Missouri River		
Temp	Weather	Wind	Temp	Weather	Wind	Rise/Fall	Feet	Inches
20a	c a s	N	31*	f	NW	R		1

Weather Diary   But little snow not enough to cover the ground

Clark   Cold windey Day Cloudy. Some little Snow last night
Ordway   cloudy a light Squawl of Snow cold air
Whitehouse   the weather continued Clear & pleasant

Wednesday, March 20

Sunrise			4 p.m.			Missouri River		
Temp	Weather	Wind	Temp	Weather	Wind	Rise/Fall	Feet	Inches
28a	c	NW	28†	f	NW	R		3

Clark   cloudy, wind hard from N
Ordway   a cloudy cool morning. The after part of the day pleasant.
Whitehouse   the weather continued Clear & pleasant

Thursday, March 21

Sunrise			4 p.m.			Missouri River		
Temp	Weather	Wind	Temp	Weather	Wind	Rise/Fall	Feet	Inches
16a	c	E	26a	s & h	S			

Weather Diary   some ducks seen to light in the river opposit the fort

Clark   a Cloudy Day, Some Snow
Ordway   Cloudy   the wind from the SE   a little Snow fell the after part of the
          day—
Whitehouse   This day we had still, pleasant Weather, nothing occur'd worth
          mentioning

---

*According to Moulton (1987a, 3: 326), Clark's Journal Codex C lists this temperature as "28 a."
†According to Moulton (1987a, 3: 326), Clark's Journal Codex C lists this temperature as "31 a."

Friday, March 22

	Sunrise			4 p.m.			Missouri River		
Temp	Weather	Wind	Temp	Weather	Wind	Rise/Fall	Feet	Inches
22a	f a s	S	36a	f	SW	F		4

Clark   Some few Drops of rain this evening for the first time this Winter, a Cloudy Day

Ordway   a clear pleasant morning.   The wind from the SE

Whitehouse   We had some Rain this morning, but it continued but a short time, the weather being cloudy & cold.

Saturday, March 23

	Sunrise			4 p.m.			Missouri River		
Temp	Weather	Wind	Temp	Weather	Wind	Rise/Fall	Feet	Inches
34a	f	W	38a	c a r	NW	F		4

Weather Diary   but little rain.

Clark   a find Day in the fore part, in the evening a little rain & the first this winter—

Ordway   clear and pleasant.   Rained a little the latter part of the day—

Whitehouse   This morning we had Snow—   towards Noon it ceased, and the weather moderated and became pleasant— in the evening it grew cold & froze during the Night.

Sunday, March 24

	Sunrise			4 p.m.			Missouri River		
Temp	Weather	Wind	Temp	Weather	Wind	Rise/Fall	Feet	Inches
28a	c a s	NE	30a	c a s	N	R		1

Weather Diary   but little Snow.

Clark   preparing to Set out, A Cloudy morning, wind from the NE, the after part of the Day fair.   Saw Swans & wild Gees flying NE this evening

Ordway   cloudy

Whitehouse   We had pleasant Weather, continued to freeze in the Evenings

MONDAY, MARCH 25

	Sunrise			4 p.m.		Missouri River		
Temp	Weather	Wind	Temp	Weather	Wind	Rise/Fall	Feet	Inches
16	f	E	32a	f	S	R		5

Weather Diary   a gang of swan return to day   the ice in the river has given way in many places and it is with some difficulty it can be passed—

Clark   a fine Day, wind SW. The ice broke up in Several places in the evenig, broke away and was nearly takeing off our new Canoes, river rise a little

Ordway   clear and pleasant.

Whitehouse   We had pleasant Weather, continued to freeze in the Evenings

TUESDAY, MARCH 26

	Sunrise			4 p.m.		Missouri River		
Temp	Weather	Wind	Temp	Weather	Wind	Rise/Fall	Feet	Inches
20	f	SE	46a	f	W	R		4 1/2

Weather Diary   the ice gave way in the river about 3 PM and came down in immense sheets very near distroying our perogues—   some gees pass today.

Clark   The river choked up with ice opposit to us and broke away in the evening, raised only 1/2 Inch, all employed prepareing to Set out

Gass   We put the canoes in the water   as the river had risen there was some water between the ice and the shore. We got three of them safe to the fort; but the ice breaking before the other three were gotten down, so filled the channel, that we were obliged to carry them the rest of the way by land.

Ordway   clear and pleasant.   About 2 oClock they returned with the perogues, but before they had landed the Ice Started So that we had to draw them out with Speed   The ice Stoped and jamed up.   Started Several times but Stoped entirely before night.

Whitehouse   This day we had moderate weather, and the Ice broke up.

WEDNESDAY, MARCH 27

	Sunrise			4 p.m.		Missouri River		
Temp	Weather	Wind	Temp	Weather	Wind	Rise/Fall	Feet	Inches
28	f	SE	60a	f	SW	R		9

Weather Diary   the first insect I have seen was a large black knat today—   the ice drifting in large quantities.—

Clark   a windey Blustering Day, wind SW. Ice running the [river] Blocked up in view for the Space of 4 hours and gave way leaveing great quantity of ice on the Shallow Sand bars.

Ordway   clear and pleasant.   The Ice kep breaking and Starting the Most of the day—

Whitehouse   This day we had pleasant weather

THURSDAY, MARCH 28

	Sunrise			4 p.m.		Missouri River		
Temp	Weather	Wind	Temp	Weather	Wind	Rise/Fall	Feet	Inches
40	f	SE	64a	f	SW	R		1

Weather Diary   it [river] raised 13 inch and fell 12.   wind hard, ice abates in quantity

Clark   the ice Stoped running owing to Some obstickle above. But few Indians visit us today, they are watching to catch the floating Buffalow which brake through the ice in Crossing

Ordway   the Ice run in the River all last night. A pleasant morning. The Ice continues to run.

Whitehouse   This day was blustering which continued the whole day—

FRIDAY, MARCH 29

	Sunrise			4 p.m.		Missouri River		
Temp	Weather	Wind	Temp	Weather	Wind	Rise/Fall	Feet	Inches
42	f	NW	52a	f	NW	F		11

Weather Diary   a variety of insects make their appearance, as flies bugs &c.   the ice ceases to run   supposed to have formed an obstruction above.—

Clark   The obstickle broke away above & the ice came down and is passing in great quantities, the river rose 13 inches the last 24 hours. I observed extrodanary dexterity of the Indians in jumping from one Cake of ice to another, for the purpose of Catching the buffalow as they float down. Maney of the Cakes of ice which they pass over are not two feet Square. The Plains are on fire in view of the fort on both Sides of the River, it is Said to be common for the Indians to burn the Plains near their villages every Spring for the benifit of ther horse, and to induce the Buffalow to come near to them.

Ordway   clear and pleasant.   The River fell 22 inches in 22 hours. We continue gitting ready to Start up the River.

Whitehouse   We had all this day high winds, but the Air was not so cold, as it had been for the several days past.

SATURDAY, MARCH 30

	Sunrise			4 p.m.			Missouri River	
Temp	Weather	Wind	Temp	Weather	Wind	Rise/Fall	Feet	Inches
28	*f*	NW	49a	*f*	NW	R	1	1

Weather Diary   ice came down in great quantities   the Mandans take Some floating Buffaloe

Clark   Cloudy Day. Several gangus of Gees and Ducks pass up the river. Not much Ice floating down today—

Ordway   clear and pleasant.   The river raised 10 Inches last night   the ice runs thick in the R. to day

Whitehouse   high winds, but not cold

SUNDAY, MARCH 31

	Sunrise			4 p.m.			Missouri River	
Temp	Weather	Wind	Temp	Weather	Wind	Rise/Fall	Feet	Inches
35	*c a r*	SE	45a	*c*	SE	R		9

Weather Diary   ducks and Gees passing   ice abates in quantity

Ordway   cloudy. Rained Some at three oclock this morning. The Ice does not run So thick in the River as it did yesterday.

Whitehouse   fine Clear warm weather—

MONDAY, APRIL 1

	Sunrise			4 p.m.			Missouri River	
Temp	Weather	Wind	Temp	Weather	Wind	Rise/Fall	Feet	Inches
33a*	*c*	NW	43a	*c a t l r & h†*	W	F		11

Weather Diary   ice ceases to run   A fine refreshing shower of rain fell about 2 PM this was the first shower of rain that we had witnessed since the fifteenth of September 1804 tho' it several times has fallen in very small quantities, and was noticed in this diary of the weather.   The cloud came from the west,

---

* According to Moulton (1987b, 4: 95), Clark's Journal Codex C lists this temperature as "38 a."

† According to Moulton (1987b, 4: 95), Clark's Journal Codex C lists this weather data as "c a t c h & r."

and was attended by hard thunder and Lightning.   I have observed that all thunderclouds in the Western part of the continent, proceed from the westerly quarter, as they do in the Atlantic States.   The air is remarkably dry and pure in this open country, very little rain or snow e[i]ther winter or summer.   The atmosphere is more transparent than I ever observed it in any country through which I have passed.

Clark   The fore part of the day haile rain with Thunder & lightning, the rain continued by intimitions all day, it is worthey of remark that this is the 1st rain which has fallen Since we have been here or Since the 15 of October last, except a fiew drops at two or three defferent times—

Gass   A considerable quantity of rain fell this day; the first of any consequence that had fallen here for six months.

Ordway   Thunder and hail & hard rain about 8 oClock this morning for about an hour. Began to rain again about 4 ock. PM rained untill 12 oC. at night & ceased.

Whitehouse   This morning we had some rain, which lasted about 2 hours & clear'd up Cool, all our party was employed in putting the Boat & Pettyaugers into the River, which they Effected—

TUESDAY, APRIL 2

	Sunrise			4 p.m.		Missouri River		
Temp	Weather	Wind	Temp	Weather	Wind	Rise/Fall	Feet	Inches
28a	c a r	NW	38a	f a c	W	F		5

Weather Diary   rained hard and without intermission last night

Clark   a Cold cloudy rain day. Rained all last night. The river falling fast.

Gass   the weather was fair bu[t] windy

Ordway   the wind blew high from NW all the later part of the night.   A cloudy morning. The wind rises from NW   the river fell 5 inches Since yesterday morning   The later part of the day plsant.

Whitehouse   This day the weather was cold & in the fore part of the day it froze. The latter part we had Blustery weather—

WEDNESDAY, APRIL 3

	Sunrise			4 p.m.		Missouri River		
Temp	Weather	Wind	Temp	Weather	Wind	Rise/Fall	Feet	Inches
24a	f	N	44a	f	W*	F		4

---

*According to Moulton (1987b, 4: 95), Clark's Journal Codex I lists this wind direction as "N."

Weather Diary   frost last night   a white frost & Some ice on the edge of the river ….

Clark   a white frost this morning, some ice on the edge of the water. A fine day.

Gass   the weather was fine and pleasant

Ordway   clear and pleasant.

Whitehouse   This day was Clear, & pleasant weather.

THURSDAY, APRIL 4

	Sunrise			4 p.m.		Missouri River		
Temp	Weather	Wind	Temp	Weather	Wind	Rise/Fall	Feet	Inches
36a	f	S	55a	f	NW	F		4

Weather Diary   Observed a flock of brant passing up the river today; the wind blew very hard as it dose frequently in this quarter; there is sarcely any timber to brake the wind from the river, & the country on both sides being level plains, wholy destitute of timber, the wind blows [over them] with astonishing violence. In this open country the winds form a great obstruction to the navigation of this river particularly with small vessels, which can neither ascend or descend should the wind be the least violent.—

Clark   a blustering windey Day.

Gass   A fine clear day.

Ordway   clear and pleasant.   The articles for St. Louis carred on board the barge ready to Set out but the wind blew high from the NW so we did not load the perogues.

Whitehouse   had fine Clear weather

FRIDAY, APRIL 5

	Sunrise			4 p.m.		Missouri River		
Temp	Weather	Wind	Temp	Weather	Wind	Rise/Fall	Feet	Inches
30a	f	NW	39a	f	N*	F		2

Clark   the wind verry high from the NW

Gass   This was a clear day and the wind blew hard and cold from the NW

Ordway   clear and pleasant. The wind high from the NW

Whitehouse   had fine Clear weather

---

*According to Moulton (1987b, 4: 95), Clark's Journal Codex C lists this wind direction as "NW."

SATURDAY, APRIL 6

	Sunrise			4 p.m.			Missouri River		
Temp	Weather	Wind	Temp	Weather	Wind	Rise/Fall	Feet	Inches
19a	f	N	48a	c	NW	F		1

Weather Diary   all the birds that we believe visit this country have now returned.—

Clark   a fine Day

Gass   The day was clear and pleasant.

Ordway   clear and pleasant. The wind Gentle from the South.

Whitehouse   had fine Clear weather

SECTION 5 To the Pacific

April 7 to December 6, 1805

On April 7, 1805, a small party returned the keelboat downriver to St. Louis, carrying journal notes and plant and animal specimens from the previous year's journey destined for President Thomas Jefferson. The permanent expedition of thirty three members set off up the Missouri River in two pirogues and canoes toward the Rocky Mountains and their ultimate goal, the Pacific Ocean. The expedition made slow progress westward as they endured strong spring winds that created sandstorms along the Missouri.

By June 2, they reached a flood-swollen fork in the Missouri. Most of the party believed the north fork was the true Missouri as it was muddy, similar to what they had seen in the previous weeks coming up the river. To be certain, Lewis and Clark took scouting parties to locate the great falls the Minetare/Hidatsa Indians had described. After an unsuccessful try, Lewis took a second party for a further investigation and discovered the Great Falls of the Missouri on June 13. It would take over a month to portage around the falls during a cool and wet early summer. At the end of July, they reached another decision point, known today as the Three Forks of the Missouri, just as the weather warmed to summer norms. The Expedition moved westward up the newly named Jefferson River in search of the Shoshone Indian Nation to obtain horses for their trek across the Continental Divide. Lewis led a small scouting party across the great divide on August 12, and found the Shoshones. Clark took a small contingent to explore Lewis's (today's Salmon) River but found it unpassable. Faced with potential failure, the captains decided they would have to use an old Indian trail to cross the mountains. After they purchased horses, their Shoshone Indian guide, Old Toby, led them over a mountain pass from Idaho back into Montana, down the Bitterroot Valley and then westward along the Lolo Trail. During this time they experienced early fall snows in the high rugged mountains. Nearly starving to death, the Expedition left the Bitterroot Mountains near the end of September and reached the Nez Perce Indian Nation east of present-day Lewiston, Idaho. Here they set up Canoe Camp to build watercraft to take them down the Clearwater, Snake, and Columbia rivers to the ocean.

With the current at their back, they set off toward the ocean from near present-day Orofino, Idaho, on October 7. After passing many rapids in the Clearwater and Snake rivers, the Expedition came to great Columbia River on October 16. They proceeded down the wide Columbia, passing many dangerous falls, chutes and rapids, and by early November they had reached the tidal waters of the Columbia. Rain began on November 4, and numerous late fall storms ravaged their camps as they progressed to the ocean. Finally, Clark declared prematurely on November 7, "*Great joy in camp we are in View of the Ocian!*"; however, they were in the great Columbia River estuary. After being pinned down by the intense storms, the Expedition reached the mouth of the river on November 16: the Pacific Ocean at last, some 4,162 miles by Clark's dead reckoning from St. Louis. Looking for winter quarters, the party moved back up the river, crossed to the south shore, and established Fort Clatsop near present-day Astoria, Oregon, on December 7, and remained there until their departure on March 23, 1806.

The systematic entries for the Lewis and Clark Expedition daily narrative journals as well as that of the army sergeants and privates were taken every day in 1805. However, not every journalist noted weather, water, or climate data each day. As in the previous section, each day's entry begins with the data from the Weather Diary's Observation Tables (when available), followed by the corresponding remarks (when available), and finally excerpts that pertain to weather and climate from the expedition members' daily narrative journals.

River rise and fall observations continued during the ascent of the Missouri River from Fort Mandan. Although not like previous recording episodes, the data is usually not for a 24-hour period unless they are encamped. In most cases, when the party stopped for the evening, following Lewis's habit from previous journal entries, it is surmised that a mark was made and measured the next morning.

Weather data reported here is from Coues (1893, 3: 1269–1291), Moulton (1987b, 1988, 1990, and 1997), and Thwaites (1904f, 6, part 2: 181–201). Different journals and notebooks were used during the expedition. For a more detailed explanation on the journals and entry practices, consult Cutright (1976) and Moulton (1986, 2: 8–48, 530–567).

Note   *On April 7, the expedition heads up the Missouri toward the Pacific in two pirogues and canoes.*

SUNDAY, APRIL 7

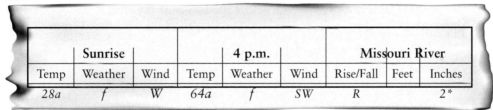

	Sunrise			4 p.m.			Missouri River	
Temp	Weather	Wind	Temp	Weather	Wind	Rise/Fall	Feet	Inches
28a	f	W	64a	f	SW	R		2*

Weather Diary   Visited by Ricara Chief wind very high. Set out on our voyage at 5 PM encampt a 4 me. S.S.

Lewis   Our vessels consisted of six small canoes, and two large perogues. This little fleet altho' not quite so rispectable as those of Columbus or Capt. Cook were still viewed by us with as much pleasure as those deservedly famed adventures ever beheld theirs; and I dare say with quite as much anxiety for their safety and preservation. We were now about to penetrate a country at least two thousand miles in width, on which the foot of civillized man had never trodden.

Clark   a windey day

Ordway   clear and pleasant. We went on verry well with a hard head wind   wind high from NW the greater part of the night

Whitehouse   This day we had fair weather—

---

*According to Moulton (1987b, 4: 95), Clark's Journal Codex C lists this river rise as "?."

MONDAY, APRIL 8

Sunrise			4 p.m.			Missouri River		
Temp	Weather	Wind	Temp	Weather	Wind	Rise/Fall	Feet	Inches
19a	f	NW	56a	f	NW	F		2

Weather Diary the Kilde, and large Hawk have returned. Buds of the Elm swolen and appear red— the only birds that I obseved during the winter at Fort Mandan was the Missouri Magpie, a bird of the Corvus genus.

Lewis Set out early this morning, the wind blew hard against us from the NW. We therefore traveled very slowly.

Clark wind hard a head from the NW.

Gass had a clear day. The wind blew hard from the NW. In the afternoon we passed very high bluffs on the South side; one of which had lately been a burning vulcano. The pumice stones lay very thick around it, and there was a strong smell of sulphur.

Ordway clear and cold. The wind high from the W we saw some Snow on the NS of the hills, and thick Ice on and under the banks of the river.

Whitehouse This day we had clear weather, the Wind blowing fresh from the Northwest.

TUESDAY, APRIL 9

Sunrise			4 p.m.			Missouri River		
Temp	Weather	Wind	Temp	Weather	Wind	Rise/Fall	Feet	Inches
38a	f	SE	70a	f	SW	F		1/2

Weather Diary the Crow has also returned saw the first today. & the corvus bird disappears the Musquitoes revisit us, saw several of them. Capt. Clark broght me a flower in full blo. it is a stranger to me.— the peroque [shakes with] is so unsteady that I can scarcely write

Clark Set out this morning verry early under a gentle breeze from the SE. I saw a Musquetor to day.   great numbers of Brant flying up the river, the Maple, & Elm has buded & Cotton and arrow wood beginning to bud. I saw flowers in the praries to day

Gass had a fine day.

Ordway clear and pleasant. A gentle breese from the South   we set off at day light. Sailed on   went Short distance further and halted for to take dinner at a bottom covered with Small cotton wood on N.S. The wind Shifted in to the West and blew Steady. We Saw a nomber of wild Geese on the river and brants flying over.   The Musquetoes begin to Suck our blood this afternoon.

Whitehouse This day Clear & pleasant weather.

WEDNESDAY, APRIL 10

	Sunrise			4 p.m.		Missouri River		
Temp	Weather	Wind	Temp	Weather	Wind	Rise/Fall	Feet	Inches
42a	f	E	74a	f	SW	R		¹/₈

Weather Diary   The prairie lark, bald Eagle, & the large plover have returned. The grass begins to Spring, and the leaf buds of the willow to appear.— Cherry birds disappear.

Lewis   The country on both sides of the missouri from the tops of the river hills, is one continued level fertile plain as far as the eye can reach, in which there is not even a solitary tree or shrub to be seen except such as from their moist situations or the steep declivities oh hills are sheltered from the ravages of the fire. About 1 ¹/₂ miles down this bluff from this point, the bluff is now on fire and throws out considerable quantities of smoke which has a strong sulphurious smell. The courant of the Missouri is but moderate, at least not greater than that of the Ohio in high tide; it's banks are falling in but little; the navigation is therefore comparitively with [its] lower portion easy and safe—

Clark   the morning cool and no wind. This day proved to be verry w[a]rm

Gass   rapid water and a great many sand-bars. But a fine pleasant day.

Ordway   a clear and pleasant. The current swift. The wind raised from West. One of our men Shot a bald Eagle. I took the quills to write.

THURSDAY, APRIL 11

	Sunrise			4 p.m.		Missouri River		
Temp	Weather	Wind	Temp	Weather	Wind	Rise/Fall	Feet	Inches
42a	f	NW	76a	f	W	F		¹/₂

Weather Diary   the lark wood pecker, with yellow wings, and a black spot on the brest common to the U' States has appeared, with sundry small birds.— many plants being to appear above the ground.— saw a large white gull today— the Eagle is now laying their eggs, and the gees have mated.— the Elm, large leafed, willow and the bush which bears a red berry, called by the engages greas de buff are in blume—   Small leaf willows in blum.

Clark   we Camped on the S.S. below a falling in bank. The river raise a little.

Gass   had a fine clear pleasant day

Ordway   Clear and pleasant.

Whitehouse   We set off at day light, this morning, the weather being Cool

Note   *The expedition arrives at the confluence of the Missouri and Little Missouri rivers in present-day western North Dakota.*

FRIDAY, APRIL 12

Sunrise			4 p.m.			Missouri River		
Temp	Weather	Wind	Temp	Weather	Wind	Rise/Fall	Feet	Inches
56a	f	NW	74a	c a r t & l	W	R		1/8

Weather Diary   small shower from the W attended with hard wind

Lewis   Canoes passed over the Lard side in order to avoid a bank which was rappidly falling in on the Stard. The night proved so cloudy I could make no further observations.

Clark   a fine morning. Set out verry early, the murcery Stood 56° above 0. The wind blew verry hard from the S all the after part of the day, at 3 oClock PM it became violent & flowey accompanied with thunder and a little rain. The water of the little Missouri is of the Same texture Colour & quality of that of the Big Missouri. The after part of the day so Cloudy that we lost the evening observation.

Gass   Another fine day. arrived at the Little Missouri (properly named), for it exactly resembles the Missouri in colour, current and taste.

Ordway   a clear and pleasant warm morning. The little River Missourie…is 120 yards wide at the mouth, but rapid and muddy like the big Missourie. About 3 oClock their came up a Squawl of verry high wind and rain. Some Thunder. The wind lasted untill after Sunsed. Then clear up   pleasant evening.

Whitehouse   This morning we had pleasant Weather,   the Little Mesouri River…its width at it mouth is 150 yards….it is Muddy, & its current runs strong

SATURDAY, APRIL 13

Sunrise			4 p.m.			Missouri River		
Temp	Weather	Wind	Temp	Weather	Wind	Rise/Fall	Feet	Inches
58a	f	SE	80a	f	SE	F		1

Weather Diary   The leaves of the Choke cherry are about half grown; the Cotton wood is in blume the flower of this tree resembles that of aspen in form, and is of a deep perple colour.—

Lewis   The wind was in our favour after 9 A.M. and continued favourable untill three 3. PM.   we therefore hoisted both the sails in the White Perogue ….which carried her at a pretty good gate, untill about 2 in the afternoon when a suddon squall of wind struck us and turned the perogue so much on the side as to allarm Sharbono who was steering at the time....the wind however abating for an instant I ordered Drewyer to the helm and the sails to be taken in,

Clark   the Missouri above the mouth of Little Missouri widens to nearly a mile

Gass   We had a pleasant day and a fair wind; but our small canoes could not bear the sail.

Ordway   clear pleasant & warm   proceeded on under a fine breeze of wind from the South.

Whitehouse   a fair wind from the Eastward, we sailed the greater part of this day

SUNDAY, APRIL 14

Sunrise			4 p.m.			Missouri River		
Temp	Weather	Wind	Temp	Weather	Wind	Rise/Fall	Feet	Inches
52a	c	SE	82a	f	SW	f		3/4

Clark   a fine morning.
Gass   had a fine morning
Ordway   clear & pleasant

MONDAY, APRIL 15

Sunrise			4 p.m.			Missouri River		
Temp	Weather	Wind	Temp	Weather	Wind	Rise/Fall	Feet	Inches
51a	f	E	78a	f	SW	F		1/2

Weather Diary   several flocks of white brant with black wings pass us today, their flight was to the NW the trees now begin to assume a green appearance, tho' the earth at the debth of about three feet is not yet thawed, which we discovered by the banks of the river, falling in [to the river] and disclosing a strata of frozen eath.—

Clark   the wind hard from the SE

Gass   We had a pleasant day and a fair wind

Ordway   a clear pleasant morning.   Sailed under a fine breeze from the SE   the river Shallow only about 8 feet deep in some places

Whitehouse   We set off this Morning, having a fresh breeze from the NE, about 8 oClock it veered round to the South East, and blew moderately—

TUESDAY, APRIL 16

	Sunrise			4 p.m.			Missouri River	
Temp	Weather	Wind	Temp	Weather	Wind	Rise/Fall	Feet	Inches
54a	f	SE	78a	f	S	F		¹/₂

Weather Diary   saw the first leather winged bat. It appeared about the same size of those common to the U' States.

Clark   Wind hard from the SE. Great numbers of Gees in the river & in the Plains feeding on the Grass

Gass   We had a clear pleasant day; and in the early part of it, a fair gentle wind. The wind became flawy (gusty), and the sailing bad.

Ordway   a clear pleasant morning.   The wind gentle from SE passed a Sand beach on the N.S. covered with Ice in Some heaps   it lay 4 feet thick where the Ice was drove in When the river broke up.   The trees are puting out Green.   The Grass begin to Grow in the bottoms & plains which look beautiful. We sailed some with a Southerly flawey (gusty) wind.   The river crooked so that we could not sail much of the time

Whitehouse   The Weather was Cool and clear; we proceeded on with all Sails set, having a fine breeze from the SE

WEDNESDAY, APRIL 17

	Sunrise			4 p.m.			Missouri River	
Temp	Weather	Wind	Temp	Weather	Wind	Rise/Fall	Feet	Inches
56a	f	NE	74a	c	SW	F		¹/₂

Weather Diary   thunder Shower passed above us from SW to NW <no> rain where we were.

Lewis   A delightfull morning.   There wase more appearance of birnt hills, furnishing large quantities of lave and pumice stone, of the latter some pieces were seen floating down the river. We had a fair wind today which enabled us to sail the greater part of the distance we have traveled

Clark   a fine morning, wind from the SE. Pumice Stone & Lava washed down to the bottoms and some Pumice Stone floating in the river. In the evening a thunder gust passed from the SW without rain

Gass   We proceeded on early as usual with a fair wind. The day was fine and we made good way.

Ordway   a clear beautiful morning.   A fair wind. We sailed on

THURSDAY, APRIL 18

Sunrise			4 p.m.			Missouri River		
Temp	Weather	Wind	Temp	Weather	Wind	Rise/Fall	Feet	Inches
52a	f	NE	64a	c	N			

Weather Diary   Wind very violent a heavy dew this morning. which is the first and only one we have seen since we passed the council bluffs last summer. there is but little dew in this open country.— saw a flock of pillecan pass from SW to NE they appear to be on a long flight.—

Lewis   a fine morning. We were detained today from one to five PM in consequence of the wind which blew so violently from N. that it was with difficulty we could keep the canoes from filling with water altho' they were along the shore. We came too on the Stard side under a boald well timbered bank which sheltered us from the wind which had abated but not yet ceased.

Clark   until near Sunset before Capt. Lewis and the party came up, they were detained by the wind, which rose Soon after I left the boat from the NW & blew verry hard untill verry late in the evening.

Gass   The morning was fine and we went on very well until 1 o'clock, when the wind blew so hard down the river, we were obliged to lie to for 3 hours, after which we continued our voyage. Encamped in a good harbour on the north on account of the wind, which blew very hard all night accompanied with some drops of rain.

Ordway   a clear pleasant morning. The wind shifted in to the NW and blew hard against us.   The wind rose so high that we could not go wit the cannoes without filling them with water. Detained us about 3 hours. The river has been verry crooked and bearing towards the South the most of the day. The Game is gitting pleantyier every day—

Whitehouse   This morning Clear pleasant weather. We set off Early, having the wind from the South   the water in the River was at a stand in rega[r]d to its depth. In the night the dew fell, which was what we had not seen for a long time—

FRIDAY, APRIL 19

Sunrise			4 p.m.			Missouri River		
Temp	Weather	Wind	Temp	Weather	Wind	Rise/Fall	Feet	Inches
45a*	c	NW	56a	c	NW			

---

*According to Moulton (1987b, 4: 95), Clark's Journal Codex I lists this temperature as "54 a."

Weather Diary   wind violent. The trees have now put forth their leaves. The goosbury, current, servisbury, and wild plumbs are in blume.

Lewis   the wind blew So hard this morning from NW that we dared not to venture our canoes on the river— the wind detained us through the course of this day, tho' we were fortunate in having placed ourselves in a safe harbour.

Clark   a blustering windey day, the wind So hard from the NW that we were fearfull of ventering our Canoes in the river. The Praries appear to green, the cotton trees bigin to leave, Saw some plumb bushes in full bloom

Gass   A cloudy morning, with high wind.

Ordway   Cloudy. The wind blew high from the Northward so that we were obleged to lay at our last nights harbour all day. The evening clear blustry & cold. Winds—

Whitehouse   This morning we had the Weather dark and Cloudy— the Wind blowing hard from the North   The water still at a stand, we remained here this day, the wind blowing so hard that we could make no headway—

SATURDAY, APRIL 20

	Sunrise			4 p.m.		Missouri River		
Temp	Weather	Wind	Temp	Weather	Wind	Rise/Fall	Feet	Inches
40a	c	NW	42a	c a s	NW			

Weather Diary   wind violent.

Lewis   the wind continued to blow tolerably hard this morning but by no means as violently as it did yesterday; we determined to set out and accordingly departed a little before seven. The wind blew so hard that I concluded it was impossible for the perogues and canoes to proceed and therefore returned and joined them about three in the evening. Capt. Clark informed me that soon after setting out, a part of the bank of the river fell in near one of the canoes and had very nearly filled [her] with water. That the wind became so hard and the waves so high that it was with infinite risk he had been able to get as far as his present station.

Clark   wind a head from the NW. we set out at 7 oClock proceeded on, Soon after we Set out a Bank fell in near one of the canoes which like to have filled her with water, the wind became hard and waves So rough that we proceeded with our little canoes with much risque, our Situation was Such after Setting out that we were obliged to pass round th 1st Point or lay exposed to the blustering winds & waves. The wind Continued So hard that we were Compelled to delay all day. Several buffalow lodged in the drift wood which had been drouned in the winter in passing the river. This morning was verry cold, some Snow about 2 oClock from flying clouds, Some frost this morning & the mud at the edge of the water was frosed.

Gass   had a cold disagreeable morning; rapid water and a strong wind. We were obliged again to lie too, on account of the wind

Ordway   cloudy. The wind is not so high as it was yesterday this morning. We found it cold polling. The air chilley. The wind rose and blew same as yesterday so that we could hardly make any head way. Delayed som time the wind abated a little. We proceeded on the wind Shortly rose again and blew so hard that the canoes were near filling they took in considerable water. The Sand blew off the sand bars & beaches so that we could hardly see, it was like a thick fogg.   High squawls of wind & flights of round Snow this day.   We took in some water in the Canoe I was in.   The water came up to my Box so that a part of my paper Got wet.

Whitehouse   wind blew so fresh from the North, that we could make no headway

Note   *The expedition passes present-day Williston, North Dakota.*

SUNDAY, APRIL 21

Sunrise			4 p.m.			Missouri River		
Temp	Weather	Wind	Temp	Weather	Wind	Rise/Fall	Feet	Inches
28a	f	NW	40a	c	NW	F		1/2

Weather Diary   wind violent white frost last night— the earth friezed along the water's edge.—

Lewis   Set out at an early hour this morning...the wind tho' a head was not violent. The wind blew so hard this evening that we were obliged to halt several hours.

Clark   Se out early, the wind gentle & from the NW. The river being verry Crooked. In the evening the wind became a verry hard a head.

Gass   a fine clear morning, but cold; there was a sharp frost. About 12 the wind again rose and was disagreeable, but we continued our voyage.

Ordway   a hard white frost last night. Froze water in the buckets setting near the fire. A clear and pleasant morning, but verry chilly & cold. About 3 oClock clouded up cold   the wind began blow as usal. A cool evening.

Whitehouse   This morning we had pleasant Weather, in the night we had a frost, we sett out early, the Wind blowing from the Northwest. The Water in the River fell one Inch.*

MONDAY, APRIL 22

Sunrise			4 p.m.			Missouri River		
Temp	Weather	Wind	Temp	Weather	Wind	Rise/Fall	Feet	Inches
34a	f a c	W	40	f	NW	R		2

Weather Diary   wind very hard greater part of the day—

Lewis   Set out at an early hour this morning; proceeded pretty well untill breakfat, when the wind became so hard a head that we proceeded with difficulty even with the

---

*According to Moulton (1997, 11: 137), Whitehouse had two entries labeled April 20 and one for April 21. By looking at the Lewis & Clark Weather Diary's river rise/fall records, these dates have been adjusted to the twenty first and twenty second.

assistance of our toe lines. The white river...the water is much clearer than that of the Missouri.

Clark   a verry cold morning some frost, we Set out...untill brackfast at which time the wind began to blow verry hard ahead, and Continued hard all day. The river riseing a little.

Gass   The wind was unfavourable to day, and the river here is very crooked.

Ordway   clear and cold. Delayed again on acct. of the high wind. Saw a buffaloe calf which had fell down the bank & could not git up again. We helped it up the bank and it followed us a short distance (the river raised 4 Inches last & white frost) we have Seen a great nomber of dead buffaloe lying on each shore all the way from the little missourie R. We suppose that they Got drowned attempting to cross on the Ice last fall before it got Strong.

Whitehouse   the Wind blew from the N East and the Water fell 2 Inches in the River

Tuesday, April 23

	Sunrise			4 p.m.			Missouri River		
Temp	Weather	Wind	Temp	Weather	Wind	Rise/Fall	Feet	Inches
34a	f	W	52	c	NW	R		2

Weather Diary   wind very hard greater part of the day—   saw the first robbin. Also the brown Curloo

Lewis   about nine AM the wind arose, and shortly after became so violent that we were unabled to proceed, in short it was with much difficulty and some risk that I was enabled to get the canoes and perogues into a place of tolerably safety. We remained untill five in the evening when the wind abating in some measure, we reloaded and proceeded. These winds being so frequently repeated, become a serious source of detention to us—

Clark   a cold morning, about 9 oClock the wind as usial rose from the NW and continued to blow verry hard untill late in the evening. The wind which had become violently hard, I joined Capt Lewis in the evening & after the winds falling which was late in the evening we proceeded on & encamped. The winds of this Countrey which blow with Some violence almost every day, has become a Serious obstruction in our progression onward, as we Cant move when the wind is high without great risque, and [if] there was no risque the winds is generally a head and often too violent to proceed.

Gass   had a fine day; but the wind was ahead, and we were obliged to lie to about three hours.

Ordway   a clear and pleasant morning. Not quite as cold as it has been for Several morning. The river verry crooked. The wind blew so hard that the large perogues Sailed in a bend where the wind came fair verry high the Small canoes took in some water. The large perogues Sailed verry fast. A short distanc we were obleged to halt the first safe place untill the wind abated which was about 3 hours. Dryed the articles which was wet. Towards evening the wind abated and we proceeded on round a point and camped

Whitehouse This morning, we had Clear weather. We proceeded on about 3 Miles, when the Wind blew so fresh, that we had to come too, it being a head Wind from the North west—

WEDNESDAY, APRIL 24

	Sunrise			4 p.m.			Missouri River	
Temp	Weather	Wind	Temp	Weather	Wind	Rise/Fall	Feet	Inches
40	f	N	56	f	N	R		1

Weather Diary wind very hard this morning

Lewis The wind blew so hard during the whole of this day, that we were unable to move. Notwithstanding that we were sheltered by high timber from the effects of the wind, such was [its] violence that it caused the waves to rise in such manner as to wet many articles. Soar eyes is a common complaint among the party. I believe it origenates from the immence quantities of sand which is driven by the wind from the sandbars of the river in such clouds that you are unable to discover the opposite bank of the river in many instances. The particles of this sand are so fine and light that they are easily supported by the air, and are carried by the wind for many miles, and at a distance exhibiting every appearance of a collumn of thick smoke. So penitrating is this sand that we cannot keep any article free from it; in short we are compelled to eat, drink, and breath it very freely.

Clark The wind rose last night and continued blowing from the N & NW, and Sometimes with great violence untill 7 oClock PM. As the wind was a head we could not move to day.

Gass This was a clear day, but the wind blew so hard down the river we could not proceed. While we lay here some of the men went to see some water at a distance which appeared like a river or small lake. In the afternoon they returned, and had found it only the water of the Missouri, which had run up a bottom.

Ordway Clear and cold. The wind high from the NW so that we had to delay here all this day. The woods got on fire.

Whitehouse This day we had Clear weather; but the Wind still blowing from the North West (ahead Wind) that we lay by, at the place we encamped the last night.

Note *The expedition sends an advance party to the confluence of the Missouri and Rochejhone (Yellowstone) rivers in present-day North Dakota. The main party camps there on April 26.*

THURSDAY, APRIL 25

Sunrise			4 p.m.			Missouri River		
Temp	Weather	Wind	Temp	Weather	Wind	Rise/Fall	Feet	Inches
36a	f	N	52a	f	NW	R		2

Weather Diary   wind very hard until 5 oClock PM

Lewis   The wind was more moderate this morning, tho' still hard; we set out at an early
hour. the water friezed on the oars this morning as the men rowed.   about 10
oclock AM the wind began to blow so violently that we were obliged to lye too. The
wind had been so unfavourable to our progress for several days past, and seeing but
little prospect of a favourable change; knowing that the river was crooked...believing
that we were at no very great distance from the Yellow stone River; proceed by
land...with a few men...to the entrance...and make the necessary observations. At
5 PM after I left him [Capt. Clark] the wind abated in some measure.

Clark   The wind was moderate & ahead this morning. The morning cold, Some flying
Clouds to be Seen, the wind from the N; ice collected on the ores this morning, the
wind increased and became So violent about 1 oClock we were obliged to lay by.
At 5 oClock the wind luled and we proceeded on and incamped.

Gass   We set out as usual and had a fine day; but about 11 were obliged to halt again the
wind so strong ahead. I remarked, as a singular circumstance, that there is now dew
in this Country, and very little rain. Can it be owing to the want of timber? At 5
o'clock in the afternoon, we renewed our voyage

Ordway   a clear cold morning.   The river rose 2 inches last night   The wind blew from the
N   Sailed some in a bend of the river.   The perogues could go no further as the
wind blew them a head so that they halted for it to abate on the N.S.

Whitehouse   this morning, having fine clear weather; about 11 oClock AM we had to come
too, on account of the Wind being a head & blowing hard. The dew at this
place never falls; and it seldom Rains, this we were told, by an Indian Women
that was with us

FRIDAY, APRIL 26

Sunrise			4 p.m.			Missouri River		
Temp	Weather	Wind	Temp	Weather	Wind	Rise/Fall	Feet	Inches
32a	f	S	63a	f	SE	R		3

Lewis   Capt. Clark measured these rivers just above their confluences; found the bed of
the Missouri 520 yards wide, the water occupying 300. [its] channel deep. The
yellowstone river including [its] sandbar, 858 yds. of which the water occupied 297
yards; the depest part 12 feet; it was falling at this time & appeard to be nearly at

[its] summer tide. …the water of this river is turbid tho' dose not possess as much sediment as that of the Missouri. The clouds this morning prevented my observing the moon.

Clark   last night was verry Cold. The Thermometer Stood at 32 above 0 this morning. I Set out at an early hour, as it was cold I walked on the bank. The river has been riseing for Several days, & 3 inches last night. Yellow Stone…it is at this time falling, the Missouri rising. I saw maney buffalow dead on the banks…those animals either drounded in attempting to Cross on the ice dureing the winter or Swiming across to bluff banks where they Could not get out & too weak to return. We Saw several in this Situation.

Gass   a fine day. The river Jaune is shallow, the Missouri is deep and rapid.

Ordway   a Clear pleasant morning (at the Yellowstone River) Capt. Clark measured these two rivers to day and found the Missourie to be 337 yards wide only the water but at high water mark 529 yards   the River Roshjone is 297 water, high water mark is 858 yards wide. The River Roshjone is not quite as rapid as the missourie

Whitehouse   This morning we had a fine Clear weather. At 12 oClock AM…arrived at the River's mouth called Roshjone (Yellowstone). At this junction…the River Mesouri was 337 Yards wide; and very deep; and the River Roshjone at its mouth, 97 yards wide…Shallow….Clear and its current rapid—

Note   *The expedition crosses the present-day North Dakota–Montana state line.*

SATURDAY, APRIL 27

	Sunrise			4 p.m.			Missouri River		
	Temp	Weather	Wind	Temp	Weather	Wind	Rise/Fall	Feet	Inches
	36a	f	SW	64a	f	NW	F		2

Weather Diary   wind very hard from 11 to 4 oClock

Lewis   at 11 AM the wind became very hard from the NW insomuch that the perogues and canoes were unable either to proceeded or pass the river to me. The wind abated about 4 PM and the party proceeded.

Clark   wind moderate & a head, at 11 oClock the wind rose and continued to blow verry hard a head from the NW untill 4 oClock PM, which blew the Sand off the Points in Such clouds as almost Covered us on the opposit bank, at 4 I Set out from my unpleasant Situation and proceeded on

Gass   About 9 o'clock in the forenoon we renewed our voyage. The day was fine, but on account of a strong wind we were obliged at 1 to halt till 4

Ordway   a clear and pleasant morning.   About 12 oC. The wind rose so high from the NW and the Sand flew so thick from the Sand bars that we halted about 1 oClock, to wait untill the wind abates,   about 4 oClock the wind abated   the current swift

Whitehouse   weather being clear and pleasant.   Stopped at One oClock to dine….shortly after the Wind blew so hard a head, from the Westward that we were delayed from starting till 4 oClock PM. River Mesouri, having had a strong current against us

SUNDAY, APRIL 28

	Sunrise			4 p.m.			Missouri River	
Temp	Weather	Wind	Temp	Weather	Wind	Rise/Fall	Feet	Inches
44a	f	SE	63a	f	SE	F		1 ½

Weather Diary   Vegetation has progressed but little since the 18th in short the change is scarcely perceptible.

Lewis   Set out this morning at an early hour; the wind was favourable and we employed our sails to advantage.

Clark   a fine day. River falling. Wind favourable from the SE and moderate

Gass   had a fine day

Ordway   clear and pleasant   the wind had shifted to SE and blew gently So that we Sailed some part of the time

Whitehouse   This day we had fine clear weather & pleasant   The River Mesouri, was not so high

MONDAY, APRIL 29

	Sunrise			4 p.m.			Missouri River	
Temp	Weather	Wind	Temp	Weather	Wind	Rise/Fall	Feet	Inches
42a	f	NE	64a	f	E	F		1 ½

Lewis   Set out this morning as the usual hour; the wind was moderate.

Clark   Set out this morning as the usial hour; the wind was moderate & from the NE.

Gass   had a clear morning. This forenoon we passed some of the highest bluffs I had ever seen

Ordway   a clear pleasant morning.

TUESDAY, APRIL 30

	Sunrise			4 p.m.			Missouri River	
Temp	Weather	Wind	Temp	Weather	Wind	Rise/Fall	Feet	Inches
50a	f	NW	58a	f	SE	F		½

Lewis   the wind blew hard all last night, and continued to blow pretty hard all day, but not so much, as to compell us to ly by.

Clark   The wind blew hard from the NE all last night, we Set out at Sunrise, the wind blew hard the greater part of the day and part of the time favourable, we did not lie by to day on account of the wind.

Gass   had a fine morning and went on very well

Ordway   clear and pleasant   we sailed a little in the bends of the River this afternoon—

Whitehouse   this morning, having fine pleasant Weather.

Note   *The expedition camps in eastern Montana between present-day Culbertson and Wolf Point.*

WEDNESDAY, MAY 1

	Sunrise			4 p.m.		Missouri River		
Temp	Weather	Wind	Temp	Weather	Wind	Rise/Fall	Feet	Inches
36a	c	E	46a	c a f	NE	F		1 ¹⁄₂

Weather Diary   wind violent form 12 0C. to 6 pM

Lewis   the wind being favourable...untill about 12 Ock. when the wind became so high that the small canoes were unable to proceed. One of them which seperated from us just before the wind became so violent, is now lying on the opposite side of the river, being unable to rejoin us in consequence of the waves, which during those gusts run several feet high. Here the wind compelled us to spend the ballance of the day.

Clark   We Set out at Sun rise under a Stiff Breeze from the East, the morning Cool & Cloudy. The wind became verry Hard and we put too on the L. side, as the wind Continued with Some degree of violence and the waves too high for the Canoes we were obliged to Stay all day.

Gass   a cool morning; and went on till 12 o'clock, when the wind rose so high, that our small canoes could not stand the waves.

Ordway   a clear pleasant morning, but cold. The wind from the East.   About 12 oClock the wind rose so high that the Small canoes could not go on without filling. We halted at a bottom covered with timber on S.S. One of the canoes lay on the opposite Shore and could not cross the water ran so high.   The wind continued so high that we delay and camped for the night.

Whitehouse   a clear pleasant morning but cold. We set off at Sun rise, the wind from the East. We Sailed Some. About 12 oClock the wind rose So high that we were obliged to halt. I and one other stopped on the other side of the river on account of the wind and ware obliged to lay out all night without any blanket. It being verry cold I Suffered verry much.

THURSDAY, MAY 2

Sunrise			4 p.m.			Missouri River		
Temp	Weather	Wind	Temp	Weather	Wind	Rise/Fall	Feet	Inches
28a	s	NE	34a	c a s	NW	F		1

Weather Diary   snow 1 inch deep the wind continued so high from 12 oClock yesterday, untill 5 this evening that we were unable to proceed. The snow which fell last night and this morning one inch deep has not yet disappeared.— it forms a singular contrast with the trees which are now in leaf.—

Lewis   the wind continued violent all night nor did it abate much of it's violence this morning, when at daylight it was attended with snow which continued to fall untill about 10 AM being about one inch deep, it formed a singular contrast with the trees and other vegitation which was considerably advanced. Some flowers had put forth in the plains, and the leaves of the cottonwood were as large as a dooler. The [water] friezed on the oars as they rowed. The wind dying at 5 PM we set out.

Clark   The wind blew verry hard all the last night, this morning about Sunrise began to Snow, (The Thermomtr. At 28 abov o) and Continued untill about 10 oClock, at which time Seased, the wind Continued hard untill about 2 PM . The Snow which fell to day was about 1 In deep, verry extroadernaley Climate, to behold the tree Green & flowers Spred on the plain, & Snow an inch deep.

Gass   At day break it began to snow; and the wind continued so high, we could not proceed until the afternoon. The snow did not fall more than an inch deep.

Ordway   at day light it began Snowing & continued Snowing & blowing so that we did not Set off. About 3 oClock it let off Snowing. The wind shifted in to the West. The Snow lay on the edge of the Sand bars & Sand beaches where the wind had blew it up one foot deep, but on the hills it was not more than half an Inch deep. The air & wind verry cold.

Whitehouse   at day light it began to Snow & the wind blow hard So that we did not set off this morning. About 3 oClock PM the Wind bated, and it quit snowing. Proceeded on our Voyage—the wind shifted and blew from the West. The Snow lay on the edge of the Sand Bars, blown by the wind, it was 12 Inches deep. The air verry cold during the whole day.

FRIDAY, MAY 3

Sunrise			4 p.m.			Missouri River		
Temp	Weather	Wind	Temp	Weather	Wind	Rise/Fall	Feet	Inches
26a	f	W	46a	c	W	F		$^1/_4$

Weather Diary hard frost last night. At four PM the snow has not yet entirely disappeared.— the new horns of Elk being to appear.

Lewis The morning being very [cold] we did not set out as early as usual; ice formed on a kettle of water $1/4$ of an inch thick. The snow has melted generally in the bottoms, but the hills still remain covered. The wind continued to blow hard from the West but not so strong as to compel us to ly by.

Clark we Set our reather later this morning than usial owing to weather being verry cold, a frost last night and the Thermt. Stood this morning at 26 above 0 which is 6 degrees blow freeseing—   The ice that was on the Kettle left near the fire last night was $1/4$ of an inch thick. The Snow is all or nearly all off the low bottoms, the Hills are entirely Covered.   the wind Continued to blow hard from the West, altho' not Sufficently So to detain us. As this Creek is 2000 miles up the Missouri we Call it the 2000 miles Creek. The greater part of the Snow is melted.

Gass though very cold and disagreeable, and a severe frost. The snow and green grass on the prairies exhibited an appearance somewhat uncommon. The cotton wood leaves are as large as dollars, notwithstanding the snow and such hard frost.

Ordway clear but verry cold for May. Saw the standing water froze over   the Ice froze to ore poles as we poled where the sun Shined on us.   A hard white frost last night. The ground covered with Snow. The wind rose high from the W about one.   The wind verry high & cold.

Whitehouse This morning we had Clear weather, but very Cold for the Season; We set out about 7 oClock AM   the Standing water froze last night, and the Water froze to our Setting pole. We had a severe white frost last night, and the ground covr'd with Snow— The wind rose & blew hard from the West. The wind continued Cold during the whole day. IN the Evening, the Snow had melted away.

SATURDAY, MAY 4

	Sunrise			4 p.m.			Missouri River	
Temp	Weather	Wind	Temp	Weather	Wind	Rise/Fall	Feet	Inches
38a	c	W	48a	f a c	W			

Weather Diary the black martin makes it's appearance. The snow has disappeared. Saw the first grasshoppers today.— there are great quantities of a small blue beatle feeding on the willows.—

Lewis We were detained this morning untill about 9 Ock. in order to repare the rudder....we then set out, the wind hard against us. I walked on shore this morning, the weather was more plesant, the snow has disappeared; the frost seems to have effected the vegetation much less than could have been expected. The leaves of the cottonwood the grass the box alder willow and the yellow flowering pea seem to be scarcely touched; the rosebushes and honeysuckle seem to have sustained the most considerable injury. At noon the sun so much obscured that I could not obtain his maridian Altitude.

Clark   detained untill 9 oClock...the wind a head from the west. The river has been falling for Several days passed; it now begins to rise a little, the rate of rise & fall is from one to 3 inches in 24 hours.

Gass   This day was more pleasant.

Ordway   clear and moderate this morning   the Snow is all melted off the hills.

Whitehouse   This morning we had clear, pleasant weather

SUNDAY, MAY 5

Sunrise			4 p.m.			Missouri River		
Temp	Weather	Wind	Temp	Weather	Wind	Rise/Fall	Feet	Inches
38a	f	NW	62a	f a r	SE	R		1

Weather Diary   a few drops of rain only

Lewis   a fine morning. The country is as yesterday beatifull in the extreme—

Clark   the river riseing & Current Strong.

Gass   The morning was fine with some white frost.

Ordway   Clear and pleasant.   We Sailed considerable in the course of the day with an East wind.

Whitehouse   Clear and pleasant weather

MONDAY, MAY 6

Sunrise			4 p.m.			Missouri River		
Temp	Weather	Wind	Temp	Weather	Wind	Rise/Fall	Feet	Inches
48a	f	E	61a	c a r	SE	R		2

Weather Diary   rain very inconsiderable as usual

Lewis   The morning being fair and pleasant and wind favourable we set sale at and early hour. The rains in the spring of the year (in a few days) suddonly melts the snow at the same time and causes for a few days a vast quantity of water which finds [its] way to the Missouri through those channels; by reference to the diary of the weather &c it will be percieved that there is scarcely any rain during the summer Autumn and winter in this open country distant from the mountains. At noon the sun being obscured by clouds I was unable to observe Altitude; it continued cloudy the ballance of the day and prevented all further observation.

Clark   a fine morning, wind from the NE. I believe those Streams to be Conveyance of the water of the heavy rains & melting Snows in the Countrey back &c. &c.

Gass   We set sail with a fair wind and pleasant weather. At 12 a few drops of rain fell, but it soon cleared up.

Ordway pleasant and warm. Sailed on under a gentle breeze from the East. Some Sprinkling rain, but did not last long.

Whitehouse clear pleasant and warm weather, a fair wind from the East, we Sailed on verry well. About 4 oclock, A light Sprinkling of rain, but did not last long.

TUESDAY, MAY 7

	Sunrise			4 p.m.			Missouri River	
Temp	Weather	Wind	Temp	Weather	Wind	Rise/Fall	Feet	Inches
42a	c	S	60a	f	NE	R		1 1/2

Lewis a fine morning. The drift wood begins to come down in consequence of the river's rising; the water is somewhat clearer than usual, a circumstance I did not expect on [its] rise. At 11 AM the wind became so hard that we were compelled to ly by for several hours. Vegitation appears to have advanced very little since the 28th Ulto— we continue to see great number of bald Eagles.

Clark A fine morning, river rose 1 1/2 inches last night, the drift wood beginning to run, the water Something Clearer than usial, the wind became verry hard,

Gass went on very well till 12 when it began to blow hard and being all under sail one of our canoes turned over. Fortunately the accident happened near the shore; and after halting three hours we were able to go again.

Ordway clear pleasant and warm. We set off cairly. The wind rose from the East. We Sailed verry fast untill about 12 oC. One of the canoes filled with water, but we got it Safe to Shore , and halted for the wind to abate

Whitehouse weather clear & pleasant. We set off eirly. The wind rose from the East, we set out sails. Sailing till about 12 oClock AM at which time the wind rose so high, that one our canoes filled with water....stopped on shore for the wind to about. About 4 oClock PM we set out again.

Note *The expedition arrives at the confluence of the Missouri and Milk rivers southeast of present-day Glasgow, Montana.*

WEDNESDAY, MAY 8

	Sunrise			4 p.m.			Missouri River	
Temp	Weather	Wind	Temp	Weather	Wind	Rise/Fall	Feet	Inches
41a	c	E	52a	c a r	E	F		1/4

Weather Diary   rain inconsiderable a mear sprinkle   the bald Eagle, of which there are great numbers, now have their young. The turtledove appears.

Lewis   Set out at an early hour under a gentle brieze from the East. A black cloud which suddonly spring up at SE, soon over shaddowed the horizon; at 8 AM it gave us a slight sprinkle of rain, the wind became much stronger but not so much so as to detain us. Examined the river [Milk River] I found it generally 150 yards wide....it is deep, gentle in [its] courant...the water of this river possesses a peculiar whiteness, being about the colour of a cup of tea with the admixture of a tablespoonfull of milk. From the colour of [its] water we call it Milk river.

Clark   a verry black Cloud to the SW. We Set out under a gentle breeze from the NW. About 8 oClock began to rain, but not Sufficient to wet. The water of this river [Milk River] will justify the belief that it has its Source at a considerable distance, and waters a great extent of Countrey— we are willing to believe that this is the river the Minitarres Call the river which Scolds at all others.

Gass   We were again very early under way in a cloudy morning; about 12 some rain fell; passed the Milk River there is a good deal of water in this river which is clear, and its banks beautiful.

Ordway   it clouded up of a Sudden, and rained Some. We Sailed on under a fine breeze from the East.

Whitehouse   Clouded up and rained shortly after we set off early. We found the current of the River to run very strong against us. Set our Sails and we proceeded on under a fine breeze from the East

THURSDAY, MAY 9

	Sunrise			4 p.m.			Missouri River		
Temp	Weather	Wind	Temp	Weather	Wind	Rise/Fall	Feet	Inches
38a	f	E	58a	f	W	R		3/4

Weather Diary   The choke Cherry is now in blume.

Lewis   Set out at an early hour; wind being favourable we used our sails and proceeded very well. The river for several days has been as wide as it is generally near [its] mouth, tho' it is much shallower. I begin to feel extreemly anxious to get in view of the rocky mountains.

Clark   a fine day, wind from the East. Came to a river...This river did not Contain one drop of running water...Those dry Streams which are also verry wide, I think is the Conveyance of the melted Snow, & heavy rains which is Said to Probable fall in from the high mountainious Countrey which is Said to be between this river & the Yellow Sone river— The Missouri keeps its width...water not So muddy & Sand finer

Gass   had a fine day.   The river more crooked

Ordway   Clear and pleasant. The game is gitting so pleanty and tame in this country that Some of the party clubbed them out of their way. Passed Big Dry creek, it is

220 yards wide at its high water mark, but at this time the water is So low that the water all Sinques in the quick Sand.

Whitehouse   This morning we had Clear pleasant Weather.

Friday, May 10

Sunrise			4 p.m.			Missouri River		
Temp	Weather	Wind	Temp	Weather	Wind	Rise/Fall	Feet	Inches
38a	f a c	WNW	62a	c a r	NW	F		3/4

Weather Diary   rain but slight a few drops

Lewis   Set out at sunrise and proceeded but a short distance ere the wind became so violent that we were obliged to come too, which we did on the Lard side in a suddon or short bend of the river where we were in a great measure sheltered from the effects of the wind. The wind continued violently all day, the clouds were thick and black, had a slight sprinkle of rain several times in the course of the day.

Clark   river fell 3/4 of an inch last night, wind from the NW we proceeded on but a short distance ere the wind became So violent we could not proceeded...the wind Continued all day. Several times in the course of the day We had some fiew drops of rain from verry black Clouds, no thunder or lightning latterly

Gass   We set out early in a fair morning; but having gone five miles were obliged to halt and lye by during the day, on account of hard wind. Some small showers of rain occasionally fell.

Ordway   a clear cold morning. The wind rose So high from the NW that obledged us to halt   The wind rose verry high

Whitehouse   We had clear and pleasant weather, the wind rose from the NW, we went about 4 miles, and halted for the Wind to abate, it blowing fresh   The wind rose considerably high, accompanied with Squalls of Rain—

Saturday, May 11

Sunrise			4 p.m.			Missouri River		
Temp	Weather	Wind	Temp	Weather	Wind	Rise/Fall	Feet	Inches
44a	f	NE	60a	c	SW			

Weather Diary   frost this morning

Lewis   the courant strong; and the river very crooked; the banks are falling in very fast. I sometimes wonder that some of our canoes or perogues are not swallowed up by means of these immence masses of earth which are eternally precipitating themselves into the river. The wind blue very hard the forepart of last night but abated toward morning; it again arose in the after part of this day and retarded our progress very much.

Clark   Wind hard fore part of last night. The latter part verry Cold a white frost this morning, the river riseing a little and verry Crooked. River rose 2 In

Gass   The morning was fine.

Ordway   a clear cool morning & white frost

Whitehouse   a clear cold morning, a white frost this morning.

Sunday, May 12

	Sunrise			4 p.m.		Missouri River		
Temp	Weather	Wind	Temp	Weather	Wind	Rise/Fall	Feet	Inches
52a	f	SE	54a	c a r	NW	R		2

Weather Diary   rain but slight

Lewis   Set out at an early hour; the weather clear and Calm. About 12 Oclock the wind veered about to the NW and blew so hard that we were obliged to Ly by the ballance of the day. About sunset it began to rain, and continued to fall a few drops at a time untill midnight; the wind blew violently all night—

Clark   the morning Clear and Calm. About 12 oClock the wind becam Strong from the E. About half past one oClock the wind Shifted round to the NW and blew verry hard all the latter part of the day, which obliged us to Lay by— about Sunset it began to rain, and rained very moderately only a fiew drops at a time for about half the night, wind Continued violent all night.

Gass   had a pleasant morning. At 1 we halted for dinner and a violent storm of wind then arose, which continued until night when some rain fell.

Ordway   a clear pleasant & warm morning   The wind rose high from the NW the detained us the remainder part of the day. Some Squawls of rain this evening—

Whitehouse   We had a clear, pleasant warm morning   The wind rose from the NW, and blew hard. The wind continuing to blow hard, detained us here the remainder of the day—   In the evening we had some Squalls of Rain.

Monday, May 13

	Sunrise			4 p.m.		Missouri River		
Temp	Weather	Wind	Temp	Weather	Wind	Rise/Fall	Feet	Inches
52a	c a r	NW	54a	f a c	NW	F		2 1/4

Weather Diary   rain but slight

Lewis   The wind continued to blow so violently this morning that we did not think it prudent to set out. At 1 PM the wind abated. The courant reather stronger than usual and the water continues to become reather clearer.

Clark   The wind Continued to blow hard untill one oClock PM to day at which time it fell a little as we Set out.

Gass   The weather continued stormy, and a few drops of rain fell.

Ordway   the wind blew verry hard all last night. Some Sprinkling rain and high wind this morning. About one oC. PM the wind abated So that we set off   The afternoon pleasant.

Whitehouse   The wind blew hard all last night, and this morning some Squalls of rain and high wind, which occasioned our not setting off Early— about 2 oClock PM the weather cleared off & became pleasant, and the wind abated— The current of the River running very swift

TUESDAY, MAY 14

Sunrise			4 p.m.			Missouri River		
Temp	Weather	Wind	Temp	Weather	Wind	Rise/Fall	Feet	Inches
32a	f	SW	52a	c	SW	F		1 ³/₄

Weather Diary   white frost this morning

Lewis   Some fog on the river this morning, which is a very rare occurrence. Surfice it so say, that the Perogue was under sail when a sudon squawl of wind struck her obliquely, and turned her considerably, the steersman allarmed, in stead of puting, her before the wind, lufted her up into it, the wind was so violent that it drew the brace of the squarsail out of the hand of the man who was attending it, and instantly upset the perogue and would have turned her completely topsaturva, had it not have been from the resistance mad by the oarning against the water.

Clark   A verry Clear Cold morning, a white frost & some fog on the river. The Thermomtr Stood at 32 above 0, wind from the SW. We proceeded on verry well untill about 6 oClock a Squawl of wind Struck our Sale broad side and turned the perogue nearly over

Gass   There was some white frost in the morning. About 12 the day became warm. Banks of snow were seen lying on the hills on the North side.   About 4…a sudden gust of wind arose, which overset one of the periogues before the sail could be got down.

Ordway   a hard white frost last night. Our mocassons froze near the fire. A clear and pleasant morning. It was verry warm or much warmer than it has been before this Spring. We saw some banks of Snow laying in the vallies at the N.S. of the hills. About 4 oClock the white peroque of the Captains was Sailing a long, there came a violent gust of wind from the NW which was to the contrary to the course they were sailing. It took the sail and before they had time to douse it turned the perogue down on one Side So that she filled with water, and would have turned over had it not been for the awning which prevented it. With much a diew they got the sail in and got the pirogue to shore

Whitehouse   a hard white frost last night, so that our Mocasins froze near the fire last night. A clear and pleasant morning. Sergt. Gass who was out hunting, Saw Some banks of Snow on the N. Side of Some hills north of the river. Hoisted Sail as the wind blew fair. Shortly after we set off a Violent Storm came from a black Cloud, which lay to the Northwest, and the Wind rose and shifted suddenly to the Northwest. This wind took the sail of the Pettyauger and had it not been for the awning, and mast She would have turned up Side down. The Pettyauger filled full of Water, and with much trouble they got her to shore—

WEDNESDAY, MAY 15

Sunrise			4 p.m.			Missouri River		
Temp	Weather	Wind	Temp	Weather	Wind	Rise/Fall	Feet	Inches
48	c a r	SW	54a	c	NW	F		3/4

Weather Diary   slight shower

Lewis   as soon as a slight shower of rain passed over this morning, we spread the articles to dry which had got wet yesterday in the white perogue; tho' the day proved so cloudy and damp that they received but little benifit from the sun or air.

Clark   Our medisons, Instruments, merchandize, Clothes, provisions &c. &c. which was nearly all wet we had put out to air and dry. The day being Cloud & rainey those articles dried but little to day— We see Buffalow on the banks dead, others floating down dead, and others mired every day, those buffalow either drown in Swiming the river or brake thro' the ice.

Gass   We remained here all day to dry our baggage that had got wet. It was cloudy and unfavourable for the purpose and some rain fell.

Ordway   cloudy. A Small Shower of rain about 11 oClock. Continued cloudy all day.

Whitehouse   This morning we had Cloudy Weather. A Shower of rain coming on, which lasted about One hour, when it cleared off—

THURSDAY, MAY 16

Sunrise			4 p.m.			Missouri River		
Temp	Weather	Wind	Temp	Weather	Wind	Rise/Fall	Feet	Inches
48	c	SW	67a	f	SW			

Lewis   The morning was fair and the day proved favourable to our operations; by 4 oClock in the evening our Instruments, Medicine, merchandize provision &c. were perfectly dryed.

Clark   a morning fair, our articles all out to Dry. At 4 oClock we had everything that was Saved dry.

Gass   This was a fine day.

Ordway   a heavy diew last night.   A clear and pleasant morning.

Whitehouse   a heavy dew fell last night, but a pleasant clear morning.

FRIDAY, MAY 17

	Sunrise			4 p.m.		Missouri River		
Temp	Weather	Wind	Temp	Weather	Wind	Rise/Fall	Feet	Inches
60a	*f*	NE	68a	*f*	SW			

Weather Diary   the Gees have their young; the Elk being to produce their young, the Antelope and deer as yet have not.— the small species of Goatsucker or whiperwill begin to cry—   the blackbirds both small and large have appeared. We have had scarcely any thunder and lightning. The clouds are generally white and accompanyed with wind only

Lewis   we were roused late at night by the Sergt. Of the guard, and warned of the danger we were in from a large tree that had taken fire and which leant immediately over our lodge. We had the loge removed, and few minutes after a large proportion of the top of the tree fell on the place the lodge had stood; had we been a few minutes later we should have been crushed to attoms. the wind blew so hard, that notwithstanding the lodge was fifty paces distant from the fire it sustained considerable injury from the burning coals which were thrown on it; the party were much harrassed also by this fire which communicated to a collection of fallen timber, and could not be extinguished.

Clark   a fine morning, wind from the NW. Mercury at 60° a 0. River falling a little. River much narrower than below from 2 to 300 yards wide.   We were roused late at night and warned of the danger of fire from a tree which had Cought... the wind blew hard and the dry wood Cought & fire flew in every direction, burnt our Lodge verry much...the whole party was much disturbed by this fire, which could not be extinguished &c.

Gass   The morning was fine. The hills...some of them, which at a distance resembled ancient steeples.

Ordway   A clear pleasant morning. Verry high hills and white knobs, which are washed by rains.

Whitehouse   a Clear pleasant morning

SATURDAY, MAY 18

	Sunrise			4 p.m.		Missouri River		
Temp	Weather	Wind	Temp	Weather	Wind	Rise/Fall	Feet	Inches
58a	*f*	W	46a	*c a r*	NW	F		1

Weather Diary   a good shower saw the wild rose blume   the brown thrush or mocking bird has appeared.— had a good shower of rain today, it continued about 2 hours; this is the first shower that deserves the appellation of rain, which we have seen since we left Fort Mandan.— no thunder or lightning

Lewis   the wind blew hard this morning from the West. There are now but few sandbars, the river is narrow and current gentle.

Clark   A windey morning, wind from the West. The after part of the day was Cloudy & at about 12 oClock it began to rain and continued moderately for about 1 1/2 hours, not Sufficient to wet a man thro' his clothes; this is the first rain Since we Set out this Spring

Gass   A cloudy morning. We had some showers of rain in the forenoon; hail in the afternoon; and a fine clear evening.

Ordway   a clear warm morning. The Missourie is gitting clear and gravelly bottom & Shore a pleasant warm afternoon

Whitehouse   a clear fine warm morning. About 10 oClock AM it clouded up and began to rain; and we had Several Small Showers. Here the Water of the Mesouri River, that had been muddy ever since we first entered it, began to get clear, and the bottom that was muddy, is gravelly—

Note   *The expedition members catch their first glimpse of the Little Rocky Mountains of Montana.*

SUNDAY, MAY 19

	Sunrise			4 p.m.			Missouri River		
Temp	Weather	Wind	Temp	Weather	Wind	Rise/Fall	Feet	Inches
38a	f	E	68a	f a c	SW			

Weather Diary   heavy fog this morning on the river

Lewis   This last night was disagreeably [cold]; we were unable to set out untill 8 oclock AM in consequence of a heavy fogg, which obscured the river in such a manner that we could not see our way; this is the first we have experienced in any thing like so great a degree; there was also a fall of due last evening, which is the second we have experienced since we have entered this extensive open country. On Capt. Clark's return…he saw a range of Mountains, bearing W. distant 40 or 50 miles. The NNE extremity of these mountains appeared abrupt. [The Little Rocky Mountains, MT] This afternoon the river was croked, rappid and containing more sawyers than we have seen in the same space since we left the enterence of the river Platte.

Clark   a verry cold night, the murckery stood at 38 at 8 oClock this morning, a heavy *dew* which is the 2d I have seen this spring. The fog (which was the first) was So thick this morning that we could not Set out untill the Sun was about 2 hours up, at which time a Small breeze Sprung up from the E. which Cleared off the fog & we proceeded on by means of the Cord. I also Saw a high mountain in a westerly direction, bearing SSW about 40 to 50 miles distant.

Gass   The morning was foggy and there was some dew.
Ordway   a heavy diew fell last night. We set off at 7 oC Clear and pleasant.
Whitehouse   a heavy dew fell last night, and this morning was clear and pleasant

---

Note   *The expedition arrives at the confluence of the Missouri and Musselshell rivers in central Montana.*

MONDAY, MAY 20

	Sunrise			4 p.m.		Missouri River		
Temp	Weather	Wind	Temp	Weather	Wind	Rise/Fall	Feet	Inches
52a	f	NE	76a	f	E	F		1

Lewis   river narrow and croked. At 11 AM we arrived at the entrance of a handsome bold river [Lewis later calls it "Muscleshell"]...is 110 yards in width...the waters is of a greenish yellow cast. The Missouri opposite to this point is deep, gentle in [its] courant, and 222 yards in width. About 5 miles above this river...a handsome river of about fifty yards in width...we called it Sah-ca-gar-me-ah or bird woman's River [Sacagawea River], after our interpreter the Snake woman.
Clark   a fine morning, wind from the NE. River falling a little.. Passed some verry swift water, river narrow and crooked ....the Missouri water is not so muddey as below, but retains nearly its usial cholour, and the sand principally confined to the points
Gass   had a fine morning. The water of the Missouri is becoming more clear. The water of the Musselshell is of a pale colour and the current is not rapid.
Ordway   a clear pleasant morning   the Missouri at the mouth of the (Mussel) Shell River is 222 yds wide with a small current....not so muddy as below but retains nearly the usal colour.
Whitehouse   A Clear pleasant morning

TUESDAY, MAY 21

	Sunrise			4 p.m.		Missouri River		
Temp	Weather	Wind	Temp	Weather	Wind	Rise/Fall	Feet	Inches
50a	f	SW	76a	f	NW			

Lewis   a delightfull morning. The wind which was moderate all the fore part of the day continued to encrease in the evening, and about dark veered about to the NW and blew a storm all night, in short we found ourselves so invelloped with clouds of dust and sand that we could neither cook, eat, nor sleep; and were finally compelled to remove our lodge about eight oClock at night to the foot of an adjacent hill where we were covered in some measure from the wind by the hills.

Clark   a butifull morning, wind from the West, river falling a little. Wind which blew moderatly all the forepart of the day increased and about Dark Shifted to the NW and Stormed all night, Several loose articles were blown over board, our lodge & Camp which was on a Sand bar on the Std. Side & opposte the lower point of the Island we were obliged to move under the hills, the dust & Sand blew in clouds.

Gass   had a fine morning; towards the middle of the day the wind blew hard

Ordway   a butiful morning. Wind from the west. River falling a little. The wind which moderately all the fore part of the day increased and about dusk Shifted to the NW and blew high & Stormed all night. Several loose articles were blown overboard   the dust & Sand blew in clouds.

Whitehouse   This morning we had clear and pleasant Weather. About 1 oClock the wind rose So high from the NW that we delayed about 2 hours and then proceeded on. Came about 15 miles....the wind rose very high and hard, Soon after we made Camp, and made the Sand fly So that it was verry disagreeable. The most of the part moved back towards the hills. Some of our party that was out hunting Yesterday reported that they had seen, a high ridge of Mountains, which lay to the West, but appeared to be a very great distance from them. (The Little Rocky Mountains)

WEDNESDAY, MAY 22

	Sunrise			4 p.m.			Missouri River		
Temp	Weather	Wind	Temp	Weather	Wind	Rise/Fall	Feet	Inches	
46a	c	NW	48a	c	NW	F		1/2	

Weather Diary   the wind excessively hard all night—   saw some particles of snow fall today it did not lye in sufficient quantity on the ground to be perceptible.—

Lewis   The wind blew so violently this morning that we did not think it prudent to set out untill it had in some measure abated; this did not happen untill 10 AM. The river continues about the same width from 200 to 250 yds wide

Clark   The wind continued to blow So violently hard we did not think it prudent to Set out untill it lulel a little, about 10 oClock we set out the morning cold. River falls about an inch a day. Maney of the Creeks which appear to have no water near ther mouths have Streams of running water higher up which rise & waste in the Sand or gravel. The water of those Creeks are So much impregnated with the Salt Substance that it cannot be Drank with pleasure.

Gass   A cloudy morning. The wind blew so hard this morning, we did not get under way until 9 o'clock. The forenoon was cold and disagreeable, but the afternoon became more pleasant.

Ordway   the wind continued to blow so violently hard we did not think it prudent to Set out untill it luled a little about 11 oClock we set out the morning cold. River falls about an Inch a day. Many of the creeks which appear to have no water near their mouths have Streams running water high up which rise & waste in the Sand or gravel the water of those creeks are so much impregnated with the Salt Substance that it cannot be drank with pleasantness—

Whitehouse   The wind blew hard all last night and continues blowing this morning till about 11 oClock AM when it abated. Cloudy. The day was chilly and Cold.

THURSDAY, MAY 23

Sunrise			4 p.m.			Missouri River		
Temp	Weather	Wind	Temp	Weather	Wind	Rise/Fall	Feet	Inches
32a	*f*	*SW*	*54a*	*f*	*SW*	*F*		*1/2*

Weather Diary   hard frost last night; ice in the eddy water along the shore, and the water friezed on the oars this morning. Strawburies in bloom. Saw the first king fisher.

Lewis   the frost was severe last night, ice appeared along the edge of the water, water also freized on the oars. The creeks...strongly impregnated with these salts that it is unfit for uce; all the wild anamals appear fond of this watcr; I have tryed it by way of experiment & find it moderately pergative, but painfull to the intestens in [its] opperation. I am astonished how this animal as it does without water [prairie dog], particularly in a country like this where there is scarcely any rain during 3/4 of the year and more rarely any due [dew]...in the Autumn when the hard frosts commence they close their burrows and do not venture out again untill spring. River more rappid. The musquetoes troublesome this evening, a circumstance I did not expect from the temperature of the morning. The Gees begin to lose the feathers of their wings and are unable to fly.

Clark   a Severe frost last night, the Thrmotr. stood at the freesing point this morning i.e. 32 a 0. wind SW. the water freeses on the oars. Ice on the edge of the river. ....the river beginning to rise, and current more rapid than yesterday. The after part of this day was w[a]rm & the Misquitors troublesome.

Gass   The morning was clear with a white frost, and ice as thick as window glass

Ordway   a Severe frost last night. the Thurmomiter Stood at the freezing point this morning. W ind SW the water freezes on the ore   Ice on the edge of the river. The after part of the day was warm & the Musquetoes troublesome. The river begining to rise & current more rapid than yesterday.

Whitehouse   We had a cold frosty morning. The standing water was froze over, and cover'd with Ice

Note   *The expedition enters what is present-day Missouri Breaks National Monument, Montana.*

FRIDAY, MAY 24

Sunrise			4 p.m.			Missouri River		
Temp	Weather	Wind	Temp	Weather	Wind	Rise/Fall	Feet	Inches
32a	f	NW	68a	f	SE	R		3 1/2

Weather Diary   frost last night ice 1/8 of an inch thick

Lewis   The water standing in the vessels freized during the night 1/8 of an inch thick, ice also appears along the verge of the river. The folage of some of the cottonwood trees have been entirely distroyed by the frost and are again puting forth other buds. About 9 AM when a fine breeze sprung up from the SE and enabled us though the ballance of the day to employ our sails to advantage; we proceed at a pretty good pace notwithstanding the courant of the river was very strong. The air is so pure in this open country that mountains and other elivated objects appear much nearer than they really are; these mountains do not appear to be further than 15 m.

Clark   a Cold night. The water in the Small vestles frosed 1/8 of an inch thick, and the thermometer Stood this morning at the freesing point. At 9 oClock we had a Breeze from the SE which Continued all day. This Breeze afforded us good Sailing. The river rising fast, Current verry rapid.

Gass   There was again some white frost this morning. The water is high, rapid and more clear.

Ordway   a cold night.   The water in the Small vessels froze 1/8 of an inch thick & the Thurmot. Stood this morning at the freezing point   At 9 oClock we had a breeze of wind from the SE which continued all day   this Breeze aforded us good Sailing the River rising fast   current very rapid   The cotton wood in this point is begining to put out a Second time the first being killed by the frost.

Whitehouse   clear & pleasant weather. After 3 oClock...The wind began to blow from the SE, and we set all Sails— At camp...the leaves of these trees were killed by the frost—

SATURDAY, MAY 25

Sunrise			4 p.m.			Missouri River		
Temp	Weather	Wind	Temp	Weather	Wind	Rise/Fall	Feet	Inches
46a	f	SW	82a	f	SW	R		2

Weather Diary   saw the kingbird, or bee martin; the grouse disappear. Killed three of the bighorned antelopes.

Clark   the morning Cool & pleasant, wind a head all day from the SW. I also think I saw a range of high mounts, at a great distance to the SSW but am not certain as the horozon was not clear enough to view it with Certianty. the Air of this quarter is pure and helthy. The water of the Missouri well tasted not quite So muddy as it is below, not withstanding the last rains has raised the river a little is less muddy than it was before the rain.

Gass   The forenoon was pleasant. These hills are very much washed in general; they appear like great heaps of clay, washing away with every shower; with scarcely any herbs or grass on any of them.

Ordway   the morning cool & pleasant   wind a head all day from the SW   the air of this country is pure & healthy the water of the Missourie fine and cool.

Whitehouse   clear pleasant weather. About 3 oClock the wind shifted to the Northwest and blew moderately— about 4 oClock...the current of the River running very swift— it being so for these several days past

SUNDAY, MAY 26

	Sunrise			4 p.m.			Missouri River	
Temp	Weather	Wind	Temp	Weather	Wind	Rise/Fall	Feet	Inches
58a	f	SW	80a	f	SW	R		1/2

Weather Diary   The last night was much the warmest we have experienced, found the covering of one blanket sufficient. The air is extremely dry and pure.

Lewis   ascended the river hills. As from this point I beheld the Rocky Mountains for the first time, I could only discover a few of the most elivated points above the horizon...these points of the Rocky Mountains were covered with snow and the sun shone on it in such manner as to give me the most plain and satisfactory view. while I viewed these mountains I felt a secret pleasure in finding myself so near the head of the heretofore conceived boundless Missouri; but when I reflected on the difficulties which this snowey barrier would most probably throw in my way to the Pacific, and the sufferings and hardships of myself and party in them, it in some measure is counterballanced the joy I had felt in the first moments I gazed on them; but as I have always held it a crime to anticipate evils I will believe it a good comfortable road untill I am compelled to believe differently. This is truly a desert barren country and I feel myself still more convinced of [its] being a continuation of the black hills.

Clark   proceeded as yesterday wind from the SW. Ascended the high countrey...I could plainly See the Mountains on either Side which I Saw yesterday...most S. Westerly of those Mountains there appeared to be Snow. We had a few drops of rain at dark— this Countrey may with propriety I think be termed the Deserts of America, as I do not Conceive any part can ever be Settled, as it is deficent in water, Timber & too Steep to be tilled.

Gass   a fine morning

Ordway   wind from SW   the waves roled for Some distance below (a rapid) we ascended it by the assistance of the chord & poles   we had a fiew drops of rain at dark. This country may with propriety be called the Deserts of North america for I do not conceive any part of it can ever be Setled as it is deficient of or in water except this river, & of timber & too Steep to be tilled.

Whitehouse   a clear pleasant morning. Stream running so strong. Came to a rapid which had considerable fall.

MONDAY, MAY 27

Sunrise			4 p.m.			Missouri River		
Temp	Weather	Wind	Temp	Weather	Wind	Rise/Fall	Feet	Inches
62a	f	SW	82a	f	SW			

Weather Diary   wind so hard we were unable to proceed in the early part of the day

Lewis   The wind blew so hard this morning that we did not sent out untill 10 AM. Great quantities of stone also lye in the river and garnish [its] borders, which appears to have tumbled from the bluffs where the water rains had washed away the sand and clay in which they were imbedded. About midday it was very warm to this the high bluffs and narrow channel of the river no doubt contributed greatly.

Clark   The wind blew hard from the SW which detained us untill about 10 oClock. This day verry w[a]rm— the river is Genly about 200 yards wide and Current very Swift to day and has a verry perceptiable fall in all its Course— it rises a little.

Gass   We have now got into a country which presents little to our view, but scenes of barrenness of desolation; and see no encouraging prospects that it will terminate. The grass is generally short on these immense natural pastures, which in the proper seasons are decorated with blossoms and flowers of various colours.   The day was fine, but the wind ahead.   The bed of the river is rocky.

Ordway   the wind blew hard from the SW which detained us untill about 10 oClock   this day verry warm.

Whitehouse   This morning pleasant weather, but the wind high, from the Northwest. The current of the River running very swift. Passed verry high Steep mountains and clifts Steep precipices. These mountains appear to be a desert part of the country. They wash by rains, but a little rain in this part. No diews like other parts but barron broken rich soil but too much of a desert to be inhabited, or cultivated. The Game became scarcer here...owing to there being no grass, or timber'd land for them to live in. [Missouri Breaks]

TUESDAY, MAY 28

Sunrise			4 p.m.			Missouri River		
Temp	Weather	Wind	Temp	Weather	Wind	Rise/Fall	Feet	Inches
62a	c	SW	72a	c & r	SW	R		1/2

Weather Diary   a slight thundershower; the air was turbid in the forenoon and appeared to be filled with smoke; we supposed it to proceed from the burning of the plains, which we are informed are frequently set on fire by the Snake Indians to compel the antelopes to resort to the woody and mountanous country which they inhabit.— saw a small white and black woodpecker with a red head; the same which is common to the Atlantic states.—

Lewis   the weather dark and cloudy, the are smokey, had a few drops of rain. At 10 AM a few drops of rain again fell and were attended with distant thunder which is the first we have heard since we left the Mandans—

Clark   a Cloudy morning Some fiew drops of rain and verry smokey, wind from the SW. at 1 oClock we had a few drops of rain and Some thunder which is the first thunder we have had since we Set out from Fort Mandan.

Gass   had a fine morning

Ordway   a cloudy morning. Some fiew drops of rain & Smokey wind from the SW. At 1 oClock we had a fiew drops of rain & Some Thunder which is the first Thunder we have had Since we Set out, from Fort Mandans

Whitehouse   clear and pleasant weather. We had some thunder, and small showers of rain which lasted about 2 hour. Had pleasant afternoon.

Note   *The party passes the confluence of the Missouri and Judith rivers in central Montana.*

WEDNESDAY, MAY 29

	Sunrise			4 p.m.		Missouri River		
Temp	Weather	Wind	Temp	Weather	Wind	Rise/Fall	Feet	Inches
62a	c a r	SW	67a	r	SW	R		1

Weather Diary   rained by little, some dew this morning.

Lewis   soon after we landed it began to blow & rain, and there was no appearance of even wood enough to make our fires for some distance. Passed a handsome river…100 yds wide…it appeared to contain much more water than the Muscle-Shell…the water is clear…Cap C. who assended this R. much higher than I did has thought proper to call it Judieths River. Gave each man a small dram. Notwithstanding the allowance of sperits we issued did not exceed $^1/_2$ pn. Man several of them were considerably effected by it, such is the effects of abstaining for some time from the uce of sperituous liquors; they were all very merry—

Clark   Soon after we Came too it began to rain & blow hard. A table Spoon full of water exposed to the air in a Saucer would avaporate in 36 hours when the mercury did not Stand higher than the temperate point [66° F or 55° F] in the heat of the day.

Gass   had a fine morning. At 12 it became cloudy and began to rain. It rained a little all the afternoon.

Ordway   passed a considerable rapid came to a bold running Stream. Soon after we came too it began to rain, and blew hard Some of the hunters who went out on the high land, Said it Snowed & hailed on the hills

Whitehouse   Cloudy weather this morning. About 3 oClock began to rain, the wind rose high & hard from the NW. Hunters returned & Said it Snowed and hailed on the hills back from the River. In the Evening we had rain.

THURSDAY, MAY 30

	Sunrise			4 p.m.		Missouri River		
Temp	Weather	Wind	Temp	Weather	Wind	Rise/Fall	Feet	Inches
56a	c a r	SW	50a	r	SW	R		5

Weather Diary the rain commenced about 4 Oclock in the evening, and continued moderately through the course of the night; more rain has now fallen than we have experienced since the 15th of September last.

Lewis The rain which commenced last evening continued with little intermission untill 11 this morning when we set out; the high wind which accompanied the rain rendered it impracticable to procede earlyer. More rain has now fallen than we have experienced since the 15th of September last. Many circumstances indicate our near approach to a country whos climate differs considerably from that in which we have been for many months. The air of the open country is asstonishingly dry as well as pure. I found by several experiments that a table spoon of water exposed to the air in a saucer would evaporate in 36 hours when the murcury did not stand higher than the temperate point at the greatest heat of the day; My inkstand so frequently becoming dry put me on this experiment. I also observed the well seasoned case of my sextant shrunk considerably and the joints opened. The water of the river still continues to become clearer and notwithstanding the rain which has fallen it is still much clearer than it was a few days past. The bluff were more steep than usual and were now rendered so slippery by the late rain that the men could scarcely walk. The wind was also hard against us. We had slight showers of rain through the course of the day, the air was [cold] and rendered more disagreeable by the rain. One of the party ascended the river hills and reported on his return that there was snow intermixed with the rain which fell on the hights

Clark The rain commenced yesterday evining, and continued moderately through the course of the night, more rain has now fallin than we have experienced Since the 15th of September last, the rain continued this morning, and the wind too high for us to proceed untill about 11 oClock at which time we were set out, and proceeded on with great labour. not with standing we proceeded on as well as we could wind hard from the NW. Some little rain at times all day. one man assended the hight Countrey and it was raining & Snowing on those hills, the day has proved to be raw and Cold.

Gass The forenoon was cloudy, with some rain. It rained a little all day.

Ordway the rain commenced yesterday evening & continued moderately through the course of the night. More rain has now fallen than we have experenced Since the 15th of September last, the rain continued this morning, and the wind too high for us to proceed, untill abt. 11 oClock at which time we set out   the men could Scarsely walk notwithstanding we proceeded as well as we Could, wind hard from the NW. Some little rain at times all day. One man ascended the high country & it was raining & Snowing on those high hills, the day has proved to be raw and cold back from the river is tollarably level.

Whitehouse   weather Cloudy and Rainey this morning, & the wind blowing high and hard from the NW. WE delayed setting off till 10 oClock, the weather still being very disagreeable working. The weather still continued Cold & Chilling with wind and rain just before 2 oClock.

FRIDAY, MAY 31

Sunrise			4 p.m.			Missouri River		
Temp	Weather	Wind	Temp	Weather	Wind	Rise/Fall	Feet	Inches
48a	c a r	W	53a	c a r	SW	R		1 ½

Weather Diary   but little rain. The Antelope now bring forth their young. from the size of the young of the bighorned Antelope I suppose they bring forth their young as early at least at the Elk.

Lewis   soon after we got under way it began to rain and continued untill meridian when it ceased but still remained cloudy through the ballance of the day. The obstructions of rocky points and riffles still continue as yesterday; at those places the men are compelled to be much in the water even to their armpits, and the water is yet very [cold], and so frequent are those point that they are one fourth of their time in the water. The hills and river Clifts which we passed today exhibit a most romantic appearance…horizontal stratas of white free-stone, on which the rains or water make nor impression. The water in the course of time in decending from those hills and plains on either side of the river trickled down the soft sand clifts and woarn it into a thousand grotesque figures, which with the help of a little immagination and an oblique view at a distance, are made to represent eligant ranges of lofty freestone buildings, haveing their pararpets well stocked with statuary; collumns of various sculpture both grooved and plain, are also seen supporting long galleries in front of those buildings; in other places on a much nearer approach and with the help of less immagination we see the remains or ruins of eligant buildings, some collumns standing and almost entire with their pedestals and capitals. As we passed on it seemed as if those seens of visionary inchantment would never have an end…So perfect indeed are those walls that I should have thought that nature had attempted here to rival the human art of masonary had I not recollected that she had first began her work. The river today has been from 150 to 250 yds. wide.

Clark   a cloudy morning. Soon found it verry laborious as the mud Stuck to my mockersons & was verry Slippery. Soon after we got under way it began to rain and continued moderately untill about 12 oClock when it ceased, & Continued Cloudy the ballance of the day. The Hills and river Clifts of this day exhibit a most romantick appearance…in many places this Sand Stone appears like antient ruins some like elegant buildings at a distance, some like Towers &c. &c. Remind us of Some of those large Stone buildings in the United States. As we passed on it Seemed as if those Seens of Visionary enchantment would never have an end; for here it is too that nature present to the view of the traveler vast ranges of walls of tolerable workmanship, so perfect indeed are those [walls] that I Should have thought that nature had attempted here to rival the human art of Masonry had I not recollected that She had first began her work. The river rises a little. The water is yet very cold. Little timber on the river to day. River less muddy than it was below.

Gass   a cloudy morning. About 11 o'clock it began to rain slowly, and continued raining two hours, when it cleared up. The rocky peaks...appear like the ruins of an ancient city.

Ordway   a Cloudy morning. It continued to rain moderately untill about 12 oClock when it ceased & continued cloudy. The Stones on the edges of the river continue to form very considerable rapids. We find them difficult to pass. The River rises a little it is from 150 to 250 yards wide

Whitehouse   cloudy weather this morning. About 11 oClock AM it began to rain, and rained moderately for some time. The current of the River rain very strong, the whole of this day. In the Evening, the weather cleared off, and became pleasant.

Note   The expedition nears the western edge of the Missouri Breaks and approaches the confluence of the Missouri and Marias rivers.

SATURDAY, JUNE 1

	Sunrise			4 p.m.			Missouri River	
Temp	Weather	Wind	Temp	Weather	Wind	Rise/Fall	Feet	Inches
50a	c	SW	62a	c		R		1 1/2

Weather Diary   rained a few drops only

Lewis   the morning was cloudy and a few drops of rain. Capt. C who waked on shore today informed me....he observed large banks of pure sand which appeared to have been driven by the SW winds from the river bluffs and there deposited. A range of high Mountains appear to the SW at a considerable distance covered with snow, they appear to run Westerly [probably the Highwood Mountains].   The river from 2 to 400 yards wide, courant more gentle and still becoming clearer. Some few drops of rain again fell this evening. The wind has been against us all day—

Clark   a Cloudy morning. Wind to day from the SW. Som fiew drops of rain in the morning and also in the evening, flying Clouds all day.

Gass   The morning was cloudy, but without rain. The water is not so rapid to day as usual, but continues high.

Ordway   a Cloudy morning. The River from 2 to 400 yards wide & current more jentle that yesterday. But fiew bad rapids points to day. The river rising a little. Wind to day from SW   Some fiew drops of rain in the morning and also in the evening. Flying clouds all day.

Whitehouse   a clear pleasant morning. About 2 oclock...the wind rose and blew from the SE.

Note   *The expedition arrives at the confluence of the Missouri and Marias rivers in Montana.*

SUNDAY, JUNE 2

	Sunrise			4 p.m.			Missouri River		
Temp	Weather	Wind	Temp	Weather	Wind	Rise/Fall	Feet	Inches
56a	c a r	SW	68a	f	SW			

Lewis   The wind blew violently last night and was attended by a slight shower of rain; the morning was fair and we set out at an early hour. The courant was strong tho' regular. The wind was hard and against us yet we proceded with infinitely more ease than the two preceding days. A small showers of rain today but it lasted only a few minutes and was very moderate. Remain here untill the morning, as the evening was favourable to make some observations.

Clark   we had a hard wind and a little rain last night, this morning fair, we Set out at an early hour, wind from the SW. Some little rain to day wind hard a head. The Current Swift but regular. A fair night. We took Some Luner observations of moon & Stears

Gass   a fine morning.

Ordway   we [had] a hard wind & a little rain last night. This morning fair. Wind from SW some little rain to day   wind hard a head. The current swift bu regular.   A fair night.

Whitehouse   a clear pleasant morning. About 12 oClock the Wind blew hard from the NW at this place, and the Sky became Cloudy— about 1 oClock PM we had some small sprinkling of rain—   the current is not So Swift yesterday & today as it has been Some time past.

MONDAY, JUNE 3

	Sunrise			4 p.m.			Missouri River		
Temp	Weather	Wind	Temp	Weather	Wind	Rise/Fall	Feet	Inches
46a	f	SW	60a	f	SW			

Weather Diary   Cought the 1st White Chub, and a fish resembling the Hickory Shad in the Clear Stream

Lewis   This morning early we passed over and formed a camp on the point formed by the junction of the two large rivers. An interesting question was now to be determined; which of these rivers was the Missouri. To mistake the stream at this period of the season, two months of the traveling season having now elapsed, and to ascend such stream to the rocky Mountain or perhaps much further before we could inform ourselves…to this end an investigation of both streams was the first thing to be done. Between the time of my AM and meridian Capt. C & myself stroled out to the hights in the fork of these rivers from whence we had an extensive and most inchanting view. The verdure perfectly clothed the ground, the weather was pleasant and fair, to the south we saw a range of lofty mountains which we supposed to be continuation of the S. Mountains, stretching themselves from SE to NW terminating abbrubtly about S. West for us; these were partially covered with snow; behind these Mountains and at a great distance, a second and more lofty range of mountains appeared…where their snowey tops lost themselves beneath the horizon. We took the width of the two rivers, found the left hand or S. fork 372 yards and the N fork 20. The north fork is deeper than the other but [its] courant not so swift; [its] water run in the same boiling and roling manner which has uniformly characterized the Missouri throughout [its] whole course so far; [its] waters are of a whitish brown colour very thick and terbid, also characteristic of the Missouri; while the South fork is perfectly transparent runds very rappid but with a smoth unriffled surface. The North fork…I am confident that this river rises in and passes a great distance through an open plain country. Convinced I am that if it penetrated the Rocky Mountains to any great distance [its] waters would be clearer unless it should run an immence distance. What astonishes us a little is that the Indians who appeared to be so well acquainted with the geography of this country should not have mentioned this river on wright hand if it be not the Missouri. I am equally astonished at their not mentioning the S fork which they must have passed in order to get to those large falls which they mention on the Missouri. The evening proved cloudy.

Clark   the after part of the day proved Cloudy. Cloudy evening—

Gass   The Rose (Teton) river is muddy and the current rapid.

Ordway   the after part of the day proved Cloudy. Clark measured each river & found the one to the Right hand 186 yards wide & the left had fork 372 yards wide and rapid.

Whitehouse   a fine fair clear morning. Some men went towards a mountain [perhaps the Highwoods] covred with Snow to the South of this place. The afternoon of this day proved Cloudy.

Note   *Lewis and Clark, each with his own scouting party, travel up separate river forks for research purposes. Lewis's party travels up the Marias River and Clark's party travels up the Missouri River.*

TUESDAY, JUNE 4

	Sunrise			4 p.m.		Missouri River		
Temp	Weather	Wind	Temp	Weather	Wind	Rise/Fall	Feet	Inches
48a	f a c	NE	61a	f	SW	F		3/4

Lewis   [Traveling up the Marias River] The Barn Mountin, a lofty mountain so called from [its] resemblance to the roof of a large Bar, is a separate Mountain [Square Butte, MT] it was not yet twelve when we arrived at the river and I was anxious to take the Meridian Altd. of the sun but the clouds prevented my obtaining the observation. The part of the river we have passed is from 40 to 60 yds. wide, is deep, has falling banks, the courant strong, the water terbind. It rained this evening and wet us to the skin; the air was extremely could.

Clark   [Traveling up the Missouri River] Some rain all the afternoon. The river is rapid

Gass   We saw a mountain to the South about 20 miles off, which appeared to run East and West, and some spots of it resembling snow.

Ordway   the day proved Cloudy. A fiew drops of rain towards evening & high cold wind from the North.

Whitehouse   remained here tody. The weather was Cloudy. A fiew drops of rain towards evening, and the Wind rose high & cold from the NE.

WEDNESDAY, JUNE 5

	Sunrise			4 p.m.			Missouri River		
	Temp	Weather	Wind	Temp	Weather	Wind	Rise/Fall	Feet	Inches
	40a	r	SW	42a	c a r	NE	F		3/4

Weather Diary   rained considerably   some Snow fell on them mounts. Great numbers of the sparrows larks, Curloos and other small birds common to praries are now laying their eggs and seting, their nests are in great abundance. The large batt, or night hawk appears. The Turkey buzzard appears.— first saw the mountain cock near the entrance of Maria's river.

Lewis   This morning was cloudy and so could that I was obleged to have recourse to a blanket coat in order to keep myself comfortable altho' walking. The rain continued during the greater part of last night. The wind hard from NW. A large creek...some timber but not water, notwithstanding the rain; it is astonishing what a quantity of water it takes to saturate the soil of this country

Clark   Some little rain & Snow last night. the mountains to our SE covered with Snow this morning.   air verry cold & raining a little. The top of which I could plainly See a mountain to the South & W. covered with Snow at a long distance, the mountains opposit to us to the SE is also Covered with Snow this morning—   some few drops of rain to day, the evening fair wind hard from the NE.

Gass   Some light showers of rain fell in the night, and the morning was cloudy. About 7 we set out along the plains again, and discovered the mountain South of us covered with snow, that had fallen last night.

Ordway   the wind blew high from the North all last night   A Cloudy Cold windy morning.

Whitehouse   The wind blew high during the last night from the NE, and we have a cold windy & cloudy morning.

Thursday, June 6

	Sunrise			4 p.m.		Missouri River		
Temp	Weather	Wind	Temp	Weather	Wind	Rise/Fall	Feet	Inches
35a	c a r	NE	42a	r a r	NE	F		1 ½

Weather Diary   rained hard the greater part of the day.—

Lewis   I now became well convinced that this branch of the Missouri [Marias] had [its] direction too much to the North for our rout to the Pacific. The forepart of the last evening was fair but in the latter part of the night clouded up and continued so with short intervals of sunshine untill a little before noon when the whole horizon was overcast, and I of course disappointed in making the observation which I much wished. The wind blew a storm from NE accompanyed for frequent showers of rain; we were wet and very [cold]. We continued our rout down the river only a few miles before the abbruptness of the clifts and their near approach to the river compelled us to take the plains and once more face the storm. It continues to rain and we have no shelter, an uncomfortable nights rest is the natural consequence—

Clark   a Cloudy Cold raw day, wind hard from the NE. At 12 oClock…it began to rain and Continued all day. My Self and party much fatigued haveing walked constantly as hard as we Could march over a Dry hard plain

Gass   Some light rain fell this afternoon.

Ordway   a Cloudy Cold morning. The wind high from the North   the hunters….saw a large mountain to the South of them covred with Snow   A light Sprinkling of rain this afternoon.

Whitehouse   a cold cloudy morning. The wind still blew cold from the NE. A light sprinkling of rain this forenoon and to day.

Friday, June 7

	Sunrise			4 p.m.		Missouri River		
Temp	Weather	Wind	Temp	Weather	Wind	Rise/Fall	Feet	Inches
40a	c a r	SW	43a	r a r	SW	F		1 ½

Weather Diary   rained moderately all day

Lewis   It continued to rain almost without intermission last night and as I expected we had a most disagreable and wrestless night. Our camp possessing no allurements, we left our watery beads at an early hour and continued our rout down the river [Marias]. It still continues to rain the wind hard from the NE and [cold]. the ground remarkably slipry, insomuch that we were unable to walk on the sides of the bluffs where we had passed as we ascended the river. Notwithstanding the rain that has now fallen the earth of these bluffs is not wet to a greater depth than 2 inches; in [its] present state it is precisely like walking over frozan grownd which is thawed to small debth and slips equally as bad.   We continued our disagreeable march through the rain, mud and water untill late in the evening.

Clark   rained moderately all the last night and Continus this morning, the wind from the SW off the mountains, the Thermometer Stood at 40° above 0. The rain Continue moderately all day. River falling

Gass   It rained all day.

Ordway   rained all last night. A rainy cold morning the wind NW   rained moderately all day. Capt Lewis and his party did not return this evening we expect the reason is owing to the badness of the weather as it is muddy & Slippery walking

Whitehouse   rained the greater part of last night. Cloudy and wet this morning. Rained moderately all day. Capt. Lewis & his party have not returned yet. We expect the reason is owing to the badness of the weather.

SATURDAY, JUNE 8

	Sunrise			4 p.m.			Missouri River		
Temp	Weather	Wind	Temp	Weather	Wind	Rise/Fall	Feet	Inches
41	r a r	SW	48a	f a r	SW	F		1 ½

Weather Diary   cleared off at 10 AM

Lewis   It continued to rain moderately all last night, this morning was cloudy untill about ten oClock when it cleared off and became a fine day. When sun began to shine today these birds appeared to be very gay and sun most inchantingly. Some of the inhabitants of the praries also take reffuge in these woods at night or from a storm. The whole of my party to a man except myself were fully peswaided that this river was the Missouri. I determined to give a named in honor of Miss Maria W– d. called it Maria's River. A noble river…one destined to become in my opinion an object of contention between the two great powers of America and Great Britin. In adition to which it passes through a rich fertile and one of the most beatifully picteresque countries that I ever beheld.

Clark   rained moderately all the last night & Some this morning untill 10 oClock. Aired and dried our Stores &c. The rivers at this point has fallen 6 Inches Sinc our arrival. At 10 oClock cleared away and became fair— the wind all the morning from the SW & hard— the water of the South fork is of a redish brown colour this morning the other river of a whitish colour as usual—   the mountains to the South Covered with Snow. Wind Shifted to the NE in the evening. Some rain in the evening. The left hard fork rose a little.

Gass   A fine cool morning. About 10 o'clock AM the water of the South river, or branch, became almost of the colour of claret (reddish brown) and remained so all day. The water of the other branch has appearance of milk. About five o'clock n the afternoon the weather became cloudy and cold, and it began to rain.   At dark the rain ceased.

Ordway   the wind blew cold from the NW about 9 oClock AM clear off pleasant. We Saw the high Mountains to the West. Our Camp covered with Snow the greater part of which has fallen within a fiew days. The South fork of the Missourie is high & of a yallow coulour. The N fork is more white than common owing as we expect to the late rain which has melted the Snow on the mountains. The wind blew from the East a light Shower of rain this evening.

Whitehouse   This morning we had cloudy weather, and the wind blew cold from the NW. Between 7 and 9 oClock the weather cleared off, and became pleasant. We saw on the weather clearing away, a high mountain, lying to the West of us, which was covered with Snow. The South fork of the Missouri is high & of a yellow Colour today, and rose to a great height, the north fork more white and riffling than it was before, the cause of which, we expect, is owing to the Rain that fell lately, and the snow melting in the Mountains. The wind blew from the E toward evening with a light Shower of Rain—

Sunday, June 9

	Sunrise			4 p.m.		Missouri River		
Temp	Weather	Wind	Temp	Weather	Wind	Rise/Fall	Feet	Inches
50	f	SW	62*	f	SW	F		1

Lewis   The Indian information also argued strongly in favour of the South fork. That the falls lay a little to the South of sunset from them. Cruzatte who had been an old Missouri navigator and who from his integrity knowledge and skill as a waterman…declared it as his opinion that the N fork was the true genuine Missouri and could be no other. Finding them so determined in this belief, and wishing that if we were in an error to be able to detect it and rectify it as so as possible it was agreed between Capt. C. and myself that one of us should set out with a small party by land up the South fork and continue our rout up it untill we found the falls or reached the snowy Mountains by which means we should be enabled to determine this question prety accurately.

Clark   a fair morning, the wind hard from the SW. The river during the night fell 1 inch. Took some Luner observations.

Gass   a fine morning. The water of the Missouri changed this morning to its former colour. The day was fine, but the wind blew hard from the northwest.

Ordway   a clear pleasant morning. The wind rose high from the west all day. We had a light Shower of rain about 11 oClock at night.

Whitehouse   a clear beautiful pleasant morning. The wind towards evening rose from the West. We had a frolick. We had a light Shower of rain late in the evening

---

*According to Moulton (1987b, 4: 348), Clark's Journal Codex I lists this temperature as "52 a."

MONDAY, JUNE 10

	Sunrise			4 p.m.			Missouri River		
Temp	Weather	Wind	Temp	Weather	Wind	Rise/Fall	Feet	Inches
52a	f	SW	68a	f a r	SW	R		2

Lewis The day being fair and fine we dryed all our baggage and merchandize. We drew up the red perogue...secured and made her fast to the trees to prevent the high floods from carrying her off. Put my brand on several trees standing near her, and covered her with brush to shelter her from the effects of the sun. At 3 PM we had a hard wind from the SW which continued about an hour attended with thunder and rain. As soon as the shower had passed over we drew out our canoes. The night was cloudy with some rain.

Clark a fine day, dry all our articles. Branded several trees to prevent the Indians injureing her, at 3 oClock we had hard wind from the SW. Thunder and rain for about an hour after which we repared & Corked the Canoes. The after noon or night Cloudy Some rain.   river riseing a little.

Gass about two it began to rain and blow we were obliged to desist. The rain continued only an hour

Ordway a beautiful pleasant morning.   About 4 oClock PM we had a light Shower of rain which lasted about an hour. High wind. The evening pleasant.

Whitehouse A beautiful pleasant morning. About 4 oClock PM we had a light shower of Rain, & in the Evening it cleared up, & we had pleasant Weather—

TUESDAY, JUNE 11

	Sunrise			4 p.m.			Missouri River		
Temp	Weather	Wind	Temp	Weather	Wind	Rise/Fall	Feet	Inches
54a	f	SW	66a	f	SW			

Weather Diary Capt. Lewis & 4 men Set out up the S. fork

Clark a fair morning, wind from the SW hard. The evening fair and fine wind from the NW. After night it became cold & the wind blew hard. Both rivers riseing fast.

Gass A fine day.

Ordway a clear pleasant morning. The wind from the S West hard.   The evening fair & fine wind from the NW   after night became cold.   High wind

Whitehouse a Clear pleasant morning. Capt. Lewis & party set out for the South Snowey Mountain.

WEDNESDAY, JUNE 12

	Sunrise			4 p.m.			Missouri River		
Temp	Weather	Wind	Temp	Weather	Wind	Rise/Fall	Feet	Inches
54a	f	SW	64	f a r	SW			

Lewis   the sun became warm, and I boar a little to the south in order to gain the river as well to obtain water to allay my thirst. From this hight we had a most beatifull and picturesk view of the Rocky mountains which wer perfectly covered with Snow and reaching from SE to the N of NW— they appear to be formed of several ranges each succeeding range riseing higher than the preceding one untill the most distant appear to loose their snowey tops in the clouds.

Clark   last night was Clear and Cold, this morning fair. Wind from the SW. The interpreters wife verry Sick So much So that I move her into the back part of our Covered part of the Perogue which is Cool, her our situation being a verry hot one in the bottom of the Canoe exposed to the Sun— water verry swift. At 2 oClock PM a fiew drops of rain.

Gass   The morning was fine. At 1 o'clock the weather became cloudy and threatened rain; at 2 there was a light shower, and the day became clear.

Ordway   a clear pleasant morning. The current verry rapid.

Whitehouse   a clear pleasant morning. The current verry Rapid.

Note   *Lewis's party arrives at the Great Falls of the Missouri in Montana.*

THURSDAY, JUNE 13

	Sunrise			4 p.m.			Missouri River*		
Temp	Weather	Wind	Temp	Weather	Wind	Rise/Fall	Feet	Inches
52a	f	SW	72	f	SW	R		3/4

Weather Diary   Some dew this morng.

Lewis   I overlooked a most beatifull and level plain of great extent or at least 50 or sixty miles; in this there were infinitely more buffaloe than I had ever before witnessed at a view. I proceeded on this course...whin my ears were saluted with the agreeable

*It appears observations were taken at the lower portage camp at Belt Creek, Montana, until the upper portage camp at the White Bear Islands was established on June 23, 1805.

sound of a fall of water and advancing a little further I saw the spray arrise above the plain like a collumn of smoke which would frequently dispear again in an instant caused I presume by the wind which blew pretty hard from the SW...soon began to make a roaring too tremendious to be mistaken for any cause short of the great falls of the Missouri. Here I arrived about 12 OClock...to gaze on this sublimely grand specticle. Immediately at the cascade the river is about 300 yds. On my right formes the grandest sight I ever beheld...some what projecting rocks below receives the water in [its] passage down and brakes it into a perfect white foam which assumes a thousand forms in a moment sometimes flying up in jets of sparkling foam to the hight of fifteen or twenty feet and are scarcely formed before large roling bodies of the same beaten and foaming water is thrown over and conceals them. From the reflection of the sun on the spray or mist which arrises from these falls there is a beatifull rainbow produced which adds not a little to the beauty of this majestically grand senery. After wrighting this imperfect discription, I again viewed the falls and was so much disgusted with the imperfect idea which it conveyed of the scene that I determined to draw my pen across it and begin agin, but then reflected that I could not perhaps succeed better than pening the first impressions of the mind. I wished...that I might be enabled to give to the enlightened world some just idea of this truly magnificent and sublimely grand object which has from the commencement of time been concealed from the view of civilized man.

Clark   a fair morning. Some dew this morning. Small stream...heads in a mountain to the SE...which at this time covered with Snow, we call this stream Snow river [Shonkin Creek]. The river verry rapid maney Sholes great nos of large Stones

Gass   a fine morning. Some dew fell last night.

Ordway   a beautiful pleasant morning. A heavy diew. Small river...about 50 yards wide....muddy coulour and verry rapid.

Whitehouse   we had a Clear & pleasant morning.

Friday, June 14

	Sunrise			4 p.m.		Missouri River		
Temp	Weather	Wind	Temp	Weather	Wind	Rise/Fall	Feet	Inches
60a	f	SW	74	f	SW	F		3/4

Weather Diary   Capt. Lewis Discover the falls & Send back Joe Fields to inform me

Lewis   proceeded up the river about SW. I arrived at a fall of about 19 feet; the river is here about 400 yds. wide [Lewis later calls it "Crooked Falls"]. I now thought that if a skillfull painter had been asked to make a beautifull cascade that he would most probably have presented the precise immage of this one; nor could I for some time determine on which of those two great cataracts to bestoe the palm, on this or that which I had discovered yesterday; at length I determine between these two great rivals for glory that this was pleasingly beautifull, while the other sublimely grand. Still pursuing the river...I arrived at another cataract of 26 feet....the river near six hundred yards wide at this place [this became known as Eagle Falls]...this fall is certainly much the greatest I ever behald except those two which I have mentioned below. It is incomparably a greater cataract and a more noble interesting object than the celibrated falls of Potomac or Soolkin &c. From hence I overlook a most beatifull and extensive plain reaching from the river to the base of the Snowclad mountains to the S and SW. I passed through the plain…to medicine river [Sun River], found it a handsome stream, about 200 yds. wide with a gentle current, apparently deep, [its] water clear…they had not the appearance of ever being overflown, a circumstance, which I did not expect so immediately in the neighbourhood of the mountains, from whence I should have supposed, that sudden and immence torrants would issue at certain seasons of the year; but the reverse is absolutely the case. I am therefore compelled to believe that the snowey mountains yeald their warters slowly, being partially effected every day by the influence of the sun only, and never suddonly melted down by haisty showers of rain— The weather being warm I had left my leather over shirt and had woarn only a yellow flannin one.

Clark   a fine morning. The Current excesevely rapid more So as we assend find great dificuelty in getting the Perogue & canoes up in safety

Gass   the morning was pleasant.

Ordway   a fare pleasant morning. The current verry rapid all day. We came 10 miles to day through a verry rapid current.

Whitehouse   a fare pleasant morning. The current verry rapid all day.

Note   *Starting on June 15, the expedition portages around the great falls to the White Bear Islands. They remain between the lower and upper portage camps at present-day Great Falls, Montana, until July 13.*

SATURDAY, JUNE 15

	Sunrise			4 p.m.			Missouri River	
Temp	Weather	Wind	Temp	Weather	Wind	Rise/Fall	Feet	Inches
60a	f	SW	76	f	SW	F		$^1/_2$

Weather Diary   The deer now begin to bring forth their young the young Magpies begin to fly. The Brown or grizzly bear begin to coppolate.

Lewis   when I awoke from my sleep [nap] today I found a large rattlesnake coiled on the leaning trunk of a tree under the shade of which I had been lying at the distance of about ten feet from him. I find a very heavy due on the grass about my camp every morning which no doubt procedes from the mist of the falls, as it takes place no where in the plains nor on the hills river except here.

Clark   a fair morning and warm. Proceeded on with great dificuelty as the river is mor rapid. We can hear the falls this morning verry distinctly— river rises a little this evening.

Gass   had the most rapid water I ever saw any craft taken through.

Ordway   a clear pleasant morning. Passed through the rapidest water I ever Saw any craft taken through. The afternoon verry warm.

Whitehouse   a clear pleasant morning. Passed the rapidest water I ever Seen any crafts taken through. In the afternoon, it became very warm.

SUNDAY, JUNE 16

	Sunrise			4 p.m.			Missouri River	
Temp	Weather	Wind	Temp	Weather	Wind	Rise/Fall	Feet	Inches
64	c a r	SW	58	f	SW	R		1/2

Weather Diary   Some rain last night

Lewis   a Sulpher Spring...this spring is situated about 200 yards from the Missouri on the NE side...the water is as transparent as possible strongly impregnated with sulpher...the water to all appearance is precisely similar to that of Bowyer's Sulpher spring in Virginia

Clark   Some rain last night, a cloudy morning, wind hard from the SW

Ordway   a Small Shower of rain and high wind from the west the fore part of last night.

Whitehouse   we had a Showers of rain & high wind the fore past of the last night. This morning it cleared away & we had pleasant weather

SUNDAY, JUNE 17

	Sunrise			4 p.m.			Missouri River	
Temp	Weather	Wind	Temp	Weather	Wind	Rise/Fall	Feet	Inches
50a	c	SW	57	c	SW	F		1/2

Weather Diary   the thermometer placed in the shade tree at the foot of the rappids.
Capt Clark sets out to survey the river & portage

Lewis   I found the Elk skins I had prepared for my boat were insufficient to compleat her, some of them having become dammaged by the water and being frequently wet...lyed in the sun to dry.

Clark   a fine morning, wind as usial. Proceeded up the river passing a Sucession of rapids & Cascades to the Falls, which we had herd for Several miles makeing a dedly Sound, I beheld those Cateracts with astonishment. The whole of the water of this great river Confined in a Channel of 280 yards and pitching over a rock of 97 feet 3/4 of an, from the foot of the falls arrises a Continued mist. We Camped for the night which was Cold. The mountains in every derection has Snow on them.

Ordway   a clear morning.

Whitehouse   a cloudy morning.

MONDAY, JUNE 18

Sunrise			4 p.m.			Missouri River		
Temp	Weather	Wind	Temp	Weather	Wind	Rise/Fall	Feet	Inches
48a	c	SW	64a	f a c	SW	F		½

Lewis   the wind blew violently this evening, as they frequently do in this open country where there is not a tree to brake or oppose their force.

Clark   arrived at the second great Cataract...this is one of the grandest views in nature and by far exceeds any thing I ever saw...the river is 473 yards wide...a continuel mist quite across the fall...proceeded...to the largest fountain or Spring I ever Saw, and doubt if it is not the largest in America Known, this water boils up from under the rocks near the edge of the river and falls imediately into the river 8 feet and keeps its Colour for ½ a mile which is emencely Clear and of a bluish Cast...a Considerable mist rises at this fall ocasionally, dined...opposite the mouth of the Medison River...and is 137 yards wide at its mouth. The Missouri above is 800 yards wide. I Saw the bear but the bushes was So thick that I could not Shoot him and it was nearly dark, the wind from the SW & Cool.

Ordway   the day pleasant. The wind high from the west.

Whitehouse   a fine pleasant day. After 12 oClock...the wind rose & blew high from the West.

TUESDAY, JUNE 19

Sunrise			4 p.m.			Missouri River		
Temp	Weather	Wind	Temp	Weather	Wind	Rise/Fall	Feet	Inches
52a	f	SW	70a	f	SW	F		½

Weather Diary   wind violent all day

Lewis   The wind blew violently the greater part of the day.

Clark   the wind all this day blew violently hard from the SW off the Snowey mountains, Cool, in my last rout I lost a part of my notes which could not be found as the wind must have blown them to a great distance.

Gass   A fine day, but the wind very high.

Ordway   a clear pleasant morning. The wind verry high from NW.

Whitehouse   a clear cool morning. The wind had blown very hard during the last night. Continues veryy high from the West and whole of this day.

THURSDAY, JUNE 20

	Sunrise			4 p.m.		Missouri River		
Temp	Weather	Wind	Temp	Weather	Wind	Rise/Fall	Feet	Inches
49a	c	SW	74a	f a r	SW	F		1/4

Weather Diary   wind still violent. Capt. Clark returns.

Clark   a Cloudy morning, a hard wind all night and this morning. Soon after we set out it began to rain and continued a short time. the wind hard from the SW. A fair after noon. The Countrey above the falls & up the Medison river is leavel, with low banks, a chain of mountains to the west some part of which particuler those to the NW & SW are Covered with Snow and appear verry high— We had a heavy dew this morning. The Clouds near those mountains rise Suddonly and discharge their Contents partially on the neighbouring Plains; the Same Cloud discharge hail alone in one part, hail and rain in another and rain only in a third all within the Space of a fiew miles; and on the Mountains to the South & SE of us Sometimes Snow. At present there is no Snow on those mountains; that which covered them a fiew days ago has all disappeared. the Mountains to the NW and West of us are Still entirely Covered are white and glitter with the reflection of the sun. I do not believe that the Clouds that pervale at this Season of the year reach the summits of those lofty mountains; and if they do the probability is that they deposit Snow which they Contain since we first saw them. I have thought it probable that these mountains might have derived their appellatoin of Shineing Mountains, from their glittering appearance when the Sun Shines in certain directions on the Snow which Cover them.

Dureing the time of my being on the Plains and above the falls I as also my party repeatedly heard a nois which proceeded from a Direction a little to the N. of West, as loud and resembling precisely the discharge of a piece of ordinance of 6 pounds at the distance of 5 or six miles. I was informed of it Several times by the men. J. Fields particularly before I paid any attention to it, thinking it was thunder most probably which they had mistaken. At length walking in the plains yesterday near the most extreem SE bend of the River above the falls I heard this nois very distinctly, it was perfectly calm clear and not a Cloud to be Seen, I halted and listend attentively about two hour dureing which time I heard two other discharges, and took the direction of the Sound with my pocket Compass which was as nearly West from me as I could estimate from the Sound. I have not doubt but if I had leasure I could find from whence it issued. I have thought it probable that it might be caused by running water in Some of the caverns of those emence mountains, on the principal of the blowing caverns [Blowing Cave, Bath County, Virginia]; but in Such case the Sounds would be periodical and regular, which is not the Case with this, being Sometimes heard once only and at other times Several discharges in quick Succession. It is heard also at different times of the day and night. I am at a great loss to account for this Phenomenon. I well recollect hereing the Minitarees Say that those Rocky Mountains make a great noise, but they could not tell me the Cause, neither Could they inform me of any remarkable substance or situation in these mountains which would autherise a conjecture of a probable cause of this noise— it is probable that the large river just above those Great falls which heads in the derection of the noise has taken [its] name Medicine River from this unaccountable rumbling Sound, which like all unacountable thing with the Indians of the Missouri is Called Medicine. The Ricaras inform us of the black mountains making a Simalar noise &c. &c. and maney other wonderfull tales of those Rocky mountains and those great falls.

Gass   A cloudy morning.

Ordway   Some cloudy & cold for the Season.   The wind continues high from the west off the mountains.   A light Sprinkling of rain about noon.   A light sprinkle of rain. The hunters…saw a chain of Mountains to the West Some of which perticular those to the NW and SW are covered with Snow, and appear to be verry high.

Whitehouse   This morning we had Cloudy cold weather; and the wind continues high from the West. We had about Noon some light Squalls of rain and wind. We had small Showers of rain this afternoon. 1 mile above the fall…is the largest fountain or Spring, as they think it is the largest in america known. They Saw a chain of mountains to the west, some of which particular those to the NW & SW are covered with Snow & appear to be verry high [probably the Lewis Range of the Rockies]. Capt, Clark lost a part of his notes which could not be found, as the wind blew high and hard & took them off.

FRIDAY, JUNE 21

Sunrise			4 p.m.			Missouri River		
Temp	Weather	Wind	Temp	Weather	Wind	Rise/Fall	Feet	Inches
49a	f	SW	70a	c	SW	F		1/4

Lewis   The wind blew violently all day.

Clark   a fine morning, wind from the SW off the mountains and hard. Cloudy afternoon.

Gass   This morning was also fine, but there was a high wind.

Ordway   a fine cool morning. the wind from the SW. off the mountains and hard.

Whitehouse   a fine morning, the wind blew hard from the SW off the mountains.

SATURDAY, JUNE 22

Sunrise			4 p.m.			Missouri River		
Temp	Weather	Wind	Temp	Weather	Wind	Rise/Fall	Feet	Inches
45a	c	SW	54a	f	SW	F		1/2

Weather Diary   wind not so violent. Thermometer removed to the head of the rappid and
place in the shade of a tree.
Clark   a fine morning. Wind from the
Ordway   a clear pleasant morning. the wind as usal. …a light Sprinkling of rain. …..we are
a little South of the Mandans but have had cold weather as yet.   it must of course
be a healthy county.
Whitehouse   a fair pleasant morning. The wind as usal. West. We are at this place a little
South of the Mandan Villages, but as yet have experienced no very warm
weather—

SUNDAY, JUNE 23

	Sunrise			4 p.m.		Missouri River*		
Temp	Weather	Wind	Temp	Weather	Wind	Rise/Fall	Feet	Inches
48a	f	SE	65a	c	SE	F		1/4

Lewis   during the late rains the buffaloe have troden up the prarie very much, which having
now become dry the sharp points of earth as hard as frozen ground stand up in such
abundance that there is no avoiding them. This is particulary severe on the feet of
the men who have not only their own w[e]ight to bear in treading on those hacklelike
points but have also the addition of the burthen which they draw
Clark   a Cloud morning, wind from the SE.
Gass   The morning was cloudy.
Ordway   a cloudy morning. The wind from East. A light Sprinkling of rain. The hard
ground in many places is So hard as to hurt our feet verry much. The emence
numbers of buffalow after the last rain has trod the flat places in Such a manner
as to leave them uneaven, and dryed as hard as frozen Ground.
Whitehouse   this morning the wind shifted to the east & became Cloudy an we had a light
sprinkling of rain—

MONDAY, JUNE 24

	Sunrise			4 p.m.		Missouri River		
Temp	Weather	Wind	Temp	Weather	Wind	Rise/Fall	Feet	Inches
49a	c a r	SE	74a	f a c	SW	F		

*It appears observations were moved from the lower portage camp to the upper portage camp at the White Bear
Islands and continued until the Expedition began its ascent once again up the Missouri River on July 13, 1805.

Weather Diary   slight rain last night & a heavy shower this evening.

Clark   a Cloudy morning. Some few drops of rain in the fore part of the day, at 6 oClock a black Cloud arose to the NW, the wind shifted from the S to that point and in a Short time the earth was entirely Covered with hail, Some rain Succeeded, which Continud for about an hour very moderately on this Side of the river, without the earths being wet $1/_2$ an inch, the riveins on the opposit or NW Side discharged emence torrents of water into the river, & Showed evidently that the rain was much heavyer on that Side, Some rain at different times in the night which was w[a]rm. Thunder without lightning accompanied the hail Cloudy.

Gass   In the evening there was a very heavy shower of rain, at night the weather cleared up

Ordway   a cloudy morning. A violent shower arose from the NW hard thunder caught us in a verry hard rain So that in a fiew minutes the ground was covered with water. So that we got a hearty drink of water in the holes & puddles &.C. The rain continued about half an hour, at dusk we arived at the upper camp all wet and much fatigued. The wind was considerable assistance to us in the course of the day, as we were drawing the canoes the wind being Sufficently hard at times to move the canoe on the Trucks. This is Saleing on dry land in every Since of the word.

Whitehouse   fair weather this morning. We proceeded across the prairie, the wind blowing steady from the SE, we hoisted a Sail in the largest canoe which helped us much. Towards evening when we were within 3 miles of the upper camp, up came a sudden and Violent thunder Shower & rained amazingly hard for about 15 to 20 minutes, in which time the water stood on the ground over our mockasons. Our water being all gone & the men very thirsty, they drank heartily, out of the puddles of water that lay in the plains—

TUESDAY, JUNE 25

	Sunrise			4 p.m.			Missouri River		
	Temp	Weather	Wind	Temp	Weather	Wind	Rise/Fall	Feet	Inches
	47a	c a r	SW	72a	f	SW			

Lewis   the river is about 800d yds. wide opposite to us above these islands, and has a very gentle current [White Bear Islands, upper portage camp]. It is worthy of remark that the winds are sometimes so strong in these plains that the men informed me that they hoisted a sail in the canoe and it had driven her along on the truck wheels. This is really sailing on dry land.

Clark   a fair w[a]rm morning, Clouded & a few drops of rain at 5 oClock AM fair. I had a little Coffee for brackfast which was to me a riarity as I had not tasted any Since last winter. The wind from the NW & w[a]rm. This Countrey has a romantick appearance river inclosed between high and Steep hills. A powerfull rain fell on the party on their rout yesterday, Wet Some fiew articles, and Caused the rout to be So bad wet & Deep thay Could with dificuelty proceed. A fair after noon— it may be worthy of remark that the Sales were hoised in the Canoes as the men were drawing them and the wind was great relief to them being Sufficently Strong to move the Canoes on the Trucks, this is Saleing on Dry land in every Sence of the word

Gass   A cloudy morning.

Ordway   a cloudy morning.   The day proved pleasant and warm the men much fatigued.

Whitehouse   a cloudy morning. The weather cleared up about 10 oClock AM and the day proved pleasant and warm. The evening was clear and pleasant.

WEDNESDAY, JUNE 26

Sunrise			4 p.m.			Missouri River		
Temp	Weather	Wind	Temp	Weather	Wind	Rise/Fall	Feet	Inches
49a	f	SW	78a	f	SW	R		1/2

Lewis   The Musquetoes are extremely troublesome to us.

Clark   Some rain last night, this morning verry Cloudy. The wind from the NW verry w[a]rm. Flying Clouds, in the evening the wind Shifted round to the East & Blew hard, which is a fair wind for the two Canoes to Sail on the Plains across the portage.

Gass   a fine morning. Captain Lewis measured the falls, found them in a distance of 17 miles 362 feet 9 inches. The first great pitch 98 feet, the second 19 feet, the third 47 feet 8 inches, the fourth 26 feet; and the number of small pitches amounting altogether to 362 feet 9 inches.

Ordway   Some rain last night. This morning cloudy. The day proved fair.

Whitehouse   We had some rain last night, and this morning verry cloudy. The weather cleared up at 9 oClock AM and the verry hot Sun beat down on us as the day proved fair. I took sick this evening I expect by drinking too much water when I was hot. I got bled &c. I had an opportunity of seeing the quantity of Buffalo as related; and I can without exaggeration say, that I saw more Buffalo feeding— at one time, than all the Animals I had ever seen before in my life time put together—

THURSDAY, JUNE 27

Sunrise			4 p.m.			Missouri River		
Temp	Weather	Wind	Temp	Weather	Wind	Rise/Fall	Feet	Inches
49a	f	SW	77 77	f a r & h t l*	SW	R		1 1/4

---

*According to Moulton (1987b, 4: 349), Clark's Journal Codex C lists this weather data as "f a r h."

Weather Diary   at 1 PM a black cloud which arose in the SW came on accompanyed with a high wind and violent Thunder and Lightning; a great quantity of hail also fell during this storm which lasted about 2 $\frac{1}{2}$ hours the hail which was generally about the size of a pigion's egg and not unlike them in form covered the ground to the debth of 1 $\frac{1}{2}$ inches.— for about 20 minutes during this storm hail fell of an innomus size driven with violence almost incredible, when they struck the ground they would bound to the hight of ten to 12 feet and pass 20 or thirty before they touched again. (during the emence Storm I was with the greater part of the men on the portage the men Saved themselves, Some by getting under a Canoe others by putting Sundery articles on their heads two was kocked down & Seven with their legs & thighs much brused. [Clark's notes] After the rain I measured and weighed many of these hail stones and found several weighing 3 ozs. and measuring 7 Inches in circumference; they were generally round and perfectly sollid. I am convinced if one of those had struck a man on the neaked head it would have knocked him down, if not fractured his skull.— Young blackbirds which are abundant in these Island are now beginning to fly

Lewis   at 1 PM a cloud arrose to the SW and shortly after came on attended with violent Thunder Lightning and hail &c. [See notes on diary of the weather for June.]   Soon after this storm was over Drewyer and J. Fields returned. They were about 4 miles above us during the storm, the hail was of no uncommon size where they were. Soon after the storm this evening the water on this side of the river became of a deep crimson colour which I pesume proceeded from some stream above and on this side. At 4 PM the party returned from the upper camp; Capt. C. gave them a drink of grog; they prepared for the labour of the next day. Soon after the party returned it began to rain accompanyed by some hail and continued a short time; a second shower fell late in the evening accompanyed by a high wind from the NW—

Clark   a fair warm morning, wind from the SE and moderate. The w[a]rmest day we have had this year, at 4 PM the party returned...Soon after it began to hail and rain hard and continued for a fiew minits & Ceased for an hour and began to rain again with a heavy wind from the NW. I refreshed the men with a drink of grog. The river beginning to rise a little the water is Coloured a reish brown, the Small Streams, discharges in great torrents, and partake of the Choler of the earth over which it passes– a great part of which is light & of a redish brown. Several Buffalow pass drowned & dashed to pices in passing over the falls. Cloudy all night, Cold.

Gass   a fine morning. In the afternoon a dreadful hail storm came on, which lasted half an hour. Some of the lumps of ice that fell weighed 3 ounces, and measured 7 inches in circumference. The ground was covered with them, as white as snow. It kept cloudy during the evening and some rain fell.

Ordway   a fair warm morning. a heavey dew last night. passed the upper falls which is a great catteract and look remarkable. passed the lower high falls which is the highest known except the falls of the Neagra. a hard Shower of rain and hail came on of a Sudden So I got under a Shelving rock on one Side of the creek where a kept dry through the hardest of it. hard thunder. large hail the creek rose So high in a fiew minutes that I had to move from the dry place and proceeded on. the wind blew So high that the hail cut verry hard against me and I could hardly keep my feet. the rain has made it So muddy and Slippery, cloudy all night.

Whitehouse   a fair clear warm morning. About 4 oClock PM we had a hard shower of rain which made the Portage so Slipperry

FRIDAY, JUNE 28

	Sunrise			4 p.m.			Missouri River		
Temp	Weather	Wind	Temp	Weather	Wind	Rise/Fall	Feet	Inches
46a	*f*	SW	75	*c a f*	SW	R		2

Weather Diary   Cat fish no higher

Lewis   the river is now about nine inches higher than is was on my arrival. Portage creek had arisen considerably and the water was of crimson colour and illy tasted. Soon after his arrival [Clark during the morning] at willow run he experienced a hard shower of rain which was succeeded by a violent wind from the SW off the snowy mountains, accompanyed with rain; the party being cold and wet, he administered the consolation of a dram to each.

Clark   a fair morning, wind from the South. Set out passed the Creek which had rose a little and the water nearly red, and bad tasted [Portage Creek]. Soon after we halted we had a Shower, and at dark we expereinced a most dredfull wind from off the Snow Mountains to the SW accompd. with rain which continued at intervales all night men wet. I refreshed them with a dram.

Gass   a fine morning.

Ordway   a fair morning. Wind from the South. The water is riseing and of a redish brown cholour.   Soon after we halted, we had a Shower and at dark we experienced a most dreedful wind from off the Snow Mountains to the SW accompanied with rain which lasted nearly all night.

Whitehouse   a fair clear morning, wind from the South, which continued the whole of this day

SATURDAY, JUNE 29

	Sunrise			4 p.m.			Missouri River		
Temp	Weather	Wind	Temp	Weather	Wind	Rise/Fall	Feet	Inches
47a	*r t & l*	SW	77	*f*	SW	R		4 1/2

Weather Diary   heavy gust of rain the morning and evening

Lewis   This morning we experienced a heavy shower of rain for about an hour after which it became fair. Not having seen the large fountain of which Cap. Clark spoke I determined to visit it today…passed through a level plain for about Six miles when I reached a brake of the river hills. Here we were overtaken by a violent gust of wind and rain from the SW attended with thunder and Litning. I expected a hail storm probably from this cloud and therefore took refuge in a little gully wher there were some broad stones with which I purposed protecting my head if we should have a repetition of the seene of the 27th but fortunately we had but little hail and that not large; I sat very composedly for about an hour without shelter and took a copious drenching of rain; afer the shower was over I continued my rout to the fountain. …nature seems to have dealt with a liberal hand for I have scarcely experienced a day since my first arrival in this quarter without experiencing some novel occurrence among the party or witnessing the appearance of some uncommon object. I think this fountain the largest I ever beheld, and the hadsome cascade which it affords over some steep and irregular rocks in [its] passage to the river adds not a little to [its] beauty…it is about 25 yds. from the river…the water of this fountain is extreemly transparent and cold…very pure and pleasent. After amusing myself about 20 minutes in examining the fountain I found myself so chilled with wet cloaths that I determined to return. I was astonished not to find the party yet arrived, but then concluded that probably the state of the praries had detained them, as in the west state in which they are at present the mud sticks to the wheels in such manner that they are obliged to halt frequently and clense them. [Lewis describes the following flash flood episode repeating it using his literary prose. It is omitted to allow the full effect from Clark's discussion below.]

Clark   a little rain verry early this morning after[ward] Clear, finding that the Prarie was So wet as to render it impossible to pass on to the end of the portage determined to Send back to the top of the hill…for remaining baggage…Soon after I arrived at the falls, I perceived a cloud which appeared black and threaten immediate rain, I looked out for a Shelter but Could see no place without being in great danger of being blown into the river if the wind Should prove as turbelant as it is at Some times. about ¼ of a mile above the falls I [observed] a Deep riveen in which was Shelveing rocks under which we took shelter near the river and placed our guns the compas &c. &c.…which was verry Secure from rain, the first Shower was moderate accompanied with a violent wind, the effects of which we did not feel, Soon after a torrent of rain and hail fell more violent than ever I saw before, the rain fell like one voley of water falling from the heavens and gave us time only to get out of the way of a torrent of water which was Poreing down the hill in the rivin with emence force tareing everything before it takeing with it large rocks & mud. I took my gun and Shot pouch in my left hand, and with the right Scrambled up the hill pushing the Interpreters wife (who had her child in her arms) before me, the Interpreter himself makeing attempts to pull up his wife by the and much Scared and nearly without motion— we at length retched the top of the hill Safe where I found my Servent in Serch of us greatly agitated, for our wellfar— before I got out of the bottom of the revein which was a flat dry rock when I entered it, the water was up to my waste & wet my watch, I Scrcely got out before it raised 10 feet deep with a torrent which [was] turrouble to behold, and by the time I reached the top of the hill, at least 15 feet water, I directed the party to return to the Camp at the run as fast as possible to get to our lode where Cloathes Could be got to Cover the Child whose Clothes were all lost, and the woman who was but just recovering from a Severe indisposition, and was wet and Cold, I was fearfull of a relaps. I caused her as also the others of the party to take a little Spirits, which my Servent had in a Canteen, which revived verry much. On arrival a the Camp on the willow run— met the party who had returned in great Confusion to the run leaveing their loads in the Plain, the hail & wind being So large and violent in the plains, and them naked, they were much brused, and Some nearly killed one knocked down three times, and other without hats or any thing on their heads bloodey & Complained verry much; I refreshed them with a little grog— Soon after the run began to rise and rose 6 feet in a few minites— I lost at the river in the torrent the large Compass, an eligant fusee [umbrella], Tomahawk Humbrallo, Shot pouch, & horn with powder & Ball, mockersons, & the woman lost her Childs Bear & Clothes bedding &c. —   The Compass is a Series loss; as we have no other large one.   The plains are So wet that we Can do nothing this evening particilarly as two deep reveins are between ourselves & Load

Gass   We had a very hard gust of wind and rain in the morning; but a fine forenoon after it. In the afternoon there was another heavy shower of rain, and after it a fine evening. Captain Lewis came to camp, but drenched with rain.

Ordway   a little rain verry eairly this morning after clear & warm. We find that the prarie is So wet as to render it impossable to pass on the end of the portage.   Saw a black cloud rise in the west which we looked for emediate rain   we made all the haste possable but had not got half way before the Shower met us and our hind extletree broke in too   we were obledged to leave the load Standing and ran in great confusion to Camp the hail being So large and the wind So high and violent in the plains, and we being naked we were much bruuzed by the large hail. Some nearly killed   one knocked down three times, and others without hats or any thing about their heads bleading and complained verry much. [He then relates Clark's story above.]   The plains are so wet that we could doe nothing this evening.

Whitehouse   a little rain verry eairly this morning which lasted but for a short time, when the weather cleared off and it became pleasant. In the afternoon, there arose a storm of hard wind & rain; accompanied with amazing large hail at the upper camp. We caught several of the hail Stones which was measured & weighed by us, there were 7 inches in Surcumference and weighed 3 ounces— Captain Lewis made a small bowl of punch out of one of them. As luck would have it, we were all...Safe...the party that was at the upper camp, were under a good shelter, but we feel concerned about the men on the road with the baggage from the lower Camp—

SUNDAY, JUNE 30

	Sunrise			4 p.m.		Missouri River		
Temp	Weather	Wind	Temp	Weather	Wind	Rise/Fall	Feet	Inches
49a	f	SW	76	f	SW	R		2 1/4

Lewis   We had a heavy dew this morning which is a remarkable event. I being to be extremely impatient to be off as the season is now waisting a pace. Nearly three months have now elapsed since we left Fort Mandan and not yet reached the Rocky Mountains. I am therefore fully preswaded that we shall not reach Fort Mandan again this season if we even return from the ocean to the Snake Indians. The men complained much today of the bruises and wounds which they had received yesterday from the hail.   Two men sent to the falls returned with the compass which they found covered in the mud…the other articles were irrecoverably lost. They found that part of the rivene in; which Capt. C. had been seting yesterday, filled with huge rocks. experienced a heavy gust of wind this evening from the SW after which it was a fair afternoon. More buffaloe than usual were seen about their camp; Capt. C. assured me that he believes he saw at least ten thousand at one view— I made several attempts to obtain Equal altitudes since my arrival here but have been uniformly defeated untill now by the flying clouds and storms in the evening—

Clark   a fair morning. At 3 oClock a Storm of wind from the SW after which we had a clear evening. Great numbers of Buffalow in every direction, I think 10,000 may be Seen in a view.

Gass   A fine day. The hail that fell on the 27th hurt some of the men very badly. Captain Clarke, the interpreter, and the squaw and child, had gone to see the spring at the falls; and when the storm began, they took shelter under a bank at the mouth of the run; but in five minutes there was seven feet water in the run; and they were very near being swept away. They lost a gun, an umbrella and a Surveyor's compass, and barely escaped with their lives.

Ordway   a fair morning. This run has fallen a little. Last evening it was up to a mans waist at the crossing place where it was dry before the Showers, and verry riley and bad tasted   at 3 oClock we had a Storm of wind from SW after which a fair evening.

Whitehouse   a fair morning & pleasant. We are at the upper camp, looking out for the arrival of Captain Clark and his party, with the baggage &c. fearing that they must have suffered much by the hail—

Note   It appears river observations were moved from the lower portage camp to the upper portage camp at the White Bear Islands and continued until the expedition began its ascent once again up the Missouri River on July 13, 1805.

Monday, July 1

	Sunrise			4 p.m.		Missouri River		
Temp	Weather	Wind	Temp	Weather	Wind	Rise/Fall	Feet	Inches
59a	f	SW	74a	f	SW	R		1/2

Weather Diary   wind hard during the grater part of the day.—

Lewis   the day has been warm and the Musquetoes troublesome of course

Clark   the Day w[a]rm and party much fatigued. The wind hard from the SW— the hail which fell at Capt. Lewis Camp 27 Ins [June 27] was 7 Inches in circumfrance & waied 3 ounces, fortunately for us it was not So large in the plains, if it had we Should most certainly fallen victims to its rage as the men were mostly naked, and but few with hats or any covering on their heads.

Ordway the day warm and party much fatigued. The wind hard from the SW. The hail which fell at Capt. Lewis Camp was 7 Inches in Surcumference and weighed 3 oucnes. Fortinately for us it was not so large in the plains where we was. If it had we Should most certainly fallen victims to its rage as the most of the men were without hats or any thing on their heads and mostly neaked.

Whitehouse This morning pleasant and warm. About 3 oClock PM Captain Clarke & party arrived…they informed us that they were detained by the wet weather, and that they were out in the hail Storm but as luck would have it, the hail was not So big as that which fell at the upper Camp. Captain Clarke was at the falls of the River, at the time the hail fell, and hunted a shelter for himself & party from the Rain & hail— This sheltering place, was in a deep Creek, without any Water in it, at the time it first began raining; he mentioned that the Creek rose so fast, they had scarecly time to get out, before the water was ten feet deep. The party that was hawling the crafts, had nearly all lost their lives, being naked and most without hats on, or any thing to cover them, they had no shelter & were Cut and bruised very much by the hail, and under went, as much as Men could possibly endure; to escape with their lives—

TUESDAY, JULY 2

Sunrise			4 p.m.			Missouri River		
Temp	Weather	Wind	Temp	Weather	Wind	Rise/Fall	Feet	Inches
60a	f a r	SW	78a	f	SW			

Weather Diary some rain just before sun rise

Lewis A shower of rain fell very early this morning. The wind hard from the SW all day. I think it possible that these almost perpetual SW winds proceede from the agency of the Snowey Mountains and the wide level and untimbered plains which streach themselves along their bases for an immence distance (ie) that the air comeing in contact with the snow is suddonly chilled and condenced, thus becoming heaver than the air beneath in the plains, where by the constant action of the sun on the face of an untimbered country there is a partial vacuum formed for [its] reception. I have observed that the winds from this quarter are always the coldest and most violent which we experience, yet I am far from giving full credit to my own hypothesis on this subject; i[f] ho[w]ever I find on the opposite side of these mountains that the winds take a contrary direction I shall then have more faith….completed my observation of Equal Altitudes today.

Clark Some rain at day light this morning, after which a fair morning. The Roreing of the falls for maney miles above us. Musquetors verry troublesom to day, day w[a]rm. Wind to day as usial from the SW and hard all the after part of the day, those winds are also Cool and generally verry hard.

Gass a fine morning.

Ordway Some rain at day light this morning after which a fair morning. The day warm.

Whitehouse Some rain at day light this morning, and it then cleared up, and we had clear pleasant fair weather.

WEDNESDAY, JULY 3

Sunrise			4 p.m.			Missouri River		
Temp	Weather	Wind	Temp	Weather	Wind	Rise/Fall	Feet	Inches
58a*	f	SW	74a	c a f & r	SW			

Weather Diary   slight rain in the evening.

Lewis   at 10 OCk AM we had a slight shower which scarcely wet the grass. The current of the river looks so gentle and inviting that the men all seem anxious to be moving upward as well as ourselves.

Clark   a fine morning. Wind from the SW. At 1 oClock began to rain. A Small Shower at 1 oClock which did Scercely wet the grass— The water tolerably clear and Soft in the river, Current jentle and bottoms riseing from the water; no appearance of the river riseing more than a few feet above the falls, as high up as we have yet explored. The winds has blown for Several days from the SW I think it possible that those almost perpetial SW winds, proceed from the agency of the Snowey mountains and the wide leavel and untimbered plains which Streach themselves along their borders for an emence distance, that the air comeing in Contact with Snow is Suddenly chilled and condensed, thus becoming heavyer than the air beneath in the plains, it glides down the Sides of those mountains and decends to the plains, where by the constant action of the Sun on the face of the untimbered country there is a partial vacuom formed for [its] reception. I have observed that the winds from this quarter is always the Coaldest and most violent which we experience, yet I am far from giveing full credit to this hypothesis on this Subject; if I find however on the opposit Side of these mountains that the winds take a contrary direction I Shall then have full faith. The winds take a contrary direction in the morning or from the mountains on the west Side.

Gass   a fine morning. We had a light shower of rain (in early afternoon)

Ordway   a clear pleasant warm morning.

Whitehouse   A clear pleasant morning. We had a light sprinkling Shower of rain in the afternoon.

THURSDAY, JULY 4

Sunrise			4 p.m.			Missouri River		
Temp	Weather	Wind	Temp	Weather	Wind	Rise/Fall	Feet	Inches
52a	f	Sw	76a	f a r	SW	F		1/4

---

*According to Moulton (1988, 5: 24), Lewis's Journal Codex P & Clark's Codex I lists this temperature as "56."

Weather Diary   heavy dew this morning.   slight sprinkle of rain at 2 PM.

Lewis   we all believe that we are now about to enter on the most perilous and difficult part of our voyage, yet I see no one repining; all appear ready to met those difficulties which wait us with resolution and becoming fortitude. We had a heavy dew this morning. The clouds near these mountains rise suddonly and discharge their contents partially on the neighbouring plains; the same cloud will discharge hail along in one part, hail and rain in another, and rain only in a third, all within the space of a few miles; and on the Mountains to the SE of us sometimes snow. At present there is no snow on those mountains; that which covered them when we first saw them and which has fallen on them several times since has all disappeared. The Mountains to the NW & W of us are still entirely covered are white and glitter with the reflection of the sun. I do not believe that the clouds which prevail at this season of the year reach the summits of those lofty mountains; and if they do the probability is that they deposit snow only for there has been no preceptible deminution of the snow which they contain since we first saw them. I have thought it probable that these mountains might have derived their appellation of shining Mountains, from their glittering appearance when the sun shines in certain directions on the snow which covers them. Since our arrival at the falls we have repeatedly witnessed a nois which proceeds from a dircction a little to the N of West as loud and resembling precisely the discharge of a piece of ordinance of 6 pounds at a distance of three miles. I was informed of it by the men several times before I paid any attention to it, thinking it was thunder most probably which they had mistaken. At length walking in the plains the other day I heard this noise very distinctly, it was perfectly calm clear and not a cloud to be seen. I halted and listened attentively about an hour during which time I heard two other discharges and took the direction of the sound with my pocket compass. I have no doubt but if I had leasure I could find from whence it issued. I have thout it probable that it might be caused by runing water in some of the caverns of those immence mountains, on the principal of the blowing caverns; but in such case the sound would be periodical & regular, which is not the case with this, being sometimes heard once only and at other times, six or seven discharges in quick succession. It is heard also at different seasons of the day and night. I am at a loss to account for this phenomenon. Our work being at an end this evening, we gave a drink of sperits, it being the last of our stock, and some of them appeared a little sensible of [its] effects. The fiddle was plyed and they danced very merrily untill 9 in the evening when a heavy shower of rain put an end to that part of the amusement tho' they continued their mirth with songs and festive jokes and were extreemly merry untill late at night.

Clark   A fine morning, a heavy dew last night. A black Cloud came up from the SW and rained a fiew drops. I employ my Self drawing a Copy of the river to be left at this place for fear of Some accident in advance. The party amused themselves danceing untill late when a Shower of rain broke up the amusement. The climate about the falls of the Missouri appears to be Singular Cloudy every day (Since our arrival near them) which rise from defferent directions and discharge themselves partially in the plains & mountains, in some places rain other rain & hail, hail alone, and on the mountains in Some parts Snow. A rumbling like Cannon at a great distance is heard to the west if us; the Cause we Can't account.

Gass   A fine day. We drank the last of our spirits in celebrating the day, and amused ourselves with dancing till 9 o'clock at night, when a shower of rain fell and we retired to rest.

Ordway   a beautiful clear pleasant warm morning. Last in the evening we had a light Shower of rain but did not last long—

Whitehouse   This morning we had Clear weather. We amused ourselves with frolicking, dancing &ca. untill 9 of 10 oClock PM in honor of the day. About that time, we had a slight shower of Rain, but it soon cleared away & we had fine weather—

FRIDAY, JULY 5

	Sunrise			4 p.m.		Missouri River		
Temp	Weather	Wind	Temp	Weather	Wind	Rise/Fall	Feet	Inches
49a	*f a h & r**	SW	72a	*f*	SW	F		1/2

Weather Diary   heavy shower of rain and hail last evening at 9 PM.   some thunder & lightning

Clark   a fine morning and but little wind, w[a]rm and Sultrey at 8 oClock—

Gass   a fine morning.

Ordway   a clear pleasant morning.

Whitehouse   a clear pleasant morning

SATURDAY, JULY 6

	Sunrise			4 p.m.		Missouri River		
Temp	Weather	Wind	Temp	Weather	Wind	Rise/Fall	Feet	Inches
47a	*c a h r t & l*	SW	74a	*f a c*	SW	F		1/4

Weather Diary   wind high all day.   A heavy wind from the SW attended with rain about he middle of last night. about day had a violent thunderstorm attended with Hail and rain.   the Hail Covered the ground and was about the Size of Musquet balls. I have Seen only one black bird killed with the hail, and am astonished that more have not Suffered in a similar manner as they are abundant, and I Should Suppose the hail Sufficiently heavy to kill them.

Lewis   In the course of last night had several showers of hail and rain attended with thunder and lightning. About day a heavy storm came on from the SW attended with hail rain and a continued roar of thunder and some lightning. The hail was as large as

---

*According to Moulton (1988, 5: 24), Lewis's Journal Codex P lists this weather data as " f h & r" & Clark's Codex I lists this as "h & r."

musket balls and covered the ground perfectly. We had some of it collected which kept very well through the day and served to cool our water. These showers and gusts keep my boat wet in dispite of my exertions. After the hail and rain was over this morning we dispatched 4 hunters

Clark   a heavy wind from the SW and Some rain about mid night last, at day light his morning a verry black Cloud from the SW, with a Contained rore of thunder & Some lightning and rained and hailed tremendiously for about ¹/₂ an hour, the hail was the Size of a musket ball and Covered the ground. This hail & rain was accompand. by a hard wind which lasted for a fiew minits. Cloudy all the forepart of the day, after Part Clear.

Ordway   verry hard Showers of rain and hail through the course of last night, hard Thunder & lightning, at day light this morning a hard shower came up of a Sudden attended with high wind & large hail one of the men Saved a Small tin kittle full of the hail which did not all disolve through the day. The morning cloudy.   A part of the day clear.   Light Showers of rain in the afternoon.   The wind high from the west.

Whitehouse   verry hard Showers of Rain & hail, through the course of last night. Hard Thunder. At day light a hard shower of rain, thunder & large hail; one of the men gethered a Small kittle full of the hail which he kept most part of the day, without melting. The morning continued cloudy. In the afternoon it cleared up with some light showers of rain.

SUNDAY, JULY 7

	Sunrise			4 p.m.			Missouri River		
Temp	Weather	Wind	Temp	Weather	Wind	Rise/Fall	Feet	Inches
54a	c a f	SW	77a	r a c	SW	F		¹/₄

Weather Diary   a Shower at 4 PM

Lewis   The weather warm and cloudy therefore unfavourable for many operations. We have no tents; the men are therefore obliged to have recourse to the sails for shelter from the weather and we have not more skins than are sufficient to cover our baggage when stoed away in bulk on land. Many of the men are engaged in dressing leather to cloath themselves. Ther leather cloathes soon become rotton as they are much exposed to the water and frequent wet. We had a light shower of rain about 4 PM attended with some thunder and lightning. The musquetoes are excessively troublesome to us. I have prepared my composition which I should have put on this evening but the rain prevented me.

Clark   A Warm day, wind from the SW. Cloudy as usial. Some rain in the after part of the day.

Gass   The morning was fine.   In the evening some few drops of rain fell

Ordway   a clear pleasant morning. The day warm. I the afternoon Some Thunder   A light shower of rain.

Whitehouse   A clear pleasant morning. They proved warm. Before 4 oClock PM we had some Thunder and a light Shower of Rain.

MONDAY, JULY 8

	Sunrise			4 p.m.		Missouri River		
Temp	Weather	Wind	Temp	Weather	Wind	Rise/Fall	Feet	Inches
60a	f	SW	78a	f a r	SW	F		1/4

Weather Diary   I finish taking the hight of the falls of the Missouri

Lewis   day being warm and fair. The mountains which ly before us from the South, to the NW still continue covered with snow. Slight rain this afternoon.

Clark   A w[a]rm morning, flying Clouds. Some rain this evening after a verry hot day— the mountains which are in view to the South & NW are Covered with Snow.

Gass   a fine morning.

Ordway   a clear pleasant morning.   Some Thunder and light Showers this afternoon. The River falling.

Whitehouse   A clear pleasant morning. We had this Evening some Thunder and light sprinkling of rain, &c.

TUESDAY, JULY 9

	Sunrise			4 p.m.		Missouri River		
Temp	Weather	Wind	Temp	Weather	Wind	Rise/Fall	Feet	Inches
56a	f	SW	76a	c a r	NW	F		1/4

Lewis   The morning was fair and plesant. Men loaded canoes…just at this moment a violent wind commenced and blew so hard that we were obliged to unload the canoes. again; a part of the baggage in several of them got wet before it could be taken out. The wind continued violent untill late in the evening. The buffaloe had principally deserted us, and the season was now advancing fast. I therefore relinquished all further hope of my favorite boat and ordered her to be sunk in the water. [Lewis's experimental iron-frame, elk-skin-covered boat. The iron frame being made at Harpers Ferry and transported all the way to the upper portage camp location.]

Clark   a clear w[a]rm morning, wind from the SW. This falire of our favourate boat was a great disapointment to us.

Gass   a fine morning, and heavy dew. In the afternoon a storm of wind, with some rain came on from the north west, and we had again to unload some of our canoes, the waves ran so high. After the storm we had a fine evening.

Ordway   a beautiful pleasant morning. Soon after we got the canoes loaded Thunder and high wind came on So that we had to unload again.

Whitehouse   a beautiful morning. In the afternoon, Soon after we got the canoes loaded, there came up a Violent Storm of wind & Thunder. The waves dashed over the canoes to such a height, so that all hands were employed to unload them as quick as possible—

WEDNESDAY, JULY 10

	Sunrise			4 p.m.		Missouri River		
Temp	Weather	Wind	Temp	Weather	Wind	Rise/Fall	Feet	Inches
52a	f a r	SW	66a	F	SW			

Weather Diary   wind hard all day.

Lewis   the wind blew very hard the greater part of the day. Sergt. Ordway proceeded up the river about 5 miles when the wind became so violent that he was obliged to ly by untill late in the evening when he again set out with the canoes.

Clark   a fair windey day, wind hard the most of the day from the SW— rained modderately all last night (by Showers). The Canoes did not arrive as I expected, owing to the hard wind which blew a head in maney places.

Gass   a fine cool morning.

Ordway   a clear morning.   Then the wind arose So high that we were obleged to lay by untill towards evening. Late in the afternoon the wind abated a little So we proceeded on within about 3 miles of the upper camp.

Whitehouse   a clear pleasant morning. We proceeded on about 8 miles when the wind rose so Great and high that we were obledged to lay by untill the evening. The wind then abated and we went on untill dark.

THURSDAY, JULY 11

	Sunrise			4 p.m.		Missouri River		
Temp	Weather	Wind	Temp	Weather	Wind	Rise/Fall	Feet	Inches
46a	f	SW	70a	f	SW			

Weather Diary   wind hard all day

Lewis   this evening a little before the sun set, I heard two other discharges of this unaccountable artillery of the Rock Mountains proceeding from the same quarter that I had before heard it. I now recollected the Minnetares making mention of the nois which they had frequently heard in the Rocky Mountains like thunder; and which they said the mountains made; but I paid no attention to the information supposing it either false or the fantom of a supersticious immagination. I have also been informed by the engages that the Panis and Ricaras give the same account of the Black mountains which lye West of them. This phenomenon the philosophy of the engages readily accounts for; they state it to be the bursting of the rick mines of silver which these mountains contain. The three other canoes did not arrive untill late in the evening in consequence of the wind and the fear of weting their loads which consisted of articles much more liable to be injured by moisture...Capt C. had the canoes unloaded and ordered them to float down in the course of the night to my camp, but the wind proved so high after the night that they were obliged to put too 8 miles above and remain untill morning.

Clark   a fair windey morning, wind SW.

Ordway   a clear morning. The wind verry high from the NW which obliged us to lay at Camp untill late in the afternoon. Towards evening the wind abated So that we went on and arrived at Capt. Clarks camp. We floated about 8 miles and the wind rose So high that drove us to Shore So we landed untill morning.

Whitehouse   a clear morning, but the high wind which oblegded us to lay at our Camp untill late in the afternoon. Towards evening the wind abated a little so that we went on. A party of 4 men set out to float back to the lower camp at night, but the wind rose so high, that they were forced to lay by till morning—

FRIDAY, JULY 12

	Sunrise			4 p.m.			Missouri River		
Temp	Weather	Wind	Temp	Weather	Wind	Rise/Fall	Feet	Inches
50a	f	SW	74a	f	SW	F		1/4

Weather Diary   wind violent all day.

Lewis   I feel excessively anxious to be moving on. The canoes were detained by the wind untill 2 PM.

Clark   a fair windey morning, wind from the SW. The wind hard all day.

Gass   a fine morning.

Ordway   a clear morning. The wind as usal. The wind rose So high that one canoe filled with water the other 2 took in water   the waves high but with difficulty we got down to Camp about noon. The wind continues verry high all day—

Whitehouse   a clear morning. The wind blew high and hard from the NW. The wind continued to blow so high and hard that one of the canoes filled and the other two took in water. The wind continued to blow high and hard during this day.

Note   *The expedition leaves the upper portage camp and continues its journey up the Missouri River.*

SATURDAY, JULY 13

	Sunrise			4 p.m.			Missouri River*	
Temp	Weather	Wind	Temp	Weather	Wind	Rise/Fall	Feet	Inches
42a	f	SW	76a	f	SW	F		1/4

Weather Diary   Wind violent in the latter part of the day

Lewis   This morning being calm and Clear I had the remainder of our baggage embarked in the six small canoes and maned them with two men each. I now bid a cheerfull adue to my camp. Proceeded up the river about 5 miles when the wind became so violent that two of the canoes shiped a considerable quantity of water and they were compelled to put too...and clense the canoes of water.   About 5 PM the wind abated. It is impossible to sleep a moment without being defended against the attacks of these most tormenting of all insects.

Clark   a fair Calm Morning, verry Cool before day— at 9 oClock, the wind rose and blew hard from the SE the greater part of the day.

Gass   A fine day, but high wind.

Ordway   clear and calm this morning.   About 5 miles...the wind rose so high that 2 of the canoes took water.

Whitehouse   a clear, calm pleasant morning. Proceeded on this morning abt. 5 miles...then the wind rose so high that obledged us halt untill the middle of the afternoon. When the wind abated we went on about 7 mls further

SUNDAY, JULY 14

	Sunrise			4 p.m.			Missouri River	
Temp	Weather	Wind	Temp	Weather	Wind	Rise/Fall	Feet	Inches
45a	f	SW	78a	c a r	SW			

Lewis   This morning was calm fair and warm. The grass and weeds in this bottom are about 2 feet high; which is much greater hight than we have seen them elsewhere this season. Had a slight shower at 4 PM this evening.

Clark   a fine morning, Calm and warm. Some rain this afternoon. All prepareing to Set out on tomorrow.

---

*River observations once again change daily as the expedition is again moving up the Missouri River. The data usually is not for a 24-hour period unless they are encamped. In most cases, when the party stopped for the evening, following Lewis's habit from previous journal entries, it is surmised that a mark was made and measured the next morning. Thus many observations were only a 8 to 10 hours long.

Gass   a fine morning. In the afternoon some rain fell but we continued to work at the canoes

Ordway   the morning clear and pleasant.   Abotu 4 oClock PM we expencerenced a Small Shower of rain. Warm.   The weeds and Grass in this bottom is as high as a mans knees but the Grass on the high plains & praries is not more than 3 Inches high not time in this Season.

Whitehouse   the morning clear, calm and pleasant. About 4 oClock PM we had a Small Shower of rain & very warm. The weeds and grass in this bottom is as high as a mans knees, but the grass on the high land is not more than 3 inches high.

Monday, July 15

	Sunrise			4 p.m.		Missouri River		
Temp	Weather	Wind	Temp	Weather	Wind	Rise/Fall	Feet	Inches
60a	f a r	SW	76a	f	SW	F		1 1/2

Weather Diary   Set out from our upper camp above the falls

Lewis   At 10 AM we once more saw ourselves fairly under way much to my joy and I believe that of every individual who compose the party. We have not passed Fort Mountain [Square Butte] on our right it appears to be about ten miles distant. From [its] figure we gave it the name of fort mountain. The prickly pear is now in full blume and forms one of the beauties as well as the greatest pests of the plains. The sunflower is also in blume and is abundant. The river is from 100 to 150 yds. wide. on the banks of the river there are many large banks of sand much elivated above the plains on which they ly and apear as if they had been collected in the course of time from the river by the almost incessant SW winds; they always appear on the sides of the river opposite to those winds—

Clark   rained all the last night, I was wet all night. This morning wind hard from the SW.

Gass   After a night of heavy rain, we had a pleasant morning.   The snow appears to have melted from all the mountains in view.

Ordway   rained the greater part of last night.   A clear morning. The wind high from the NW. The current verry gentle & river Smoth Since we left the falls.

Whitehouse   rained the greater part of last night. This morning was clear, but the wind blowing high and hard from the NW. The current verry gentle since we came above the falls and clear

Tuesday, July 16

	Sunrise			4 p.m.		Missouri River		
Temp	Weather	Wind	Temp	Weather	Wind	Rise/Fall	Feet	Inches
53a	f	SW	80a	f	SW	F		3/4

Lewis   We had a heavy dew last night. Early this morning we passed about 40 little booths formed of willow brushes to shelter them from the sun; they appeared to have been deserted about 10 days; we supposed that they were snake Indians. Came to a rappid [probably the later-named Half-Breed or Lone Pine Rapids]...the current of the missouri below these rappids is strong for several miles, tho' just above there is scarcely any current, the river very narrow and deep, about 70 yds. wide. The musquetoes are extreemly troublesome this evening and I have left my bier, of course suffered considerably, and promised in my wrath that I never will be guily of a similar peice of negligence while on this voyage—

Clark   a fair morning after a verry cold night, heavy dew   the river is not So wide as below from 100 to 150 yards wide & Deep Crouded with Islands & Crooked. the current of the river from the Medison river to the Mountain is gentle, bottoms low and extensive.

Gass   a fine morning. The water became more rapid; but the current not so swift as below the falls.

Ordway   a clear pleasant morning. The current Swift towards evening.

Whitehouse   a clear pleasant morning. The current begin to git Swifter.

WEDNESDAY, JULY 17

Sunrise			4 p.m.			Missouri River		
Temp	Weather	Wind	Temp	Weather	Wind	Rise/Fall	Feet	Inches
58a	f	SW	81a	f	SW	F		1 ½

Clark   took a Medn. Altitude & we took Some Luner Observations &c. The river confined in maney places in a verry narrow Chanel from 70 to 120 yards wide.

Gass   the morning was fine and pleasant.

Ordway   a clear morning. Came to a verry bad rapid. The River is about 100 yards wide.

Whitehouse   a clear morning. The current verry rapid, and river Crooked, and only about 100 yard wide.

THURSDAY, JULY 18

Sunrise			4 p.m.			Missouri River*		
Temp	Weather	Wind	Temp	Weather	Wind	Rise/Fall	Feet	Inches
60a	f	SW	84a	f	SW			

*According to Moulton (1988, 5: 24), Clark's Journal Codex I lists that the river fell one-half inch.

Lewis   passed the entrance of a considerable river…[its] current is rapid and water extreamly transparent…in honour of the Secretary of war calling it Deaborn's River. The river [Missouri] somewhat wider than yesterday and the mountains more distant from the river and not so high.

Clark   a fine morning.

Gass   The morning was fair

Ordway   a clear pleasant morning.

Whitehouse   a clear pleasant morning.

Note   *The expedition enters what Lewis names "The Gates of the Rocky Mountains," in present-day Montana.*

FRIDAY, JULY 19

	Sunrise			4 p.m.			Missouri River		
Temp	Weather	Wind	Temp	Weather	Wind	Rise/Fall	Feet	Inches
62a	*f*	SW	68a	*c a h & r*	SW	F		1/2

Weather Diary   Thunder Storm 1/2 after 3 PM

Lewis   the current has been strong all day. The river deep and from 100 to 150 yds,. wide. This evening we entered much the most remarkable clifts that we have yet seen. Every object here wears a dark and gloomy aspect. Several fine springs burst out at the waters edge from the interstices of the rocks. From the singular appearance of this place, I called it the gates of the rocky mountains. Musquetoes less troublesome than usual. We had a thundershower today about 1 PM which continued about an hour and was attended with som hail.

Clark   a find morning.

Gass   a fine morning. About 1 o'clock we had thunder, lightning and rain, which continued an hour or two, and then the weather became clear.

Ordway   a clear pleasant morning. This curious looking place (we entered) we call the gates of the Rocky Mountains.   About one oClock PM we had a Thunder Shower which lasted about one hour   a little hail attended it

Whitehouse   a clear pleasant morning. The current swift and water clear. About 1 oClock PM we had a Thunder Shower which lasted 1 hour. Shortly after we camped we had a light Sprinkling Shower of rain this evening.

SATURDAY, JULY 20

	Sunrise			4 p.m.			Missouri River		
Temp	Weather	Wind	Temp	Weather	Wind	Rise/Fall	Feet	Inches
59a	*f a r*	SW	60a	*f*	NW			

Lewis currant strong. Having lost my post Meridian Observation for Eql. Altitudes in consequence of a cloud which obscured the sun for several minutes about that time, I had recourse to two altitudes of the sun with Sextant.

Clark a fine morning.

Gass a fine morning. About 2 o'clock came to a level plain on the north side, from which we saw a strong smoke rising.   The river is very crooked in general

Ordway   a clear morning

Whitehouse   a clear morning. The current verry rapid.

Note   *The main party passes just east of present-day Montana state capitol, Helena.*

SUNDAY, JULY 21

	Sunrise			4 p.m.			Missouri River		
Temp	Weather	Wind	Temp	Weather	Wind	Rise/Fall	Feet	Inches
60a	f	NW	67a	f	NW	F		¹/₂

Lewis   the current strong…the river is not now so deep but reather wide and much more rapid. The grass near the river is lofty and green that of the hill sides and high open grounds is perfectly dry and appears to be scorched by the heat of the sun. This valley is bounded by two nearly parallel ranges of high mountains which have their summits partially covered with snow. The musquetoes were equally as troublesome to them as to ourselves this evening; tho' some hours after dark the air becomes so cold that these insects disappear.

Clark   a fine morning. I observed on the highest pinecals of Some of the mountains to the West Snow lying in Spots Some Still further North are covered with Snow and cant be Seen from this point. The Winds in those mountains are not Settled generally with the river, to day the wind blow hard from the West at the Camp.

Gass   a pleasant morning. At noon….the wind blew very hard and some drops of rain fell.

Ordway   a clear morning. The grass in the valley and on the hills look dry and pearched up. The River divides in many channels and full of Islands and Spreads about a mile wide. The current swift.

Whitehouse   a clear morning. The wind high from the NW.

MONDAY, JULY 22

	Sunrise			4 p.m.			Missouri River		
Temp	Weather	Wind	Temp	Weather	Wind	Rise/Fall	Feet	Inches
52a	f	NW	80a	f	NE			

Lewis   Onions...this appears to be a valuable plant inasmuch as it produces a large quantity to the squar foot and bears with ease the rigor of this climate. Killed an otter which sunk..The water was about 8 feet deep yet so clear that I could see it at the bottom; I swam in and obtained it by diving. I placed my thermometer in a good shade as was my custom about 4 PM and after dinner set out without it and had proceeded near a mile before I recollected it I sent Sergt. Ordway back for it, he found it and brought it on. The murcury stood at 80 a. 0. This is the warmest day except one which we have experienced this summer.

Clark   a fine morning, wind from the SE. The last night verry cold, my blanket being Small I lay on the grass & Covered with it.

Gass   We embarked early, the weather being pleasant. We saw to day several banks of snow on a mountain west of us [Elkhorn Mountains].

Ordway   a clear morning. Capt. Lewis forgot his Thurmometer where we dined. I went back for it. It Stood in the heat of the day at 80 degrees abo. 0, which has only been up to that point but once before this Season as yet.

Whitehouse   A clear pleasant morning. Capt. Lewis forgot his Thurmometer which he had hung in a Shade. It Stood this day at 80 degrees above 0. The current verry rapid and a pleasant country.

TUESDAY, JULY 23

Sunrise			4 p.m.			Missouri River		
Temp	Weather	Wind	Temp	Weather	Wind	Rise/Fall	Feet	Inches
54a	f	SW	80a	c	SW	F		1/2

Clark   a fair morning, wind from the South.
Gass   A cloudy morning.
Ordway   a little cloudy   the current swift
Whitehouse   This morning cloudy. The current verry rapid.

WEDNESDAY, JULY 24

Sunrise			4 p.m.			Missouri River		
Temp	Weather	Wind	Temp	Weather	Wind	Rise/Fall	Feet	Inches
60a	f	SW	90a	f	SW	F		3/4

Lewis   the current very strong. The mountains still continue high and seem to rise in some places like an amphatheater one rang above another as they receede from the river untill the most distant and lofty have their tops clad with snow...I fear every day that we shall meet some considerable falls or obstruction in the river notwithstanding the information of the Indian woman to the contrary who assures us that the river continues much as we see it. I can scarcely form an idea of a river runing to great extent through such a rough mountainous country without having [its] stream intersepted by some difficult and gangerous rappids or falls.

Clark   a fine day, wind from the NW. The river much like it was yesterday. The mountains on either Side appear like the hills had fallen half down & turned Side upwards [Lombard thrust fault crosses here] the bottoms narrow and no timber a fiew bushes only.

Gass   The morning was fine

Ordway   a clear pleasant morning. The current swift. The Swift water continues   some bad rapids which it is with difficulty we passed over them.

Whitehouse   a clear pleasant morning. The current still verry rapid and strong all day.

Note   *The expedition spends July 25–30 at the Three Forks of the Missouri River.*

THURSDAY, JULY 25

	Sunrise			4 p.m.		Missouri River		
Temp	Weather	Wind	Temp	Weather	Wind	Rise/Fall	Feet	Inches
60a	f	SW	86a	f	SW	F		¹/₂

Weather Diary   Snow appears on the mountains ahead.

Lewis   two rapids near the large spring we passed this evening were the worst we have seen since that we passed on entering the rocky mountains.

Clark   a fine morning we proceeded on a fiew miles to the three forks of the Missouri. Those three forks are nearly of a Size, that North appears to have the most water and must be Considered as the one best calculated for us to assend. Middle fork is quit as large about 90 yds. wide. The South fork is about 70 yds wide. The forks appear to be verry rapid. On the North Side the Indians have latterly Set the Prarie on fire, the Cause I Can't account for. The day verry hot. About 6 to 8 miles up the North fork a Small rapid river falls in on the Lard Side which affords a great Deel of water and appears to head in the Snow mountains to the SW. musquetors verry trouble Som untill the mountain breeze Sprung up, which was a little after night.

Gass   a fine morning.

Ordway   a clear morning. We discover Mountains a head which have Spots of Snow on them. Passed a large dry plain on S side. Bad rapids.

Whitehouse   A clear pleasant morning. Discovered mountains lying ahead of us, which appear to have Snow on them, if not Snow it must be verry white Clay or rocks. Several bad rockey rapids.

FRIDAY, JULY 26

Sunrise			4 p.m.			Missouri River		
Temp	Weather	Wind	Temp	Weather	Wind	Rise/Fall	Feet	Inches
60a	f	SW	82a	c a r	SW	F		3/4

Lewis   current strong with frequent riffles on entering this open valley I saw the snowclad tops of distant mountains before us. This morning Capt. Clark…proceeded up the river about 12 miles to the top of a mountain …The day being warm and the road unshaded by timber he [Clark] suffered excessively with heat and the want of water. He returned down the mountain…here Charbono was very near being swept away by the current and cannot swim, Capt. C. however risqued him and saved his life.

Clark   proceeded to top of a mountain…much fatigue…we came to a Spring of excessively Cold water, which we drank reather freely of as we were already famished; not with Standing the precautions of wetting my face, hands & feet, I Soon felt the effects of the water. We Contind. thro a Deep Vallie without a Tree to Shade us Scorching with heat. a fiew drops of rain this evening

Gass   the morning was fine. Before 4 o'clock….while we remained here it became cloudy and some rain fell

Ordway   a clear morning. The current verry Swift. We can discover high mountains a head, with Snow on them. The River wide and full of islands. Passed over several bad rapids.

Whitehouse   a clear morning. We find that we have not entered the 2nd chain of Mountains but can discover verry high white toped mountains. The wind blew hard since 10 AM and at 2 oClock & a light Sprinkling of rain.

SATURDAY, JULY 27

Sunrise			4 p.m.			Missouri River		
Temp	Weather	Wind	Temp	Weather	Wind	Rise/Fall	Feet	Inches
52a	c	SW	80a	c a r	SW	F		3/4

Weather Diary   a considerable fall of rain unattended with Lightning. ….

Lewis   We set out at an early hour and proceeded on but slowly the current still so rapid that the men are in continual state of their utmost exertion to get on, and they begin to weaken fast from this continual state of violent exertion. The limestone appears to be of an excellent quality of deep blue colour when fractured and a light led colour when exposed to the weather. We arrived at 9 AM at the junction of the SE fork of the Missouri and the country opens suddonly to extensive and beatifull plains and meadows which appear to be surrounded in every direction with distant and lofty mountains. From the E to S between the SE and middle forks a distant range of lofty mountains rose their snow-clad tops above the irregular and broken mountains which lie adjacent to this beautifull spot. A range of high mountains at a considerable distance appear to reach from South to West and are partially covered with snow. My principal consolations are that from our present position it is impossible that the SW fork can head with the waters of any other river but the Columbia

Gass   had a pleasant morning. There is very little difference in the size of the 3 branches. About 9 o'clock…we halted here, it began to rain and continued 3 hours.   In the evening the weather became clear and we had a fine night.

Ordway   a clear morning.   The current Swift as usal. About 9 oClock we arived at the three forks of the Missourie, which is in open view of the high Mountains covered in Some places with Snow.   We had a Shower of rain this afternoon.

Whitehouse   a clear morning. Current rapid as yesterday. At 9 oClock AM we arrived at the Three Forks of the Mesouri River which is in a wide valley in open view of high Mountains which has white Spots on it which has the appearance of Snow. We had showers of rain that continued till the evening.

SUNDAY, JULY 28

	Sunrise			4 p.m.		Missouri River		
Temp	Weather	Wind	Temp	Weather	Wind	Rise/Fall	Feet	Inches
49a	f a r	SW	90a	f	SW	F		1/2

Lewis   the day proved warm…our leather lodge when exposed to the sun is excessively hot. In the evening about 4 O'Ck the wind blew hard from the South West and after some little time brought on a Cloud attended with thunder and Lightning from which we had a fine refreshing shower which cooled the air considerably; the showers continued with short intervals untill after dark.

Clark   a very w[a]rm day untill 4 oClock when the wind rose & blew hard from the SW, and was Cloudy,   The Thermometr. Stood at 90° above 0 in the evening a heavy thunder Shower from the SW which continud at intervales untill after dark

Gass   As this was a fine day, the men were employed in airing the baggage.   From this valley we can discover a large mountain with snow on it, towards the southwest.

Ordway   a foggy morning, but clear after.   Towards evening we had a fine Shower of rain Some Thunder attended it which cooled the air verry much.

Whitehouse   We has some fog early this morning, but it cleared away at Sun rise, & the weather was pleasant. In the evening we had a fine shower of rain, accompanied with Thunder, which cooled the Air, & made it very pleasant.

MONDAY, JULY 29

	Sunrise			4 p.m.			Missouri River		
Temp	Weather	Wind	Temp	Weather	Wind	Rise/Fall	Feet	Inches
54a	*f a r*	N	82a	*f*	NE	R		¹/₂

Lewis   we see a great abundance of fish in the stream some of which we take to be trout, but they will not bite at any bate we can offer them.

Clark   a fair morning, wind from the North

Gass   a fine day.

Ordway   a clear pleasant morning. The day warm. The wind from the East. the width of the three Rivers at the forks we alow the North fork to be about 60 yards, the middle fork the Same, the South forks not So wide nor large.   All appear rapid but not verry deep.

Whitehouse   a clear pleasant morning. The day proved verry warm since 9 oClock AM the wind from the East.

TUESDAY, JULY 30

	Sunrise			4 p.m.			Missouri River		
Temp	Weather	Wind	Temp	Weather	Wind	Rise/Fall	Feet	Inches
50a	*f*	SE	80a	*f*	SE			

Weather Diary   Set out from 3 forks

Lewis   the night was cool but I felt very little inconvenience from it as I had a large fire all night.

Clark   the river very rapid & Sholey

Gass   This branch [Jefferson] is about 60 yards wide and 6 feet deep, with a rapid current.

Ordway   a fine pleasant morning. The River crooked rapid and full of islands &.C.

Whitehouse   a clear pleasant morning. They day warm, but verry pleasant. The current verry swift & rapids common.

Note   *The expedition embarks up the Jefferson River (West Fork of the Missouri River) in Montana.*

WEDNESDAY, JULY 31

Sunrise			4 p.m.			Missouri River		
Temp	Weather	Wind	Temp	Weather	Wind	Rise/Fall	Feet	Inches
48a	f	SW	92a	f	SW			

Lewis   Capt. Clark and party...their detention had been caused by the rapidity of the water and the circuitous route of the river. The river...from 90 to 120 yd. wide

Clark   a fair Morning.

Gass   a fine cool morning with dew.

Ordway   a fine morning. the current swift. We dined under a handsom Shady grove of cotton timber under the hills of the Mountains to o[u]r left which has heaps of Snow on the tops & sides of it. The day very warm

Whitehouse   a fine morning, the current rapid. Wed dined about 1 oC, under a delightful Grove of cotton timber on L. Side under the mountain which has large heaps of Snow on it. The day verry warm.

Note   *The expedition is just east of present-day Whitehall, Montana. River observations are taken on the Jefferson Fork of the Missouri.*

THURSDAY, AUGUST 1

Sunrise			4 p.m.			Jefferson River		
Temp	Weather	Wind	Temp	Weather	Wind	Rise/Fall	Feet	Inches
54a	f	SW	91a	f	SW	F		½

Lewis   about 2 PM much exhausted by the heat of the day. Our rout lay through the steep and narrow valleys of the mountains exposed to the intense heat of the midday sun without shade and scarcely a breath of air.

Clark   a fine day. The water Swift & very Sholey. The river so rapid that the greatest exertion is required by all to get the boats on. Wind SW. Murckery at sun rise 50° Ab. 0

Gass   a fine morning.

Ordway   a fine morning.   The current swift.

Whitehouse   a clear pleasant morning. We also saw Snow on the Mountains, a short distance to the South of our Camp—

FRIDAY, AUGUST 2

Sunrise			4 p.m.			Jefferson River		
Temp	Weather	Wind	Temp	Weather	Wind	Rise/Fall	Feet	Inches
48a	f	NW	81a	f	NW	F		1/2

Lewis   we resumed our march this morning at sunrise, the weather was fair and wind from NW. Found the current verry rapid about 90 yards wide and wait deep, this is the first time that I ever dared to make the attempt to wade the river. The tops of these mountains are yet covered partially with snow, while we in the valley are nearly suffocated with the intense heat of the mid-day sun; the nights are so cold that two blankets are not more than sufficient covering.

Clark   a fine day. The wind from the SW. We proceeded on with great dificuelty from the rapidity of the current & rapids.

Gass   The morning was fine. In the middle of the day it was very warm in the valley, and at night very cold; so much so that two blankets were scarce a sufficient covering. On each side of the valley there is a high range of mountains, which run nearly parallel, with some spots of snow on their tops.

Ordway   a fine pleasant morning. The River is now Small crooked Shallow and rapid.   The day warm.

Whitehouse   a fine pleasant morning. The river is now small crooked Shallow and rapid.

SATURDAY, AUGUST 3

Sunrise			4 p.m.			Jefferson River		
Temp	Weather	Wind	Temp	Weather	Wind	Rise/Fall	Feet	Inches
50a	f	NE	86a	f	NE	F		1/2

Lewis   …they passed a handsome little stream on Lard. which is form of several large springs which rise in the bottoms and along the base of the mountains with some little rivulets form the melting snow. In the evening they passed a very bad rappid where the bed of the river is formed entirely of solid rock and encamped on an island just above.

Clark   a fine morning, wind from the NE. The river more rapid and Sholey than yesterday. Passed a bold Stream which heads in the mountains to our right and the drean of the minting Snow in the Montn. on that side are in View— the Greater portion of the Snow on this mountain is melted…no wood being near the Snow.

Gass   A fine cool morning. The night was disagreeably cold.

Ordway   a clear morning. Passed verry rapid water. The mountains a Short distance South of us   Some Spots of Snow on it. The day pleasant and warm. Passed a large Spring run which is made by the Snow on the Mountains and runs from the foot of the Mo. through a Smooth plain.

Whitehouse   a clear morning. The day proved pleasant & warm.

SUNDAY, AUGUST 4

	Sunrise			4 p.m.			Jefferson River	
Temp	Weather	Wind	Temp	Weather	Wind	Rise/Fall	Feet	Inches
48a	f	SW	92a	f	S	F		1/2

Lewis   passed a bould runing creek about 12 yard wide and the water [cold] and remarkably clear. The middle fork…its water is much warmer than that of the rappid fork and somewhat turbid, from which I concluded that it had [its] source at a greater distance in the mountains and passed through an opener country than the other [forks of the Jefferson River].

Clark   a fine morning cool. Murcury at Sun rise 39 a. 0

Gass   a fine morning.

Ordway   a clear morning. The rapids continue.   Some of the mountains near the River has been burned by the natives Some time ago. The pine timber killed. The cotten timber in some of the R. bottoms killd. & dry also.

Whitehouse   a clear cool morning. Some of the mountains on the South side of the River has had the Grass burned off from them & the Timber killed on them, some time ago.

Note   *On August 5, they record the last entry of the Missouri River rise and fall in the Weather Diary. No explanation is given as to why they stopped. The expedition is located at the Big Hole River near the present-day town of Twin Bridges. They had followed that river up a short distance before returning to the Jefferson River. Since the river was becoming shallow and observation changes were minimal, it could be assumed that they decided the remarks were no longer necessary.*

MONDAY, AUGUST 5

	Sunrise			4 p.m.			Jefferson River	
Temp	Weather	Wind	Temp	Weather	Wind	Rise/Fall	Feet	Inches
49a	f	SE	79a	f	SE	F		1/2

Weather Diary

Lewis   the mountains put in close on both sides and arrose to great hight, partically covered with snow.

Clark   a Cold Clear morning, the wind from the SE. The river Streight & much more rapid than yesterday. 4 oClock PM Murcury 49 ab. 0

Ordway   a clear cool morning. passd. rapids as usal. the wind cold from the South. Passed over rapids covred with Slippery Small Stone and gravel. Passed over rapids worse than ever   it is with difficulty & hard fatigue we git up them Some of which are allmost perpinticular 3 or 4 feet in a Short distance. About 7 oC. PM cloud up high wind   it appears this little Stream is verry high, but has been high by the Snow melting off the Mountains.   It is now falling a little.

Whitehouse   a clear cool morning. The wind blew cold from the South. At 1 PM clouded up. Wind high. We expect this little Stream is high from the Snow melting on the mountains. It appears it has lately been higher, but is now falling a little.

TUESDAY, AUGUST 6

	Sunrise			4 p.m.			River	
Temp	Weather	Wind	Temp	Weather	Wind	Rise/Fall	Feet	Inches
52a	f	SW	71a	c	SW			

Lewis   We believe that the NW or rapid fork is the dane [drain] of the melting snows of the mountains.

Clark   a Clear morning, Cool, wind from the SW. This evening Cool...a Violent wind from the NW accompanied with rain which lasted half an hour. Wind NW.

Ordway   a clear morning.

Whitehouse   a clear morning. After 4 oClock PM...and some time after we had a small Shower of rain.

WEDNESDAY, AUGUST 7

	Sunrise			4 p.m.			River	
Temp	Weather	Wind	Temp	Weather	Wind	Rise/Fall	Feet	Inches
54a	c a r	SW	80a	c	SW			

Weather Diary   Thunder shower last evening form the NW. The river which we are now ascending is so inconsiderable and the curant So much on a Stand that I relinquished paying further attention to [its] State.

Lewis the morning being fair we spread our stores to dry at an early hour. We had a shower of rain which continued about 40 minutes attended with thunder and lightning. This shower wet me perfectly before I reached the camp. The clouds continued during the night in such manner that I was unable to obtain any lunar observations.

Clark a fine morning. At 5 oClock a thunder Storm from the NW accompanied with rain which lasted about 40 minits—

Gass We remained here during the forenoon, which was fair and clear. In the evening a heavy cloud came up, and we had hard thunder with lightening and rain. The weather cleared (early evening), and we had a fine night.

Ordway a clear morning. The morning cool, but the day warm. We had Thunder Showers this afternoon, attended with high winds.

Whitehouse a clear cool morning. This day was very warm & the party was much troubled with large horse flies— the rapids not So bad. We had Thunder Showers & high winds this afternoon.

THURSDAY, AUGUST 8

	Sunrise			4 p.m.			River	
Temp	Weather	Wind	Temp	Weather	Wind	Rise/Fall	Feet	Inches
54a	f a r	SW	82a	c a f	SW			

Weather Diary a thunder Shower last evening.

Lewis We had a heavy dew this morning. The evening again proved cloudy much to my mortification and prevented my making any lunar observations. The Indian woman recognized the point of a high plain to our right...this hill she says here nation calls the beaver's head from a conceived remblance of [its] figure to the head of that animal.

Clark wind from the SW. The Thermometer at 52 a 0 at Sunrise.

Gass a pleasant morning. The river is very crooked in this valley.

Ordway a clear cold morning. A heavy diew. The prarie is covred with grass which is high in places. The day warm. On the River, which is verry crooked but not So rapid as below, and only 25 yards wide, and verry crooked

Whitehouse a clear cold morning. Saw a little Snow on the knobs & mountains which lay but a short distance from us. The day proved warm & pleasant.

FRIDAY, AUGUST 9

	Sunrise			4 p.m.			River	
Temp	Weather	Wind	Temp	Weather	Wind	Rise/Fall	Feet	Inches
58a	f	NE	78a	c	SW			

Weather Diary   Encamped below the Forks Jeffersons River   Set out on a part of discovery

Lewis   the morning was fair and fine. The current of the river increasing in rapidity towards evening. In the evening it clouded up and we experienced a slight rain attended with some thunder and lightning. The musquetoes very troublesome this evening.

Clark   a fine morning,   wind from the NE. in the evening Clouded up and a fiew drops of rain

Gass   a fine morning with some dew. The river is narrow and very crooked

Ordway   a clear cool morning. The wind high from SE. Some Thunder. Saw Snow on the Mountains Some distance a head.   Some showers passed over.

Whitehouse   a clear cool morning. The wind high from the SE. Some Thunder in the afternoon. Saw Snow on the Mountains Some distance a head. In the evening we had some Showers of rain accompanied with Thunder with passed round or over. We all expect that we are near the head Waters or source of the Mesouri River, as the River, here is growing much narrower that it was.

Note   *The expedition passes present-day Dillon, Montana.*

SATURDAY, AUGUST 10

	Sunrise			4 p.m.			River	
Temp	Weather	Wind	Temp	Weather	Wind	Rise/Fall	Feet	Inches
60a	c a r f & l	SW	68a	t l & r	SW			

Weather Diary   rain Commenced at 6 PM and continued Showery through out the night. Musqueters bad.

Lewis   The mountains do not appear very high in any direction tho' the tops of some of them are partially covered with snow. This convinces me that we have ascended to a great hight since we have entered the rocky Mountains, yet the ascent has been so gradual along the vallies that it was scarcely perceptable by land. I do not believe that the world can furnish an example of a river runing to the extent which the Missouri and Jefferson's rivers do through such a mountainous country and at the same time so navigable as they are.

Clark   Some rain this morning at Sun rise and Cloudy. At 4 oClock a hard rain from the SW accompanied with Hail Continued half an hour, all wet, the men Sheltered themselves from the hail with bushes. River narrow, & Sholey but not rapid.

Gass   a fine morning. At 1 o'clock we halted t[o] dine, when a shower of rain came on with thunder and lightening, and continued an hour, during which some hail fell.

Ordway   a clear pleasant morning. A hard Thunder Show arose of rain and large hail which lasted nearly an hour.

Whitehouse   a clear pleasant morning. About 1 oClock…had a hard Thunder Shower of large hail and rain.

SUNDAY, AUGUST 11

	Sunrise			4 p.m.			River	
Temp	Weather	Wind	Temp	Weather	Wind	Rise/Fall	Feet	Inches
58a	c a r & j	NE	70a	f	SW			

Weather Diary   heavy Dew last evening

Lewis   Discovered an Indian on horseback…he suddonly turned his horse about, gave him a whip leaped the creek and disapeared.…we now set out on the track of the horse…halted…cooked breakfast.…before we had finised our meal, a heavy shower of rain came on with some hail which continued about 20 minutes and wet us to the skin, after this shower we pursued the track of the horse but as the rain had raised the grass which he had trodden down it was with difficult that we follow it.

Clark   A Shower of rain this morning at Sun rise, Cloudy all the morning wind from the SW.   the river shallow and rapid. Passed a large Island which I call the 3000 miles Island.

Gass   This morning was cloudy and we did not set out untill after breakfast.   About 2 some rain fell.

Ordway   a wet rainy morning. The day warm.   We Saw high Mountains a head some distance large Spots of Snow on them.

Whitehouse   a cool cloudy morning & some Rain. They turned warm & large flies became very troublesome. We see mountains, lying a head of us some Short distance; which appear very high, and large spots of snow on them.

Note   *Lewis's advance party crosses the Continental Divide at Lemhi Pass, the present-day Montana–Idaho state line.*

MONDAY, AUGUST 12

	Sunrise			4 p.m.			River	
Temp	Weather	Wind	Temp	Weather	Wind	Rise/Fall	Feet	Inches
58a	f a r & j	W	72a	f a r a h	NW			

Lewis   two miles below McNeal had exultingly stood with a foot on each side of this little rivulet and thanked his god that he had lived to bestride the mighty & heretofore deemed endless Missouri. After refreshing ourselves we proceeded on to the top of the dividing ridge from which I discovered immence ranges of high mountains still to the West of us with their tops partially covered with snow. Here I first tasted the water of the great Columbia river.

This morning Capt. Clark set out early. Found the river shoally, rapid shallow, and extreemly difficult.   At noon they had a thunderstorm which continued about half an hour.

Clark   We Set out early (Wind NE). The weather Cool

Gass   A few drops of rain fell to day.

Ordway   a clear morning. The current verry rapid.   Some of these rapids is deep and dangerous to pass up. We had a hard Thunder Shower   rained some time.

Whitehouse   A clear morning. The current verry rapid. About 2 oClock PM a hard Thunder Shower arose rained hard a Short time.

TUESDAY, AUGUST 13

	Sunrise			4 p.m.			River		
Temp	Weather	Wind	Temp	Weather	Wind	Rise/Fall	Feet	Inches
52a	c a f	NW	70a	f a r	NW			

Weather Diary   very cold last night. Passed the dividing ridge to the waters of the Columbia river.

Lewis   [Meets the Shoshones] the sun was verry warm and no water at hand. This was the first salmon I had seen and perfectly convinced me that we were on the waters of the Pacific Ocean.

Clark   a verry Cool morning. The Thermometer Stood at 52 a 0. All the fore part of the day Cloudy. At 8 oClock a mist of rain

Gass   A cloudy morning.   The weather was cold during the whole of this day.

Ordway   Cloudy. The current rapid.   In the afternoon the current more gentle.

Whitehouse   Cloudy. The current of the river running very rapid

WEDNESDAY, AUGUST 14

	Sunrise			4 p.m.			River		
Temp	Weather	Wind	Temp	Weather	Wind	Rise/Fall	Feet	Inches
51a	f a r	NW	76a	f	NW			

Lewis   [Indian Chief describes route to Clearwater River, Lewis notes the following] the next part of the journey of the rout was about 10 days through a dry and parched sandy desert...the sun had now dryed up the little pools of water which exist through this desert plain in the spring season and had also scorched all the grass.

Clark   a Cold morning, wind from the SW. Thermometer Stood at 51° a 0, at Sunrise the morning being cold and men Stiff. I deturmind to dealy & take brackfast...we set out at 7 oClock. River verry Crooked and rapid. Stream on the Stard. Side which head in a mountain to the North on which there is Snow.

Gass   The morning was clear and cold.

Ordway   a clear cold morning.   The water is verry cold.   The wind high from SW the current continued rapid all day.

Whitehouse   a clear cold morning. Water in the river is clear and Cold.

THURSDAY, AUGUST 15

	Sunrise			4 p.m.			River		
Temp	Weather	Wind	Temp	Weather	Wind	Rise/Fall	Feet	Inches
43a*	f	SE	74a	f	SW			

Weather Diary   remarkably cold this morning

Lewis   [On Horse Prairie Creek near present-day Grant, Montana] the Cove is called Shoshone Cove. The grass being birned on the North side of the river we passed over...as I came up this cove.

Clark   a Cool windey morning, wind from the SW.

Gass   a fine morning. The river meanders   The water is very cold and severe and disagreeable to the men, who are frequently obliged to wade and drag the canoes

Ordway   clear & cold this morning. We passed Several bad rapids.

Whitehouse   a cold clear morning. The river shallow.

Note   *The main party establishes Camp Fortunate at present-day Clark Reservoir, Montana.*

FRIDAY, AUGUST 16

	Sunrise			4 p.m.			River		
Temp	Weather	Wind	Temp	Weather	Wind	Rise/Fall	Feet	Inches
48a	f	SW	70a	f	SW			

*According to Moulton (1988, 5: 182), Clark's Journal Codex I lists this temperature as "52 a."

Weather Diary   Capt Lewis Join with the Snake Indians at the forks

Clark   as this morning was cold and the men fatigued Stiff and Chilled…detained till 7 oClock. The Thmtr. Stood at 48° a 0 at Sunrise, wind SW. The water excessively cold.

Gass   we proceed through rapid water, the river is very narrow, crooked and shallow.

Ordway   a clear morning but verry cold. The Thurmometer Stood at 47°. The water So cold that we delayed untill after we took breakfast.   We find the current Swift the river Shallow.   The water not So Swift above the bad rapid.

Whitehouse   a clear but verry cold morning. The thermometer stood at 47° the river water So cold that we delayed untill after breakfast

SATURDAY, AUGUST 17

Sunrise			4 p.m.			River		
Temp	Weather	Wind	Temp	Weather	Wind	Rise/Fall	Feet	Inches
42a	f	NE	76a	f	SW			

Lewis   the nights are very cold and the sun excessively hot in the day.

Clark   a fair Cold morning, wind SW. The Thermometer at 42 a. 0 at Sunrise, set out at 7 oClock

Gass   a fine morning.   The water so shallow

Ordway   a clear cold morning. We have been cold this Several nights under 2 blankets or Robes, over us. A little white frost. The air chilley & cold.

Whitehouse   a clear cold morning. The weather was so cold last night, that our party had to lay under 2 buffalo robes each in order to keep themselves warm. Some frost this morning.

Note   *Clark takes at least twelve men in an advanced party to the Salmon River to determine if it is navigatable and to build canoes. They return back up the Salmon from August 18 to 29 and meet the main party near Tendoy, Idaho. It appears Clark copied the Weather Diary information when he returned from the scouting trip, but his daily narrative remarks, as well as Sergeant Gass's, are separated from Lewis and the main group.*

SUNDAY, AUGUST 18

Sunrise			4 p.m.			River		
Temp	Weather	Wind	Temp	Weather	Wind	Rise/Fall	Feet	Inches
45a	c	SW	78a	r	SW			

Lewis   [his birthday] …resolved in future, to redouble my exertions and at least indeavour to promote those two primary objects of human existence, by giving them the aid of that portion of talents which nature and fortune have bestoed on me; or in future, to live for mankind, as I have heretofore lived for myself—

Ordway   a clear morning.   We had Showers of Rain this afternoon.

Whitehouse   a clear morning.   We had some Showers of rain this afternoon—

Clark Scouting Party

Clark   the fore part of the day w[a]rm, at 12 oClock it became hasey with a mist of rain, wind hard from the SW and Cold which increased untill night.   The rain Seased in about two hours.

Gass   a fine morning.   At three o'clock this afternoon there was a violent gust of wind, and some rain fell.   In about an hour the weather became clear, and very cold, and continued cold all night.

MONDAY, AUGUST 19

Sunrise			4 p.m.			River		
Temp	Weather	Wind	Temp	Weather	Wind	Rise/Fall	Feet	Inches
30a	f a r	SW	71a	f a r	SW			

Weather Diary   ice on Standing water $^1/_8$ of an inch thick.

Lewis   the frost which perfectly whitened the grass this morning had a singular appearance to me at this season.

Ordway   a clear cold morning.   A white frost & the grass Stiff with frost it being disagreeably cold.   The day pleasant & warm.   Light showers of rain this evening.

Whitehouse   a cold morning. A white frost this morning. A clear pleasant day. We had a shower of rain about 3 oClock PM which lasted but a few minutes.

Clark Scouting Party

Clark   A verry Cold morning, Frost to be seen

Gass   A fine morning, but cold.   At 1 o'clock we dined at the head spring of the Missouri and Jefferson river.   A bout 5 miles south of us we saw snow on top of a mountain, and in the morning there was a severe white frost, but the sun shines very warm where we now are.   It is not more than a mile from the head spring of the Missouri to the head of one of the branches of the Columbia.

TUESDAY, AUGUST 20

Sunrise			4 p.m.			River		
Temp	Weather	Wind	Temp	Weather	Wind	Rise/Fall	Feet	Inches
32a	f a r *	SW	74a	f	SW			

---

*According to Moulton (1988, 5: 182), Clark's Journal Codex C list this weather data as "f."

Weather Diary   hard frost last night.
Ordway   a clear cold morning.
Whitehouse   a clear cold morning. A white frost.

Clark Scouting Party

Clark   Frost last night
Gass   A fine cool frosty morning.

WEDNESDAY, AUGUST 21

Sunrise			4 p.m.			River		
Temp	Weather	Wind	Temp	Weather	Wind	Rise/Fall	Feet	Inches
19a	f	SE	78a	f	E			

Weather Diary   ice $1/2$ an inch thick on standing water. Most astonishing difference between the hight of the Murcury at Sunrise and at 4 PM today
there was 59° and this in the Space of 8 hours, yet we experience this wonderfull transicion without feeling it near so Sensibly as I should have expected.

Lewis   This morning was very cold. The ice $1/4$ of an inch thick on the water which stood in the vessels exposed to the air. Some wet deerskins that had been spread on the grass last evening are stiffly frozen. The ink feizes in my pen. The bottoms are perfectly covered with frost, insomuch that they appear to be covered with snow. notwithstanding the coldness of the last night the day has proved excessively warm.

Ordway   the ground is covered with a hard white frost. The water which stood in the Small vessells froze $1/4$ of an Inch thick, a little. Some Deer Skins which was spread out we last night are froze Stiff this morning. The Ink freezes in my pen now   the Sun just ariseing   clear and pleasant this morning

Whitehouse   We had a hard white frost this morning. The water that stood in small Vessells froze, and some Deer Skins which was spread out wet last night, froze Stiff this morning, & the Ink froze in the pen at Sun rise; the morning was clear & got pleasant. At 8 oClock AM some of the party found Ice in some standing water $1/4$ of an inch thick

Clark Scouting Party

Clark   Frost last night. I shall in justice to Capt. Lewis who was the first white man ever on this fork of the Columbia Call this Louis's river [Salmon River]. The Westerley fork of the Columbia River is double the Size of the Easterley fork & below those forks the river is about the Size Jeffersons River near its mouth or 100 yards wide, it is verry rapid & clear

Gass   at this place the river is about 70 yard wide [Salmon River near Carmen, Idaho]

THURSDAY, AUGUST 22

	Sunrise			4 p.m.		River		
Temp	Weather	Wind	Temp	Weather	Wind	Rise/Fall	Feet	Inches
22a	f	E	70a	f	E			

Weather Diary   Snow yet appears on the summits of the mountains.

Ordway   a white frost and cold as usal.

Whitehouse   a white frost & cold as usual in the morning. The morning clear and got pleasant.

Clark Scouting Party

Clark   mountains verry Steap high & rockey, the assent of three was So Steap that it is incredible to describe the rocks in maney places.

Gass   The morning was fine, with a great white frost. We proceeded down the river; but with a great deal of difficultly; the mountains being so close, steep and rocky. The river here is about 80 yards wide, and a continual rapid but not deep [Salmon River above Shoup, Idaho]

FRIDAY, AUGUST 23

	Sunrise			4 p.m.		River		
Temp	Weather	Wind	Temp	Weather	Wind	Rise/Fall	Feet	Inches
35a	f	E	72a	f	SE			

Weather Diary   white frost this morning

Lewis   the season is now far advanced to remain in these mountains as the Indians inform us we shall shortly have snow; the salmon have so far declined that they are themselves haistening from the country. The bends of the river are short and the currant beats from side to side against the rocks with great violence. The river is about 100 yds. wide and so deep that it cannot be foarded but in a few places. The sides of the mountains are very steep, and the torrents of water which roll down their sides at certain seasons appear to carry with them vast quantities of the loose stone into the river. This view was terminated by one of the most lofty mountains, Capt. C. informed me, he had ever seen which was perfectly covered with snow.

Ordway   a clear pleasant morning.

Whitehouse   a clear pleasant morning.

Clark Scouting Party

Clark   [Salmon River] Current So Strong that is dangerous crossing the river. 100 yards wide. The mountains Cloe and is perpendicular Clift on each Side, and Continues for a great distance and that the water runs with great violence from one rock to the other on each Side foaming & roreing thro rocks in every direction, So as to render the passage of any thing impossible. The Hills or mountains were not like those I had Seen but like the Side of a tree Streight up— The Torrents of water which came down after a rain carries with it emence numbers of those Stone into the river.

Gass   We proceeded down the river through dreadful narrows, where the rocks were in some places breast high, and no path or trial of any kind.

SATURDAY, AUGUST 24

Sunrise			4 p.m.			River		
Temp	Weather	Wind	Temp	Weather	Wind	Rise/Fall	Feet	Inches
40a	f	SE	76a	f a r	SE			

Weather Diary   Set out with the Indians and pack horses for the Columbia river

Lewis   I had now the inexpressible satisfaction to find myself once more under way with all my baggage and party. after we encamped we had a slight shower of rain.

Ordway   a clear cold morning. We had a Small Shower of rain

Whitehouse   a clear cold morning.

Clark Scouting Party

Clark   marked my name on a pine Tree.   Every man appeared disheartened from the prospects of the river.

Gass   had a pleasant morning. The river at this place is so confined by the mountains that it is not more than 20 yard wide and very rapid. The water is so rapid and the bed of the river so rocky, that going by water appeared impracticable; and the mountains so amazingly high, steep and rocky, that it seemed impossible to go along the river by land.

SUNDAY, AUGUST 25

Sunrise			4 p.m.			River		
Temp	Weather	Wind	Temp	Weather	Wind	Rise/Fall	Feet	Inches
32a	f a r	SE	65a	c	SE			

Weather Diary   white frost this morning,

Lewis   This morning while passing through the Shone cove Frazier fired his musquet at some ducks in a little pond at the distance of about 60 yards from me; the ball rebounded from the water and pased within a very few feet of me. This part of the cove on the NE side of the Creek has lately been birned by the Indians as a signal on some occasion.

Ordway   a clear morning.   Some frost.
Whitehouse   a clear morning a little light frost last night.

Clark Scouting Party

Clark   Mountain to our right...Pine timber which is thick on that side
Gass   a fine morning.

Note   *The main party crosses the Continental Divide at Lemhi Pass, at today's Montana–Idaho state line, proceeds down the Lemhi and Salmon rivers, and then up the north fork of the Salmon River.*

MONDAY, AUGUST 26

Sunrise			4 p.m.			River		
Temp	Weather	Wind	Temp	Weather	Wind	Rise/Fall	Feet	Inches
31a	f	SE	45a	f	SE			

Weather Diary   hard white frost and some ice on standing water this morning   arrived with baggage and party on the Columbia river at 5 PM

Lewis   This morning was excessively cold; there was ice on the vessels of water which stood exposed to the air nearly a quarter of an inch thick. We soon arrived at the extreem source of the Missouri; here I halted a few minutes, the men drank of the water and consoled themselves with the idea of having at length arrived at this long wished for point. From hence we proceeded to a fine spring on the side of the mountain [west side]...here I halted to dine and graize our horses, there being fine green grass on that part of the hillside which was moistened by the water of the spring while the grass on the other parts was perfectly dry and parched with the sun. [...Lewis ponders the Native American birthing process...] — if a pure and dry air, an elivated and cold country is unfavourable to childbirth, we might expect every difficult incident to that operation of nature in this part of the continent. — The tops of the high and irregular mountains which present themselves to our view on the opposite side of this branch of the Columbia are yet perfectly covered with snow; the air which proceeds from those mountains has an agreeable coolness and renders these parched and South hillsides much more supportable at this time of the day it being now about noon.

Ordway   a clear cold morning. The water in the Small vessells froze. We Set out at Sunrise and proceeded on with out big coats on and our fingers ackd with the Cold. Saw considerable of Snow on the mountain near us which appear but little higher than we are. It lies in heaps and a cold breeze always comes from these mountains.

Whitehouse   a clear morning, we find it verry cold and a heavy frost every morning. The water froze a little in the Small Vessells. Crossed a high ridge between the Mesouri and Calumbian River. Saw high mountains to the SW with Some Spots of Snow on them.

Clark Scouting Party

Clark   a fine morning.
Gass   a pleasant morning.

Tuesday, August 27

	Sunrise			4 p.m.			River	
Temp	Weather	Wind	Temp	Weather	Wind	Rise/Fall	Feet	Inches
32a	f	SE	56a	f	SE			

Weather Diary   hard frost white this morning.   on the Columbian waters (Clark).

Ordway   a beautiful pleasant morning. Snow now lying a Short distance to the South of us on the broken mountains.

Whitehouse   a beautiful pleasant morning.

Clark Scouting Party

Clark   Some frost this morning.
Gass   A fine morning with frost

Wednesday, August 28

	Sunrise			4 p.m.			River	
Temp	Weather	Wind	Temp	Weather	Wind	Rise/Fall	Feet	Inches
35a	f	SW	66a	f	SW			

Ordway   a clear pleasant morning.

Whitehouse   a clear pleasant morning. Some Spots of Snow continues to lay on the mountains a fiew miles to the South of us.

Clark Scouting Party

Clark   a frost this morning.
Gass   The morning again was pleasant. I found the weather very cold for the season.

Thursday, August 29

	Sunrise			4 p.m.			River	
Temp	Weather	Wind	Temp	Weather	Wind	Rise/Fall	Feet	Inches
32a	f	SW	68a	f	SW			

Ordway   a clear pleasant morning. They find that the mountains are So bad that we cannot follow the river by land and the river So rapid and full of rocks that it is impossable for crafts to pass down.

Whitehouse   a clear pleasant morning.

Clark Scouting Party

Clark   a Cold morning Some frost. The Wind from the South.

Gass   There was a severe white frost this morning.

FRIDAY, AUGUST 30

	Sunrise			4 p.m.			River	
Temp	Weather	Wind	Temp	Weather	Wind	Rise/Fall	Feet	Inches
34a	c	NE	59a	c	NE			

Weather Diary   Set out with the party by land at 2 PM.

Clark   a fine Morning.

Ordway   a fine morning.

Whitehouse   a clear pleasant morning.

SATURDAY, AUGUST 31

	Sunrise			4 p.m.			River	
Temp	Weather	Wind	Temp	Weather	Wind	Rise/Fall	Feet	Inches
38a	c a r	NE	58a	c a r & h	NE			

Clark   A fine morning. The wind hard from the SW. This day warm and Sultrey. Praries or open Valies on fire in Several places— The Countery is Set on fire for the purpose of Collecting the different bands

Ordway   a fare morning. The River bottoms narrow and verry much dryed up.

Whitehouse   a fine morning

Note   *The expedition starts toward the Lost Trail Pass along the North Fork of the Salmon River, near the future mining town of Gibbonsville, Idaho.*

SUNDAY, SEPTEMBER 1

	Sunrise			4 p.m.			River		
Temp	Weather	Wind	Temp	Weather	Wind	Rise/Fall	Feet	Inches
38a	c	NW	67a	c	NW			

Weather Diary   Service berries dried on the bushes abundant and very fine.   black colour.

Clark   a fine morning. Some rain to day at 12 and in the evening which obliges us to Continu all night.

Gass   a fine morning.   At noon some rain fell, and the day continued cloudy. At 3 o'clock….we halted the weather became cloudy, and considerable quantity of rain fell.

Ordway   Cloudy. In the afternoon we had Several Shower of rain and a little hail.   Several Small Showers of rain this evening—

Whitehouse   a fine clear morning. We stopped and camped about 3 hours before night on account of its raining. We passed across several large Creeks, the water of which was very Cold. During this afternoon we had several small Showers of rain—

MONDAY, SEPTEMBER 2

	Sunrise			4 p.m.			River		
Temp	Weather	Wind	Temp	Weather	Wind	Rise/Fall	Feet	Inches
36a	c a r	NE	60a	c a r h	NE			

Clark   a Cloudy Morning, rained Some last night. Camped. Some rain at night

Gass   The morning was cloudy. In the afternoon we had a good deal of rain, and the worst road (if road it can be called) that was ever travelled.

Ordway   a cloudy wet morning. This is a verry lonesome place.

Whitehouse   a wet cloudy morning.

Note   *The expedition crosses the Lost Trail Pass at the Idaho–Montana state line. The journalists note that the expedition's last thermometer was broken during an accident as they proceeded up the Lost Trail pass on September 3. However, this is not recorded in the Weather Dairy remarks section until September 6. Temperature observations are noted for two additional days in the Weather Diary, but the readings may be suspect. No further temperature observations are recorded for the rest of the journey.*

TUESDAY, SEPTEMBER 3

	Sunrise			4 p.m.			River	
Temp	Weather	Wind	Temp	Weather	Wind	Rise/Fall	Feet	Inches
34a	c a r	NE	52a	c a r	NE			

Weather Diary   Choke Cherries ripe and abundant.

Clark   a Cloudy morning. we assended after crossing Several Steep points & one mountain, but little to eate at dusk it began to Snow, at 3 oClock Some Rain. The last mountains we had passed to the East Covered with Snow.   We met with a great misfortune, in haveing our last Th[er]mometer broken by accident. This day we passed over emence hils and Some of the worst roads that ever horses passed, our horses frequently fell. Snow about 2 inches deep when it began to rain which termonated in a Sleet

Gass   The morning of this day was cloudy and cool. We halted for dinner….we staid here about two hours, during which time some rain fell, and the weather was extremely cold for the season. This was not the creek our guide wished to have come upon; and to add to our misfortunes we had a cold evening with rain.

Ordway   Several Small Showers of rain. So we lay down wet hungry and cold.

Whitehouse   We had a cloudy morning & set out as usual. We crossed a dividing ridge [Lost Trail Pass]. At dark it began to rain hard, We lay down to sleep being wet, hungry & Cold. Saw Snow on the tops of Some of these mountains this day.

WEDNESDAY, SEPTEMBER 4

	Sunrise			4 p.m.			River	
Temp	Weather	Wind	Temp	Weather	Wind	Rise/Fall	Feet	Inches
19a	r a s	NE	34a	c a r	NE			

Weather Diary   ice one inch thick.

Clark   a verry cold morning every thing wet and frosed, we detained untill 8 oClock to thaw the covering for the baggage &c &c. Ground covered with Snow. I was the first white man who ever wer on the water of this river [Bitterroot River]

Gass   A considerable quantity of snow fell last night, and the morning was cloudy.

Ordway   the morning clear, but very cold.   The ground covred with frost.   Our moccasins froze. The mountains covred with Snow. the Snow over our mockasons in places. The air on the mountains verry chilley and cold. Our fingers aked with the cold.

Whitehouse   the morning clear but verry cold. Our mockersons froze hard. The mountains covred with Snow. The Snow lay on the mountain So that it stuck to our mockisons   The air verry cold   our fingers aked with the cold.

THURSDAY, SEPTEMBER 5

Sunrise			4 p.m.			River		
Temp	Weather	Wind	Temp	Weather	Wind	Rise/Fall	Feet	Inches
17a	c a s	NE	29a	c a r & s	NE			

Weather Diary   Ground Covered with Snow.

Clark   a Cloudy morning

Gass   This was a fine morning with a great white frost.

Ordway   a clear cold morning. The Standing water froze a little.   A hard white frost this morning.

Whitehouse   a clear cold morning. The Standing water we had in our small Vessells froze a little last night.

FRIDAY, SEPTEMBER 6

Sunrise			4 p.m.			River		
Temp	Weather	Wind	Temp	Weather	Wind	Rise/Fall	Feet	Inches
	c a r	NE		r	NE			

Weather Diary   Thermometer broke by the Box strikeing against a tree in the Rocky Mountains

Clark   Some little rain. Rained contd. untill 12 oClock. rained this evening

Gass   A cloudy morning. About 12 o'clock some rain fell

Ordway   a clear cold morning. Light Sprinkling of rain, through the course of this day—

Whitehouse   a clear cold morning. Light Sprinklings of rain through the course of the day.

SATURDAY, SEPTEMBER 7

Sunrise			4 p.m.			River		
Temp	Weather	Wind	Temp	Weather	Wind	Rise/Fall	Feet	Inches
	c a r	NE		c a r	NE			

Clark   A Cloudy & rainie Day. The Vallie from 1 to 3 miles wide the Snow top mountains to our left, open hilley Countrey on the right.

Gass   We set out early in a cloudy cool morning. 12 o'clock...some rain fell.   Some rain fell in the afternoon

Ordway   a cloudy cold morning. High mountains a little to the Lard. Side which is covred thick with Snow.   We had Several Showers of rain.

Whitehouse   a cloudy cold morning. The high mountains...Several Small Showers of rain in the course of the day.

SUNDAY, SEPTEMBER 8

Sunrise			4 p.m.			River		
Temp	Weather	Wind	Temp	Weather	Wind	Rise/Fall	Feet	Inches
	c	NE		*c a r*	NE			

Weather Diary   Mountains Covered with Snow to the SW a singular kind of Prickly Pears.

Clark   a Cloudy morning. The wind from the NW & Cold. The foot of the Snow mountains approach the River on the left Side. Some Snow on the mountain to the right also. A hard rain all the evening we are all Cold and wet.

Gass   The morning was wet   had a cold, wet disagreeable afternoon

Ordway   cloudy and cold. Saw Snow on the mount to our left. High barron hills to our right.   The wind cold from the NW & Showers of rain, and a little hail. The mountains are rough on each side and are covred with pin[e] and on the tops of which are covred. with Snow.

Whitehouse   cloudy and verry chilley and cold. Saw Snow on the Mountains to our left. The wind from NW & the Air chilley and cold. The Snow lays thick on the mountains a little to our left....the tops of which are covered with Snow and at places appear to lay thick.

Note   *The expedition arrives at Travelers Rest, near present-day Lolo, Montana.*

MONDAY, SEPTEMBER 9

Sunrise			4 p.m.			River		
Temp	Weather	Wind	Temp	Weather	Wind	Rise/Fall	Feet	Inches
	c a r	NE		*f a r*	NE			

Weather Diary   arrived at travelers rest Creek

Lewis   as our guide informes that we should leave the river at this place and the weather appearing settled and fair I determined to halt the next day rest our horses and take some scelestial Observations. We called this Creek Travellers rest.

Clark a fair morning. Day fair, wind NW. The foot of the Snow toped mountains approach near the river.

Gass The morning was fair, but cool. The nigh snow-topped mountains are still in view on our left. The Flathead River [Bitterroot River] is 100 yards wide.

Ordway Cloudy. The Snow continues on the Mont. each side of this valley.

Whitehouse a cloudy cold morning, wind blew from the NW. The Snow continues on the Mountains on both sides of the valley. The afternoon pleasant, but the Snow Still continues on the Mountains as usul.

TUESDAY, SEPTEMBER 10

	Sunrise			4 p.m.			River	
Temp	Weather	Wind	Temp	Weather	Wind	Rise/Fall	Feet	Inches
	f	*NW*		*f*	*NW*			

Lewis The morning being fair I sent out all the hunters

Clark a fair morning. The day proved fair

Gass We remained here all this day, which was clear and pleasant.

Ordway a fair morning. The day warm.

Whitehouse a clear pleasant morning, and the weather moderate, not So cold as usal. As our road now leads over a mountain to our left, our Captains conclude to Stay here this day to take observations, and for the hunters to kill meat to last us across the mountains and for our horses to rest, etc….the day continued to grow warm, but the Snow does not melt on the mt. a Short distance from us…The snow on the Mountains have the appearance of the Middle of winter.

Note   *The Expedition crosses the Bitterroot Mountains on the Lolo Trail from September 11 to 22.*

WEDNESDAY, SEPTEMBER 11

	Sunrise			4 p.m.			River	
Temp	Weather	Wind	Temp	Weather	Wind	Rise/Fall	Feet	Inches
	f	*NW*		*f*	*NW*			

Clark a fair morning, wind from the NW. nothing killed this evening hills on the right high & ruged, the mountains on the left high & Covered with Snow. The day Verry w[a]rm.

Gass   This was a fine morning.

Ordway   a clear pleasant morning.

Whitehouse   a beautiful pleasant morning. The snow on the mountains about 1 miles to the SW of us does not melt but verry little.

THURSDAY, SEPTEMBER 12

	Sunrise			4 p.m.			River	
Temp	Weather	Wind	Temp	Weather	Wind	Rise/Fall	Feet	Inches
	f	NW		*f*	NE			

Weather Diary   Mounts to our left Covered with Snow

Clark   a white frost, Set out at 7 oClock. Creek of fine clear water. The road through this hilley Countrey is verry bad passing over hills & thro' Steep hollows, over falling timber &c &c.

Gass   a fine morning.

Ordway   a fair morning.

Whitehouse   a white frost, and clear pleasant morning. Saw high Mountains to the South of us covred with Snow, which appears to lay their all the year round.

Note   *The expedition crosses the Lolo Pass at the present-day Montana–Idaho state line.*

FRIDAY, SEPTEMBER 13

	Sunrise			4 p.m.			River	
Temp	Weather	Wind	Temp	Weather	Wind	Rise/Fall	Feet	Inches
	c	NE		*r*	NE			

Weather Diary   a hot Spring

Clark   a cloudy morning. The after part of the day Cloudy. Passed Several Springs [Lolo Hot Springs]...I tasted this water and found it hot & not bad tasted. I found this water nearly boiling hot at the places it Spouted from rocks. I put my finger in the water, at first could not bare it in a Second— Those springs come out in maney places. Some mountains in view to the SE & SW Coverd with Snow.

Gass   A cloudy morning. We came to a most beautiful warm spring, the water of which is considerably above blood-heat; and I could not bear my hand in it without uneasiness. [At Lolo Hot Springs, Montana]

Ordway   cloudy.

Whitehouse   Cloudy weather. The day proved very pleasant.

SATURDAY, SEPTEMBER 14

	Sunrise			4 p.m.		River		
Temp	Weather	Wind	Temp	Weather	Wind	Rise/Fall	Feet	Inches
	c a r	*SW*		*c a r & s**	*SW*			

Weather Diary   killed and eat a colt, Snowed rained & hailed to day.

Clark   a Cloudy day in the Valies it rained and hailed, on the top of the mountains Some Snow fell. Rained, Snowed & hailed the greater part of the day all wet and Cold. Crossed a verry high Steep mountain for 9 miles to a large fork from the left which appears to head in the Snow toped mountains Southerly and SE. Camped. The rain

Gass   We set out early in a cloudy morning.   I saw service-berry bushes hanging full of fruit; but not yet ripe, owing to the coldness of the climate on these mountains. Encamped for the night, as it rained and was disagreeable travelling.

Ordway   a little Thunder hail and rain. Saw high Mountains covred with Snow and timber—

Whitehouse   A cloudy morning. We had towards evening Several light small Showers of rain and a little hail. Several severe claps of Thunder. Saw high mountains. A little to the South of us, which are covred with Snow.

SUNDAY, SEPTEMBER 15

	Sunrise			4 p.m.		River		
Temp	Weather	Wind	Temp	Weather	Wind	Rise/Fall	Feet	Inches
	c a l & s†	*SW*		*s*	*SW*			

Weather Diary   no water   we are obliged to Substitute the coald Snow   to boil our Colt.

Clark   ….here the road leaves the river to the left and assends a mountain winding in every direction to get up the Steep assents & to pass the emence quantity of falling timber which had [been] falling from dift. Causes I e fire & wind and has deprived the greater part of the Southerly Sides of this mountain of its green timber. Several horses slip…my portable desk broken…Some others verry much hurt, from this point I observed a range of high mountains Covered with Snow from SE to SW with Their top bald or void of timber. We could find no water and Concluded to Camp and make use of the Snow [w]e found on the top to cook the remns. Of our Colt & make our Supe. evening verry Cold and Cloudy. From this mountain I could observe high ruged mountains in every direction as far as I could see. Encamped on the top of the

---

* According to Moulton (1988, 5: 243), Lewis's Journal Codex P & Clark's Codex I lists this weather data as "c a r."

† According to Moulton (1988, 5: 243), Lewis's Journal Codex P & Clark's Codex I lists this weather data as "c a s."

mountain near a Bank of old Snow about 3 feet deep lying on the Northern side of the mountain. We melted the Snow to drink

Gass   Encamped on a high mountain...there was no water, but a bank of snow answered as a substitute.

Ordway   cloudy. We found Some Spots of Snow so we camped on the top of the Mountain and melted Some Snow. This Snow appears to lay all year on this Mount.

Whitehouse   This morning we had Cold weather & cloudy. We followed on the ridge of the Mountain & went over several high knobs on it, where the Wind had blown down most of the timber on them. On top of the mountain...some Spots of Snow. We encamped on a top ridge of the Mountain, where we found plenty of Snow, which appear to have lain here all year.

MONDAY, SEPTEMBER 16

	Sunrise			4 p.m.			River	
Temp	Weather	Wind	Temp	Weather	Wind	Rise/Fall	Feet	Inches
	c a s	SW		f	SW			

Weather Diary   Snow commenced about 4 oClock AM and continued untill night. It is about 7 inches deep. Ice one inch thick.   The Snow fell on the old Snow 4 inches deep last night.

Clark   began to Snow about 3 hours before Day and continued all day   The Snow in the morning 4 inches deep on the old Snow, and by night we found it from 6 to 8 inches deep. I walked in front to keep the road and found great dificuelty in keeping maney places the Snow had entirely filled up the track and obliged me to hunt Several minits for the track. As 12 oClock we halted on the top of the mountain to w[a]rm & dry our Selves. The pine, which are So covered with Snow, that is passing thro them we are continually covered with Snow, I have been wet and as cold in every part as I ever was in my life, indeed I was at one time fearfull my feet would freeze in the thin Mockirsons which I wore. Men all wet cold and hungary. To describe the road of this day would be a repitition of yesterday expt the Snow which made it much wors to proseed as we had in maney places to derect our Selves by the appearence of the rubbings of the Packs against the trees which have limbs quiet low and bending downwards

Gass   Last night about 12 o'clock it began to snow. We renewed our march early, though the morning was very disagreeable, and proceed over the most terrible mountains I ever beheld. It continued snowing untill 3'oclock PM.   The snow fell so thick, and the day was so dark, that a person could not see to a distance of 200 yards. In the night and during the day the snow fell about 10 inches deep.

Ordway   when we a woke this morning to our great Surpize we were covred with Snow, which had fell about 2 Inches deep the later part of last night, & continues a cold Snowey morning.   Could Scarsely keep the old trail for the Snow.   About one oClock finding no water we halted and melted Some snow and eat or drank a little more soup.   Saw considerable of old snow   The Snow is now about 4 inches deep on a levl.   The clouds So low on the Mount that we could not See any distance no way. It appeared as if we have been in the clouds all this day.

Whitehouse   When we awoke this morning to our great Surprize we were covred with Snow, which had fallen about 2 Inches the latter part of last night, & [it] continues a verry cold Snow Storm. Could hardly See the old trail for the Snow. The Snow fell so fast that it is now in common 5 or 6 Inches deep & where old Snow remained it was considerably deeper. We mended up our mockasons. Some of the men without Socks, wrapped rags on their feet, and loaded up our horses and Set out without anything to eat, and proceeded on.   It has quit Snowing this evening, but continues very chilley and cold.

Tuesday, September 17

	Sunrise			4 p.m.			River	
Temp	Weather	Wind	Temp	Weather	Wind	Rise/Fall	Feet	Inches
	f	SW		f	SW			

Clark   Cloudy morning. One oClock PM at which time we Set out the falling Snow & snow falling from the trees which kept us wet all the after noon. Snow on the Knobs, no Snow in the vallies. The after part of the day fare. Road emencely bad as usial, no Snow in the hollers   all the high knobs of the mounts Covered.

Gass   It was a fine day with warm sunshine, which melted the snow very fast on the south sides of the hills, and made the travelling very fatiguing and uncomfortable. We continued over high desert mountains.

Ordway   Cloudy and cold. The afternoon clear and pleasant & warm. The Snow melted fast.

Whitehouse   cloudy and cold. Set out about Noon, the Snow lay heavy on the timber. In the afternoon the weather cleared away & then it became clear & warm. The snow melted so that the water Stood in the trail over our mockasons in Some places & in some places it was very Slippy

Wednesday, September 18

	Sunrise			4 p.m.			River	
Temp	Weather	Wind	Temp	Weather	Wind	Rise/Fall	Feet	Inches
	f	SW		f	SW			

Weather Diary   hard black frost this morning

Lewis   used the snow for cooking—

Clark   The want of provisions together with the dificuely of passing those emence mountains damped the Spirits of the party which induced us to resort to Some plan of reviving ther Sperits. A fair morning, cold. From the top of a high part of the mountain at 20 miles I had a view of an emence Plain and leavel Countrey to the SW & West at a great distance [on top of Sherman Peak looking at prairies toward Grangeville & Lewiston Idaho]

Gass This was a clear cold frosty morning. About 12 we passed a part where the snow was off, and no appearance that much had lately fallen. At 3 we came to snow again, and halted to take some soup, which we made with snow water, as no other could be found. We see now prospect of getting off these desert mountains yet, except the appearance of a deep cove on each side of the ridge we are passing along.

Ordway a clear pleasant morning.

Whitehouse a clear pleasant morning. Melt a little Snow as we found not water to make a little Port. Soup. The weather moderated, & the snow melted a little. The Mountains appear a head of us as far as we can see & continue much further than we expected—

THURSDAY, SEPTEMBER 19

Sunrise			4 p.m.			River		
Temp	Weather	Wind	Temp	Weather	Wind	Rise/Fall	Feet	Inches
	f	SW		f	SW			

Weather Diary rose raspberry ripe and abundant. Snow is about 4 Inches deep.

Lewis this plain appeared to be about 60 miles distant.

Clark passed…two high mountains, ridges and through much falling timber. As we decend the mountain the heat becomes more proseptable every mile.

Gass last night was disagreeably cold. About 8 this morning….the sun shining warm and pleasant. The snow is chiefly gone except on the north points of the high mountains.

Ordway a clear morning.

Whitehouse a clear pleasant morning. The ground was covered with Snow & froze.

FRIDAY, SEPTEMBER 20

Sunrise			4 p.m.			River		
Temp	Weather	Wind	Temp	Weather	Wind	Rise/Fall	Feet	Inches
	f	Sw		f	SW			

Lewis our road was much obstructed by fallen timber particularly in the evening. Saw the hucklebury.

Gass day was fine

Ordway a cold frosty morning

Whitehouse a cold frosty morning.

SATURDAY, SEPTEMBER 21

	Sunrise			4 p.m.			River	
Temp	Weather	Wind	Temp	Weather	Wind	Rise/Fall	Feet	Inches
	f	SE		*f*	SW			

Weather Diary   I arrive at the Flat head Camp of 200 lodges in a Small prarie

Clark   a fine morning. The day proved warm. The weather verry w[a]rm after decending into the low Countrey—

Gass   The morning was pleasant

Ordway   a clear pleasant morning.

Whitehouse   a Clear pleasant morning. On Some of the ridges the timber has been killed by fire and fell across the trail So that we had Some difficulty to git a long the trail.

Note   *The expedition leaves the Bitterroot Mountains and travels down onto Idaho's Weippe Prairie.*

SUNDAY, SEPTEMBER 22

	Sunrise			4 p.m.			River	
Temp	Weather	Wind	Temp	Weather	Wind	Rise/Fall	Feet	Inches
	f	SW		*f*	SW			

Clark   a fine morning. A verry w[a]rm day. Some few drops of rain this evening.

Gass   This was a fine warm day. in the evening arrived in a fine large valley

Ordway   a clear pleasant morning.   and white frost.

Whitehouse   a clear pleasant morning. A white frost.

MONDAY, SEPTEMBER 23

	Sunrise			4 p.m.			River	
Temp	Weather	Wind	Temp	Weather	Wind	Rise/Fall	Feet	Inches
	f	SW		*f*	SW			

Clark   Hot day. At dark a hard wind from the SW accompanied with rain which lasted half
        an hour.
Gass   The morning was warm and pleasant. About dark a shower of rain fell.
Ordway   a fair morning. Had a Thunder Shower this evening.
Whitehouse   a clear pleasant morning. We had a Shower of rain attended with Thunder
        this evening.

TUESDAY, SEPTEMBER 24

	Sunrise			4 p.m.			River	
Temp	Weather	Wind	Temp	Weather	Wind	Rise/Fall	Feet	Inches
	f a r t & *l**	*SE*		*f a r*†	*SE*			

Weather Diary   a thunder cloud last evening.

Clark   a fine morning. Hot day.
Gass   The morning was fine. In the evening we arrived at the camp of our hunters on a
        river about 100 yards broad, a branch of the Columbia River [Clearwater River].
Ordway   a clear morning. We Set out and proceeded on the day warm. Towards evening
        we came down on a fork of Columbia River [Clearwater].
Whitehouse   a clear pleasant morning. The day warm.

WEDNESDAY, SEPTEMBER 25

	Sunrise			4 p.m.			River	
Temp	Weather	Wind	Temp	Weather	Wind	Rise/Fall	Feet	Inches
	f	*E*		*f*	*SW*			

Weather Diary   I proceed to the forks   w[a]rm day

Clark   a verry hot day. Calculated to build Canoes, as we had previously deturmined to
        proceed on by water.
Gass   A fine, pleasant warm morning. The climate here is warm; and the heat to day was
        as great as we had experienced at any time during the summer. The water also is
        soft and warm, and perhaps causes our indisposition more than any thing else.
Ordway   a fair morning. This river is about 60 yards wide.
Whitehouse   a fine morning. The fork of the Columbia we are now [on] is about 60 yards
        wide, and generally very deep.

---

* According to Moulton (1988, 5: 243), Lewis's Journal Codex P lists this weather data as "f."

† According to Moulton (1988, 5: 243), Lewis's Journal Codex P and Clark's Codex I lists this weather data as "f."

Note *The expedition stops to build canoes for their journey to the Pacific Ocean and establishes Canoe Camp, near Orofino, Idaho.*

THURSDAY, SEPTEMBER 26

	Sunrise			4 p.m.			River	
Temp	Weather	Wind	Temp	Weather	Wind	Rise/Fall	Feet	Inches
	f	E		*f*	SW			

Weather Diary Form a Camp at the forks

Clark this day proved verry hot.

Gass The morning was fine.

Ordway a clear pleasant morning. Several of the party Sick with a relax by a Sudden change of diet and water as well as the change of climate also.

Whitehouse clear and pleasant morning. Several of the men Sick with the relax, caused by a Suddin change of diet and water as well as the Climate Changed a little also.

FRIDAY, SEPTEMBER 27

	Sunrise			4 p.m.			River	
Temp	Weather	Wind	Temp	Weather	Wind	Rise/Fall	Feet	Inches
	f	E		*f*	SW			

Weather Diary day very warm

Clark the day verry hot.

Gass A fine warm morning. The river below the fork is about 200 yards wide, the water is clear as crystal, from 2 to 5 feet deep, and abounding with salmon of an excellent quality.

Ordway a fair morning.

Whitehouse a fine fair pleasant morning.

SATURDAY, SEPTEMBER 28

	Sunrise			4 p.m.			River	
Temp	Weather	Wind	Temp	Weather	Wind	Rise/Fall	Feet	Inches
	f	E		*f*	SW			

Clark this day proved verry w[a]rm and Sultery

Gass   We had a pleasant morning
Ordway   a clear morning.
Whitehouse   This morning we had clear fine pleasant weather.

SUNDAY, SEPTEMBER 29

	Sunrise			4 p.m.			River	
Temp	Weather	Wind	Temp	Weather	Wind	Rise/Fall	Feet	Inches
	f	E		*f*	SW			

Weather Diary   ³/₄ of the party Sick. Day very hot

Clark   a Cool morning,   wind from the SW. The after part of the day w[a]rm.
Gass   A fine day
Ordway   a fair morning.
Whitehouse   A fair and pleasant morning.

MONDAY, SEPTEMBER 30

	Sunrise			4 p.m.			River	
Temp	Weather	Wind	Temp	Weather	Wind	Rise/Fall	Feet	Inches
	f	E		*f*	SW			

Weather Diary   Great numbers of Small Ducks pass down the river.   hot day

Clark   a fine morning. Cool
Gass   The weather continued pleasant.
Ordway   a fair morning.
Whitehouse   Fair and pleasant weather this morning.

Note   *Lewis kept no weather data in October. Clark wrote a combined table for October, November, and December of 1805. No river observations were made during October, November, or December.*

TUESDAY, OCTOBER 1

	Sunrise			4 p.m.			River	
Temp	Weather	Wind	Temp	Weather	Wind	Rise/Fall	Feet	Inches
	f	E						

Weather Diary   from the 1st to 7th of octr. we were at the month of the Chopunnuish river makeing Canoes to Decend the Kooskooske.

Clark   a cool morning, wind from the NE and East. W[a]rm evening

Gass   This was a fine pleasant warm day.

Ordway   a clear pleasant morning.

Whitehouse   a fair fine clear morning.

WEDNESDAY, OCTOBER 2

	Sunrise			4 p.m.			River	
Temp	Weather	Wind	Temp	Weather	Wind	Rise/Fall	Feet	Inches
	f	N						

Clark   day excesively hot in the river bottom, wind North

Ordway   a fair morning.

Whitehouse   a fair and pleasant morning.

THURSDAY, OCTOBER 3

	Sunrise			4 p.m.			River	
Temp	Weather	Wind	Temp	Weather	Wind	Rise/Fall	Feet	Inches
	f	E						

Weather Diary   The easterly winds which blow imediately off the mountains are very cool untill 10 a.m. when the day becomes verry w[a]rm and the winds Shift about

Clark   a fair cool morning, wind from the East.

Ordway   a clear morning.

Whitehouse   a fair and very pleasant morning.

FRIDAY, OCTOBER 4

	Sunrise			4 p.m.			River	
Temp	Weather	Wind	Temp	Weather	Wind	Rise/Fall	Feet	Inches
	f	E						

Clark   This morning is a little cool, wind from off the Eastern mountains. The after part of this day verry warm.
Gass   a white frost, afterwards a fine day.
Ordway   a fair morning.
Whitehouse   a fair morning.

SATURDAY, OCTOBER 5

Sunrise			4 p.m.			River		
Temp	Weather	Wind	Temp	Weather	Wind	Rise/Fall	Feet	Inches
	f	E						

Clark   a cool morning, wind East for a Short time, wind is always a Cool wind
Ordway   a clear cool morning. A little white frost.
Whitehouse   a fair clear cool frosty morning.

SUNDAY, OCTOBER 6

Sunrise			4 p.m.			River		
Temp	Weather	Wind	Temp	Weather	Wind	Rise/Fall	Feet	Inches
	f	E						

Clark   a cool Easterly wind which Spring up in the latter part of the night Continues untill about 7 or 8 oClock AM. The winds blow cold from a little before day untill the Suns gets to Some hight from the Mountains East as they did from the mountains at the time we lay at the falls of the Missouri from the West. The river below this forks is Called Kos kos kee, it is Clear rapid with Shoals or Swift places—
Ordway   a pleasant morning.
Whitehouse   a clear pleasant morning. A raft Seen floating down the River with Several Indians on it.

Note   *The expedition resumes its journey and uses canoes to proceed via the river system. It begins by traveling along the Kooskooskee River (present-day Clearwater River) in Idaho.*

MONDAY, OCTOBER 7

	Sunrise			4 p.m.			River	
Temp	Weather	Wind	Temp	Weather	Wind	Rise/Fall	Feet	Inches
	f	E						

Note from the 7th to the 16th octr. We were decending Kooskooske & Lewises river, the 17th 18 at the mouth of Lewis River.

Clark The after part of the day cloudy

Gass a pleasant morning. About 3 o'clock in the afternoon we began our voyage down the river, and found the rapids in some places very dangerous.

Ordway a clear morning. Some part of the River is deep and current gentle &.C.

Whitehouse a fair clear weather and a pleasant morning. Some places the water is deep & and the current is gentle. The Evening proved cloudy.

TUESDAY, OCTOBER 8

	Sunrise			4 p.m.			River	
Temp	Weather	Wind	Temp	Weather	Wind	Rise/Fall	Feet	Inches
	f	E						

Clark a Cloudy morning

Gass a fine morning. Several rapids. Water was not more than waist-deep.

Ordway a fair morning. Passed Some clifts of rocks and barron hills on each side. The waves roared over the rocks.

Whitehouse a fair day. The day proved warm.

WEDNESDAY, OCTOBER 9

	Sunrise			4 p.m.			River	
Temp	Weather	Wind	Temp	Weather	Wind	Rise/Fall	Feet	Inches
	c	SW						

Clark The morning Cool as usial the greater part of the day proved to be Cloudy, which was unfavourable for drying our things &c. which got wet yesterday. The wet articles not Sufficiently dried to pack up obliged us to delay another night

Gass We stayed here during the whole of this day, which was very pleasant.
Ordway a fair morning, and warm.
Whitehouse a fair and pleasant morning.

Note *The expedition arrives at the confluence of the Kooskooskee and Lewis's rivers (today known as the Clearwater and Snake rivers), at present-day Clarkston, Washington / Lewiston, Idaho. It crosses the present-day Idaho–Washington state line and proceeds into Washginton.*

THURSDAY, OCTOBER 10

	Sunrise			4 p.m.			River	
Temp	Weather	Wind	Temp	Weather	Wind	Rise/Fall	Feet	Inches
	f	NW						

Clark a fine morning. a verry w[a]rm day. W[a]rm night— I think Lewis's [Snake] River is about 205 yards wide, the Koos koos kc River about 150 yards wide and the river below the forks about 300 yards wide. The water of the South fork is a greenish blue, the north as clear as cristial. We came to on the Stard. Side below with a view to make some luner observations, the night proved Cloudy and we were disapointed— . The Indians...winter hunting the dear on Snow Shoes in the plains

Gass We had a fine morning. the southwest branch very large, and of a goslin-green colour. The wind blew so hard we could not proceed.

Ordway a clear morning. We Set out eairly and procccd on down passed over a number of bad rapids. About 4 oClock PM we came to the Columbia River   this great Columbia River is about 400 yards wide and afords a large body of water and of a greenish coulour. We went down it a short distance and the wind blew so high from NW

Whitehouse a fair and pleasant morning. About 5 oClock PM we arived at the forks of the Columbian river. The wind blew So high from the west that we Camped on the north side. This river is about 400 yards wide, and a greenish coulour.

FRIDAY, OCTOBER 11

	Sunrise			4 p.m.			River	
Temp	Weather	Wind	Temp	Weather	Wind	Rise/Fall	Feet	Inches
	c	E & SW						

Clark a cloudy morning, wind from the East. The after part of the day the wind from the SW and hard. The day w[a]rm.

Gass   a fine morning; proceeded on

Ordway   a clear morning. The country is barron and broken. Some high plains.   No timber.   Passed over Some rapids where the waves roled high.

Whitehouse   fair and clear weather a pleasant morning. Some rapid water but the current mostly gentle.

Saturday, October 12

	Sunrise			4 p.m.			River	
Temp	Weather	Wind	Temp	Weather	Wind	Rise/Fall	Feet	Inches
	f	*E & SW*						

Clark   a fair cool morning, wind from the East. In the afternoon the wind Shifted to the SW and blew hard. the wind blew hard this evening—

Gass   a fine morning.

Ordway   a fair morning. Bad rockey rapid.

Whitehouse   a clear pleasant morning. The current swift in Some places, but gentle in general. After 12 oClock…the wind rose hard and blew from the west.

Sunday, October 13

	Sunrise			4 p.m.			River	
Temp	Weather	Wind	Temp	Weather	Wind	Rise/Fall	Feet	Inches
	f a r	*SW*						

Weather Diary   rained moderately from 4 to 11 AM to day.

Clark   rained a little before day. A windey dark raney all the morning, a hard wind from the SW untill 9 oclock, rain continued moderately untill near 12 oClock. the rained Seased & wind luled. This must be a verry bad place in high water. Wind hard from the SW in the evening and not very cold.

Gass   This was a cloudy wet morning, and we did not set out till 11 o'clock. In the afternoon the weather cleared and we had a fine evening.

Ordway   a rainy morning. High wind. Rapids   The current swift

Whitehouse   a rainy wet morning which delayed us untill about 10 oClock AM. The wind hard a head. Between 2 and 3 oClock PM the weather became clear and pleasant.

MONDAY, OCTOBER 14

	Sunrise			4 p.m.			River	
Temp	Weather	Wind	Temp	Weather	Wind	Rise/Fall	Feet	Inches
	f	SW						

Clark a verry Cold morning, Set out at 8 oClock, wind from the West and Cool untill about 12 oClock When it Shifted to the SW. All wet we had every articles exposed the Sun to dry on the Island, our loss in provisions is verry Considerable. The wind this after noon from the SW as usial and hard.

Gass a fine clear cool morning.

Ordway a clear cold morning. The wind high NW the current rapid. Bad rockey rapid The country continues barron and broken in places &.C.

Whitehouse A clear cold morning, the wind blowing hard and high ahead of us from the West. Very bad Rockey rapid, it being the worst rapid that we had passed in this River.

TUESDAY, OCTOBER 15

	Sunrise			4 p.m.			River	
Temp	Weather	Wind	Temp	Weather	Wind	Rise/Fall	Feet	Inches
	f	SW						

Clark a fair morning after a Cold night. Some frost this morning and Ice. A high point to the west. Plain wavering.

Gass This day was fine, clear and pleasant. This river in general is very handsome, except at the rapids, where it is risking both life and property to pass; and these rapids, when the bare view or prospect is considered distinct from advantages of navigation, may add to its beauty, by interposing variety and scenes of romantick grandeur where there is so much uniformity in the appearance of the county.

Ordway a clear cold morning. The current very rapid.

Whitehouse clear cool weather this morning.

Note *The expedition arrives at the confluence of Lewis's River (present-day Snake River) and the Columbia River, near present-day Pasco (Tri-Cities), Washington.*

WEDNESDAY, OCTOBER 16

	Sunrise			4 p.m.			River	
Temp	Weather	Wind	Temp	Weather	Wind	Rise/Fall	Feet	Inches
	f	SW						

Clark   a cool morning. haveing taken Diner Set out and proceeded on Seven miles to the junction of this river and the Columbia which joins from the NW.

Gass   a fine morning. Having gone 21 miles we arrived at the great Columbia river, which comes in from the northwest.

Ordway   towards evening we arived at the big forks. The large River which is wider than the Columbia River comes in from a northerly direction.   No timber. Not a tree to be Seen as far as our Eyes could extend.

Whitehouse   a pleasant morning. Towards evening we arived at the forks of the river, which came from a northly direction and is larger than this Columa. R.

THURSDAY, OCTOBER 17

	Sunrise			4 p.m.			River	
Temp	Weather	Wind	Temp	Weather	Wind	Rise/Fall	Feet	Inches
	f	SE						

Clark   This morning after the Luner observations, the old chief came down. This river is remarkably Clear and Crouded with Salmon in maney places. The number of dead Salmon on the Shores & floating in the river is incrediable to say. The Cause of the emence numbers of dead Salmon I can't account for. – at this Season they have only to collect the fish Split them open and dry them on their Scaffolds on which they have great numbers. No wood to be Seen in any direction— Those people as also those of the flat heads which we had passed on the koskoske and Lewis's rivers are Subject to Sore eyes, and maney are blind of one and some of both eyes. This misfortune must be owing to the reflections of the Sun &c. on the waters in which they are continually fishing during the Spring Summer & fall, & the Snows dureing the, winter Seasons, in this open countrey where the eye has no rest. The roughfs are nearly flat, which proves to me that rains are not common in this open countrey.

Gass   The Columbia here is 860 yards wide, the Kimooeenum [Snake River] is 475 yards wide at the junction.

Ordway   a clear pleasant morning. Saw a great quantity of sammon….a great number lay dead on the Shores which the Indians had giged.

Whitehouse   We had a clear and pleasant morning. The columbia River is more Smooth and the current gentle

Note *The expedition begins the final leg of its journey by proceeding down the Columbia River from its confluence with the Snake River. On October 18, the expedition begins its journey parallel to the border of present-day Oregon and Washington (the Columbia River).*

FRIDAY, OCTOBER 18

Sunrise			4 p.m.			River		
Temp	Weather	Wind	Temp	Weather	Wind	Rise/Fall	Feet	Inches
	f	*SE*						

Note from the eighteenth to the twenty-second of October descending the Great Columbia to the falls.

Clark a cold morning faire & wind from SE. Distance across the Columbia 960 3/4 yrds...Ki-moo-e im 575 yrds. At 4 Oclock we Set out down the Great Columbia. No timber in view. Saw a mountain bearing SW Conocal form, Covered with Snow [probably Mount Hood]

Gass a fine day. At one we proceeded on down the Great Columbia, which is a very beautiful river.

Ordway a clear pleasant morning. Capt. Clark measured Columbian River and the Ki mo e nem [Snake] Rivers and found the Columbia to be 860 yards wide, and the [Snake] to be 475 yard wide at the forks. Proceeded down the Columbia....we passed over Several rapids.

Whitehouse a clear pleasant morning. Columbia 860 yards wide and ki-moo-e-nem River 475 yards wide at the forks. Columbia...verry wide from a half a mile to three forths wide and verry Smooth & pleasant.

SATURDAY, OCTOBER 19

Sunrise			4 p.m.			River		
Temp	Weather	Wind	Temp	Weather	Wind	Rise/Fall	Feet	Inches
	f	*SE*						

Clark I assended a high clift about 200 feet above the water...from this place I descovered a high mountain of emence hight covered with Snow, this must be one of the mountains laid down by Vancouver, as seen from the mouth of the Columbia River, from the Course which it bears which is West I take it to be Mt. St. Helens [probably Mt. Adams], destant about 156 miles, a range of mountains in the Derection crossing a conical mountain SW toped with snow.

Gass   The morning was clear and pleasant, with some white frost.

Ordway   a clear cold morning. We discovered a verry high round mountain a long distance down the River which appears to have Snow on the top of it.

Whitehouse   A clear cold morning. We discovered a high hill or mountn a long distance down the River which appears to have Snow on it. We found the day pleasant

SUNDAY, OCTOBER 20

	Sunrise			4 p.m.			River	
Temp	Weather	Wind	Temp	Weather	Wind	Rise/Fall	Feet	Inches
	f	SW						

Clark   a very cold morning, wind SW. The river to day is about $^1/_4$ of a mile in width. The current much more uniform than yesterday or the day before.

Gass   A fine clear frosty morning.

Ordway   a clear frosty morning. The River Smooth. Passed many rapid places of water.

Whitehouse   a clear frosty morning.

MONDAY, OCTOBER 21

	Sunrise			4 p.m.			River	
Temp	Weather	Wind	Temp	Weather	Wind	Rise/Fall	Feet	Inches
	f	SW						

Clark   a verry Cold morning, we Set out early, wind from the SW. The Conocil mountain is SW....high and the top covered with Snow.

Gass   a fine morning.

Ordway   a clear cold morning. Passed a number of bad rockey rapids where the River is nearly filled with high dark couloured rocks

Whitehouse   A clear cold morning.

Note   *The expedition portages around what its members called "The Great Falls of the Columbia," Celilo Falls, near Wishram, along today's Oregon–Washington state line.*

TUESDAY, OCTOBER 22

	Sunrise			4 p.m.			River	
Temp	Weather	Wind	Temp	Weather	Wind	Rise/Fall	Feet	Inches
	f	SW						

Note   from the twenty-second to the twenty-ninth about the Great Falls of the Columbia river.

Clark   a fine morning, calm and fare. Portage of 457 yards & down a Slide…of the Great falls…of the Columbia

Gass   The morning was fine.   came to the first falls or great rapids

Ordway   a fair morning.   A short distance [Ordway earlier indicates that they are near a large river, today known as Deschutes River] below we came to the first falls of the Columbia River.

Whitehouse   A clear pleasant morning. A mist rises contiuually from the falls…the first falls of the Columbia River. the water falling in such an immense quantity, makes a roaring that can be heard several miles below it.

WEDNESDAY, OCTOBER 23

	Sunrise			4 p.m.			River	
Temp	Weather	Wind	Temp	Weather	Wind	Rise/Fall	Feet	Inches
	f	SW						

Clark   a fine morning.

Gass   A pleasant day.   The high water mark below the falls is 48 feet,   The reason of this rise in the water below the falls is, that for three miles down, the river is so confined by rocks (being not more than 70 yards wide) that it cannot discharge the water, as fast as it comes over the falls,

Ordway   a clear pleasant morning. The height of these falls…in all is 37 feet eight Inches

Whitehouse   a clear pleasant morning.

Note   *The expedition portages around the Long and Short Narrows at present-day The Dalles, Oregon.*

THURSDAY, OCTOBER 24

Sunrise			4 p.m.			River		
Temp	Weather	Wind	Temp	Weather	Wind	Rise/Fall	Feet	Inches
	f	W						

Clark   a fine fare morning, after a beautiful night. Below those falls are Salmon trout and great numbers of the heads of a Species of trout Smaller than the Salmon. Narrow chanel of 45 yards wide. Those narrows the water was agitated in a most Shocking manner boils Swell & whorl pools, we passed with great risque... Those narrows was the whorls and Swills arriseing from the Compression of th water...notwithstanding the horrid appearance of this agitated gut Swelling, boiling & whorling in every direction.

Gass   a fine morning.

Ordway   a clear pleasant morning. The current rapid

Whitehouse   a clear cool morning. The current verry rapid.

FRIDAY, OCTOBER 25

Sunrise			4 p.m.			River		
Temp	Weather	Wind	Temp	Weather	Wind	Rise/Fall	Feet	Inches
	f	W						

Clark   A Cold morning. Came to...a Creek which falls in on the Lard Side and head up towards the high Snow mountain to the SW.

Ordway   a fair morning.   The River gitting Smoth. The country timbred back a little from the River.

Whitehouse   a clear & pleasant morning.

SATURDAY, OCTOBER 26

Sunrise			4 p.m.			River		
Temp	Weather	Wind	Temp	Weather	Wind	Rise/Fall	Feet	Inches
	f	W						

Clark   a fine morning. The river has rose nearly 8 Inches to day and has every appearance of a tide, from what Cause I can't say— as the tides cannot effect the river here as there is a falls below, I conjecture that the rise is owing to the wind which has Set up the river for 24 hours past. All our articles we have exposed to the Sun to Dry

Gass   a fine morning.

Ordway   a clear pleasant morning. the River raised considerable this afternoon.

Whitehouse   a clear pleasant morning. The river began to raise about 4 oClock PM and raised Several Inches, the cause of which we think that the tide Swels a little up to this place. Called the long narrows of Columbia River.

SUNDAY, OCTOBER 27

	Sunrise			4 p.m.			River	
Temp	Weather	Wind	Temp	Weather	Wind	Rise/Fall	Feet	Inches
	f	W						

Clark   a verry windy night and morning from the west and hard. The wind verry high. The wind increased in the evening and blew verry hard from the Same point W. Day fair and Cold—

Gass   This was a fine clear morning, but the wind blew very hard up the river, and we remained here all day. The wind blew hard all this day.

Ordway   fair morning. The wind high from the west. The waves roled verry high—

Whitehouse   a clear morning, but the wind blew high from the west., which continued so the whole of this day.

MONDAY, OCTOBER 28

	Sunrise			4 p.m.			River	
Temp	Weather	Wind	Temp	Weather	Wind	Rise/Fall	Feet	Inches
	r a f	NW						

Weather Diary   a Violent wind   a moderate rain commenced at 4 oClock PM and continued untill 8 PM.

Clark   A cool windey morning. Wind from the West. Set out at 9 oClock AM.   The wind rose and we were obliged to lie by about 1 mile below. The wind which is the cause of our delay, does not retard the motions of those people at all, as their canoes are calculated to ride the highest waves. Wind blew hard accompanied with rain all the evening, our Situation not verry good one for an encampment, but such as it is we are obliged to put up with, the harbor is a Safe one, we encamped on the Sand, wet and disagreeable.

Gass   Just before day light there was a shower of ra[i]n; but at sunrise the morning was fine and clear. Went about 4 miles…here we stayed about an hour and proceeded again for about a mile, when we were compelled to stop on account if the wind, which blew so hard ahead that we were unable to continue our voyage. In the course of the day there were some showers of rain.

Ordway   rained hard the later part of last night. Cleared up this morning. We then went on a Short distance further   the wind rose so high NW that obledged us to halt on the Lard. Side under Some clifts of rocks. A little rain this evening.

Whitehouse   the wind Seased the later part of last night; when it began to rain and rained moderately untill morning, then cleared off. About 9 oClock AM the wind rose again & blew westward. We continued on our way a short distance further down the river, when the Wind rose so high from the Westward, & the waves ran also so high, that our officers thought it dangerous and not safe to proceed. We had Several Squalls of wind which were high during this day.

TUESDAY, OCTOBER 29

	Sunrise			4 p.m.			River	
Temp	Weather	Wind	Temp	Weather	Wind	Rise/Fall	Feet	Inches
	f a r	W						

Weather Diary   rained moderately all day. Saw the first large Buzzard or Voultur of the Columbia.

Note   from the 29th of Octr. to the 3rd of Novr. in passing through the western mountains below falls.

Note   the balance of Novr. and December between the Mountains & Pacific Ocean.

Clark   a Cloudy morning wind from the West but not hard. A good Situation for winter quarters if game can be had is just below Sepulchar rock on the Lard Side. The falls mountain [Mount Hood] covered with Snow is South. Here the mountains are high on each Side, those to the Lard. Side has Some Snow on them at this time, more timber than above and of greater variety.

Gass   We embarked early in a cloudy morning. In the evening we discovered a high mountain to the south, not more than five miles off, covered with snow [Mt. Hood]. We have here still water; and the breadth of the river is from three quarters to a mile.

Ordway   a cloudy morning. The current gentle. Saw Snow on a mountain on the Lard Side—

Whitehouse   a cloudy cool morning. We proceeded on in the gentle current. We saw mountains lying on the South side of the River; a distance back from it; Covered with timber, which had Snow lying on them.

WEDNESDAY, OCTOBER 30

	Sunrise			4 p.m.			River	
Temp	Weather	Wind	Temp	Weather	Wind	Rise/Fall	Feet	Inches
	r a r	SE						

Weather Diary   rained moderately all day. arrived at the Grand rapids. Saw a different Species of ash.

Clark   A cool Cloudy morning.   moderate rain all the last night. Rained moderately all day we are wet and cold. The day proved Cloudy dark and disagreeable with Some rain all day which kept us wet. Rained all the evening, a wet disagreeable evening. This part of the river resembles a pond partly dreaned leaving many Stumps bare both in & out of the water, current about 1 mil pr. Hour. Saw 4 Cascades caused by Small Stream falling from the mountains on the Lard. Side, a remarkable circumstance in this part of the river is, the Stumps of pine trees are in maney places are at some distance in the river, and gives every appearance of the rivers b[e]ing damed up below from Some cause which I am not at this time acquainted with, the current of the river is also verry jentle and about 3/4 of a mile in width. Examine the Shute and river below.

Gass   The morning was cloudy. It rained hard all day.

Ordway   a cloudy morning. The River wide and Strait the current gentle.   The after part of the day rainy and foggy. A number of the Savages came to our Camp and Signed to us that they were Surprized to See us   they thought we had rained down out of the clouds.

Whitehouse   cool and cloudy morning. The river verry Strait and wide current gentle. The after part of the day rainy and foggey. These savages were Surprized to See us they Signed to us that they thought we had rained down out of the clouds. Continued raining.   The rain continued the greater part of this night—

Note   *The expedition portages past the Cascades of the Columbia.*

THURSDAY, OCTOBER 31

	Sunrise			4 p.m.			River	
Temp	Weather	Wind	Temp	Weather	Wind	Rise/Fall	Feet	Inches
	f a r	SW						

Weather Diary   Som rain last night and this morning.

Clark   A cloudy rainey disagreeable morning. At a mile lower is a verry Considerable rapid at which place the waves are remarkably high. I could not See any rapids below for in the extent of my view which was for a long distance down the river, which from the last rapids widened and had everry appearance of being effected by the tide— This great Shute or falls is about $1/2$ mile with the water of this great river Compressed within the Space of 150 paces…water passing with great velocity forming & boiling in the most horriable manner. The Shute, must be the Cause of the rivers daming up to Such a distance above, where it Shows Such evidant marks of the Common current of the river being much lower than at the present day.

Gass   The morning was cloudy.

Ordway   Cloudy.   This Shoote is full of rocks and roles verry high waves &.C. The after part of the day pleasant.

Whitehouse   This morning was cool & Cloudy. About 9 oClock AM the weather cleared off and became pleasant.

Note   *The expedition camps near the Cascades of the Columbia River (The Great Shute) near present-day Bonneville Dam.*

FRIDAY, NOVEMBER 1

	Sunrise			4 p.m.			River	
Temp	Weather	Wind	Temp	Weather	Wind	Rise/Fall	Feet	Inches
	f	NE						

Clark   A verry Cool morning, wind hard from the NE

Gass   We had a cool frosty morning. We carried down our baggage…we could not go into the water, without uneasiness on account of the cold.

Ordway   a fair morning.   The wind high from the NE and cold.

Whitehouse   a clear morning. The wind rose high from the NE and cold.

Note   *The expedition proceeds into the tidal water of the lower Columbia River system.*

SATURDAY, NOVEMBER 2

	Sunrise			4 p.m.			River	
Temp	Weather	Wind	Temp	Weather	Wind	Rise/Fall	Feet	Inches
	f	SW						

Clark   The river wider and bottoms more extencive. River about 2 miles wide. The ebb tide rose here about 9 Inches, the flood tide must rise here much higher.

Ordway   a fair morning.   The river got more Smooth the current gentle wide and Strait. Saw a number of Spring runs flowing from the high clifts and Mountains. Some of which falls off about 100 feet perpinticular.

Whitehouse   a clear morning but cool. The Riv. got Smooth the current verry gentle &c. The river wide and Strait the remdr. of the day. Great number of springs runs, and Springs flowing from the high clifts and mountains and fell off down 100 feet or more.*

Note   *On November 3 and 4, the party passes present-day Portland, Oregon, but does not see the confluence of the present-day Willamette and Columbia rivers.*

SUNDAY, NOVEMBER 3

Sunrise			4 p.m.			River		
Temp	Weather	Wind	Temp	Weather	Wind	Rise/Fall	Feet	Inches
	f a fog	*NE*						

Weather Diary   a thick fog which continud untill meridian   cleared off and was fair the remainder of the day.

Clark   The fog so thick this morning we did not think it prudent to Set out...we could not see a man 50 Steps off, this fog detained us untill 10 oClock at which time we Set out. The water rose 2 Inches last night the effects of tide. The water Shallow for a great distance from shore. The fog continued thick untill 12 o'clock

Gass   The morning was foggy. At 9 we proceed on ,but could not see the country we were passing, on account of the fog, which was very thick till noon when it disappeared, and we had a beautiful day. We see the high point of a mountain covered with snow, in about a southeast direction from us. (Mt. Hood)

Ordway   a foggy morning.   We perceive the tide rise and fall a little at this place.   We saw the round mountain some distance a head which we expect is the Same which was discovred by Lieut. Hood and is called Hoods Mountain.

Whitehouse   a foggy morning & we delayed setting out till about 9 oClock AM. The fog So thick this morning that we cannot See more than one hundred yards distance. Agreeable to all calculations it cannot be more than two hundred miles from this to the ocean. We then set out abt. 9oC. and proceeded on the fog continued So thick that we could Scarsely See the Shores or Islands. We Saw a high round mountain on the Lard Side which we expect is the Same we Saw abo. the great falls and the Same that Lieut. Hood gave an account off. (it is nearly covd. with Snow). River verry wide better than a mile in general. About 2 oClock PM...the weather now got clear & pleasant.

*These falls may be the Multnomah and other falls in Multnomah County, Oregon.

MONDAY, NOVEMBER 4

Sunrise			4 p.m.			River		
Temp	Weather	Wind	Temp	Weather	Wind	Rise/Fall	Feet	Inches
	c a r	W						

Clark   A cloudy cool morning, wind from the west. Tide rose last night 18 inches perpndicular at Camp. High tide at 6 o'clock pm. Verry w[a]rm. Saw Mount Helien [Mt. St. Helens]…it is emensely high and covered with Snow, riseing in a kind of Cone perhaps the highest pinecal from the common leavel in america.   The river here is 1 ¹/₂ miles wide, and current jentle.

Gass   a fine morning. The tide raised the water last night 2 feet. In the evening we saw Mount Rainy [Mt. St. Helens]   It is a handsome point of a mountain, with little or no timber on it, very high, and a considerable distance off this place.

Ordway   cloudy. The tide Ebbs and floes abt. 3 feet at this place   discovered a high round mountain some distance back from the River on the Stard side which is called mount rainy— [probably Mt. St. Helens]

Whitehouse   This morning was cold, foggy & cloudy. We are now tide way, the tide fell during last night 2 feet perpendicular, and is on the rise this morning. The river wider.

TUESDAY, NOVEMBER 5

Sunrise			4 p.m.			River		
Temp	Weather	Wind	Temp	Weather	Wind	Rise/Fall	Feet	Inches
	r c r	SW						

Weather Diary   Commenced raining at 2 PM and continued to rain with intervales throughout the day.   Saw 14 Garter Snakes.

Clark   Rained all the after part of the last night, rain continues this morning. We are all wet Cold and disagreeable, rain Continues & encreases. My feet and legs cold. I saw 17 Snakes to day on a Island, but little appearance of Frost at this place. Here the river is about one and a half miles wide The day proved Cloudy with rain the greater part of it, we are all wet cold and disagreeable. I Saw but little appearance of frost in this valley which we call Wap-pa-too Columbia

Gass   Some rain fell last night about 2 o'clock, and the morning was cloudy. We proceed on in the afternoon, during which some rain and a little hail fell. Here the tide rises and falls 4 feet.

Ordway   hard rain the later part of last night.   We had several small showers of rain

Whitehouse   it began to rain about one oClock last night and continued to rain till day light. This morning was Cloudy. Found the river run very strait, & grew wider. We had frequent small showers of rain. This evening continued Rainey. The river was about 1 miles wide at this place.

WEDNESDAY, NOVEMBER 6

	Sunrise			4 p.m.			River	
Temp	Weather	Wind	Temp	Weather	Wind	Rise/Fall	Feet	Inches
	r a r	SW						

Weather Diary   rained the greater part of the day moderately.

Clark   a cold [Clark later adds "cool" here] wet ra[i]ny morning. Wind high a head. Cloudy with rain all day, we are all wet and disagreeable

Gass   We set out early in a cloudy morning after a disagreeable night of rain.

Ordway   Several Showers of rain in the course of last night. The wind rose from the west towards evening so that the waves run high.

Whitehouse   Several Showers of rain in the course of last night. The tide Ebbs & flows abt. 3 feet pertular. This morning was cloudy & wet. We also passed several small springs. Towards evening we had the Wind blowing hard from the Westward & the Waves ran very high.

Note   *Between November 7 and December 6, the expedition moves into the Columbia River estuary and proceeds to the river's mouth at the Pacific Ocean.*

THURSDAY, NOVEMBER 7

	Sunrise			4 p.m.			River	
Temp	Weather	Wind	Temp	Weather	Wind	Rise/Fall	Feet	Inches
	r a fog[*]	SW						

Weather Diary   Thick fog this morning which Continued untill 11 AM. Cleared off and was fair until meridian,   Several havy Showers dureing the evening

Clark   a Cloudy fogey morning, a little rain. We delayed 1 $^1/_2$ hour & set out the tide being up   one of our Canoes Seperated from us this morning in the fog.   The rain continued untill 9 oclock moderately.   Several marshey Islands towards the Lard Side

---

[*]According to Moulton (1990, 6: 101), Clark's Journal Codex I lists this weather data as "r a r fog."

the Shape of them I can't See as the river is wide and day foggey. The fog So thick we could not See across the river. rain Continud. moderately all day   our Small Canoe which got Seperated in the fog this morning joined us this evening. We are in view of the opening of the Ocian, which Creates great joy. Some high mountains to the S.W. on the top of one is Snow. *Ocian in View! O! The joy.* Great joy in camp we are in View of the Ocian, this great Pacific Octean which we been So long anxious to See. and the roreing or noise made by the waves brakeing on the rockey Showers (as I Suppose) may be heard distinctly

Gass   a foggy morning. At this place the river is about 3 miles wide

Ordway   a foggy cool morning.

Whitehouse   a cool foggy morning. These Islands were Marshy & were covered with Grass, & had Water laying in different parts of them.

FRIDAY, NOVEMBER 8

	Sunrise			4 p.m.			River	
Temp	Weather	Wind	Temp	Weather	Wind	Rise/Fall	Feet	Inches
	f a r	SW						

Weather Diary   rained moderately

Clark   a cloudy morning Some rain and wind   Cloudy and disagreeable all the day. After dinner we took the advantage of the returning tide & proceeded on to the 2nd point, at which place we found the Swells too high to proceed. The Swells Continued high all the evening   river wide & at this place too Salt to be used for drink.   here we found the Swells or Waves so high that we thought it imprudent to proceed; we landed unloaded and drew up our Canoes. Some fine rain all day at intervals, we are all wet and disagreeable, as we have been for Several days past, and our present Situation a verry disagreeable one in as much as we have not leavel land Sufficent for an encampment and for our baggage to lie cleare of the tide. The Seas roled and tossed the Canoes in such a manner this evening that Several of our party were Sea sick. Ocian 4142 Miles from the Mouth of Missouri R.

Gass   The morning was cloudy, and there was a hard wind from the east. We had to coast round [a bay], as the wind raised the waves so high we could go no other way.   The waves ran so high we were obliged to lie to, and let the tide leave our canoes on dry ground. In crossing the bay when the tide was out, some of our men got sea sick, the swells were so great. The whole of this day was wet and disagreeable

Ordway   a Cloudy morning. The waves high tossed us abt.   We can see along distance a head   We expect we can See the mo. of the Columbian River.   We bet it appears a long distance off. The waves roled So high that we were obledged to land on the Same Shower Stard. Side and took great pains to keep the canoes from filling with water.   The river water is gitting so brackish that we cannot drink of it at full tide.   The evening rainy—

Whitehouse This morning we had cool cloudy weather. Shortly after the wind rose & blew from the SE very hard, & the River got so rough, that we were tossed very much in our Canoes. Entered into a Bay, or wide place about 7 Miles wide, which continued as far as our Eyes could descern. Found the waves running so high, that we were obliged to land about 3 oClock. We found the River water at this place brackish.

SATURDAY, NOVEMBER 9

	Sunrise			4 p.m.			River		
Temp	Weather	Wind	Temp	Weather	Wind	Rise/Fall	Feet	Inches
	r	S						

Weather Diary   rained all day with wind

Clark The tide of last night did not rise Sufficiently high to come into our camp, but the Canoes which was exposed to the mercy of the waves &c. which accompanined the returning tide, they all filled and with great attention we Saved them untill the tide left them dry. wind Hard from the South, and rained hard all the fore part of the day, at 2 oClock P M the flood tide came in accompanied with emence waves and heavy winds, floated the tree and Drift which was on the point on which we Camped and tossed them about in such a manner as to endanger the canoes verry much. Our camp entirely under water dureing the hight of the tide, every man as wet as water could make them all the last night and to day all day as the rain continued all day, at 4 oClock P M the wind Shifted about to the SW and blew with great violence imediately from the Ocean for about two hours, notwithstanding the disagreeable Situation of our party all wet and cold (and one which they have experienced for Several days past) they are chearfull and anxious to See further into the Ocian, The water of the river being too Salt to use we are obliged to make use of rain water. At this dismal point we must Spend another night as the wind & waves are too high to proceed. The rain Continud all day—

Gass The morning was windy, rainy and disagreeable, and we were obliged to remain at Cape Swell [Grays Bay] all day. We had no fresh water, except what rain we caught by the putting out our vessels. We remained here all night, and the rain continued.

Ordway rained hard the greater part of last night, and the wind rose so high NW that we had to unload the canoes in the night. This morning wet   rained the most part of the day and the wind So high up the River that Caused the tide to raise much higher than common So we had to move our loads and some of the Camps further from Shore.

Whitehouse It rained the greater part of last night, & the Wind blew very hard from the SE, which caused the Waves to run so very high, that all hands...employed unloading canoes...The morning we had wet weather & rainey, & it rained the most part of this day—   The wind continued high, which caused the tide to rise, much higher than a common tides.

SUNDAY, NOVEMBER 10

	Sunrise			4 p.m.			River	
Temp	Weather	Wind	Temp	Weather	Wind	Rise/Fall	Feet	Inches
	r a r	*NW*						

Weather Diary   rained all day with wind

Clark   rained verry hard the greater part of the last night & continues this morning, the wind has luled and the waves are not high. We loaded our canoes and proceeded on. About 12 o'clock the wind rose from the NW and the swells become so high, we were compelled to return about 2 miles to a place where we could unload our canoes. We are all wet also our bedding and maney other articles. We are all employed untill late drying our bedding. We are all wet the rain haveing continued all day

Gass   We had a rainy morning, but the wind was not so high as it had been yesterday. Proceeded on….where we found swells so high, the wind having risen, that we could not proceed any further. After we had been here about 2 hours, it became calm, and we loaded our canoes again, but could not get around the point, the swells were still so high. It rained hard all night, and was very disagreeable. The water is become very salt.

Ordway   considerable of rain fell last night. A rainy morning. The waves not So high as yesterday.   Continued raining hard   abt. noon the wind rose So high that obledged us to turn back from a point of rocks and roe about 2 miles back into a cove.   The wind contind. So high that we could not proceed.

Whitehouse   We had rain the greater part of last night & a wet rainey morning, but the Waves did not run so high as they did Yesterday.   We continued on our way, it raining hard on us, 'till about noon; when it ceased. The wave then ran so high that we had to turn back and went up river about 2 miles.

MONDAY, NOVEMBER 11

	Sunrise			4 p.m.			River	
Temp	Weather	Wind	Temp	Weather	Wind	Rise/Fall	Feet	Inches
	r	*SW*						

Weather Diary   rained all day with wind

Clark   A hard rain all the last night, dureing the last tide the logs on which we lay was all on float, ….about 12 oClock 5 Indians came down in a canoe, the wind verry high from the SW with most tremendious waves brakeing with great violence against the Shores, rain falling in torrents, we are all wet as usial – and our Situation is truly disagreeable one; the great quantities of rain which has loosened the Stones on the hill Sides and the Small ones fall on us.   Rained all day. The tide was 3 hours later to day than yesterday and rose much higher.

Gass   The morning was wet and the wind still blowing, so that we could not proceed. still had enough, as we have not tents, or coverings to defend us, except our blankets and some mats we got from the Indians, which we punt on poles to keep off the rain. It continued raining and blowing all day; and at 4 o'clock in the afternoon the tide was so high that we had to leave our lodges, until it got lower in the evening.

Ordway   rained hard the greater part of last night. A rainy wet morning. Our Robes all wet as we have no Shelter that will keep the rain from us.   The wind continued so high that we did not attempt to move this day.   These Savages went in their canoe across the River in the high waves.

Whitehouse   It rained hard the greater part of last night, which made it very disagreeable to us all. The greater part of our Men had nothing to Shelter them from the rain, & were obliged to lay down in it, & their Cloathes were wet through. This morning continued wet & rainey, the wind was high, & the swell in the river ran verry high, & we did not attempt to move from this place—

TUESDAY, NOVEMBER 12

	Sunrise			4 p.m.			River	
Temp	Weather	Wind	Temp	Weather	Wind	Rise/Fall	Feet	Inches
	h r t & l	SW						

Weather Diary   violent wind from the SW acompanied with Hail thunder and lightning, the Claps of Thunder excessively loud and Continued form 3 to 6 AM. Cleared off a Short time & raind untill 12 oClock   Cleared off an hour and rained again. (Lewis)   the rain has been pretty generally falling Since the 7th inst. (Clark)

Clark   A Tremendious wind from the SW about 3 oClock this morning with Lightineng and hard claps of Thunder, and Hail which Continued untill 6 oClock a.m. in intervals when it became light for a Short time, then the heavens became sudenly darkened by a black cloud from the SW and rained with great violence untill 12 oClock, the waves tremendious brakeing with great fury against the rocks and trees on which we are encamped. our Situation became Seriously dangerous.

It would be distressing to a feeling person to See our Situation, at this time all wet and colde with our bedding &c. also wet, in a Cove Scercely large enough to Contain us, our Baggage in a Small holler about 1/2 mile from us, and Canoes at the mercey of the waves & drift wood. It was clear at 12 for a short time, I observed the Mountains on the opposit Side was covered with snow— I observe great numbers of Sea guls, flying in every direction— Rain Continued

Gass   A cloudy wet morning, after a terrible night of rain, hail, thunder and lightning. The rain still continued, and the river remained very rough.

Ordway   a hard Storm continued all last night, and hard Thunder lightning and hail this morning.   We Saw a mountain on opposite Shore covered with Snow. The rain continued hard all day.

Whitehouse   We had a hard storm the greater part of last night, & hard thunder, lightning, & hail this morning. We saw a high mountain which lay on the opposite to where we were encamped covered with snow. The rain continued hard during the most part of this day.

Wednesday, November 13

Sunrise			4 p.m.			River		
Temp	Weather	Wind	Temp	Weather	Wind	Rise/Fall	Feet	Inches
	r	*SW*						

Clark   Some intervales of fair weather last night, rain and wind continue this morning. …trees verry high & thick Cannot determine the procise course of the winds. Rain all day moderately, I am wet &c.&c. The Hail which fell 2 nights past is yet to be Seen on the mountains. The rain continuing and weather proved So cloudy that I could not See any distance. The rain continue all day. The tides at every flud come in with great swells brakeing against the rocks & drift trees with great fury— if we were to have cold weather to accompany the rain which we have had for this 6 or 8 days passed we must eneviatilbly Suffer verry much as Clothes are Scerce with us.

Gass   This was another disagreeable rainy day   at 9 o'clock in the forenoon it became a little more calm than usual

Ordway   hard rain continued all last night a rainy morning.   As the wind continues So high that obledges us to stay—

Whitehouse   The storm continued & hard rain during last night, and this morning rainey disagreeable weather. Our Buffalo robes are getting rotten, and the most part of our baggage were wet. We have a very disagreeable time of it, the most part of our Men having slept in the rain, ever since this storm began, & are continually wet. The waves continued high & the Storm continued during the whole of this day.

Thursday, November 14

Sunrise			4 p.m.			River		
Temp	Weather	Wind	Temp	Weather	Wind	Rise/Fall	Feet	Inches
	f							

Weather Diary   a blustery rainey day

Clark   rained all the last night without intermission, and this morning wind blows verry hard, but our Situation is Such that we cannot tell from what point it comes. one of our canoes is much broken by the waves dashing it against the rocks in high tide. The rain &c. which has continued without a longer intermition than 2 hours at a time for ten days past has distroyd. the robes and rotted nearly one half of the fiew clothes the party has, perticularley the leather clothes   if we have cold weather before we can kill & Dress Skins for clothing the bulk of the party will Suffer verry much. The rain Continue all day. Rained as usial all the evening, all wet and disagreeable Situated.

Gass We expected last night to have been able to proceed on this morning, but the rain continued, and the river still remained rough, and we are therefore obliged to lie by.   3 men returned...as the swells ran so high that they could not possibly get the canoe along. The weather continued wet, and the most disagreeable I had ever seen.

Ordway the Storm continues, and obledges us to Stay in this disagreeable harbour with nothing but pounded Sammon to Eat.

Whitehouse We had considerable quantity of rain during last night, & this morning we had wet rainey weather. About 10 oClock AM the weather cleared off, & in the afternoon it became tolerable calm weather.

Friday, November 15

	Sunrise			4 p.m.			River	
Temp	Weather	Wind	Temp	Weather	Wind	Rise/Fall	Feet	Inches
	f a r	SE						

Weather Diary The after part of this day fair and calm for the first time since the 5th instant. and no rain move our encampment.

Clark Rained all the last night at intervales of sometimes of 2 hours, This morning it became calm & fair, I prepared to set out at which time the wind sprung up from the SE. and blew down the River & in a fiew minits raised such swells and waves brakeing on the Rocks at the Point as to render it unsafe to proceed. The sun shown untill 1 oClock P.M. which gave an oppertunity for use to dry some of our bedding & examine our baggage.

The rainey weather continued without a longer intermition than 2 hours at a time, from the 5th in the morng. untill the 16th is eleven days rains, and the most disagreeable time I have experenced confined on a tempiest coast wet, Scerce Provisions, and torrents of rain poreing on us all the time— , where I can neither git out to hunt, return to a better situation, or proceed on; in this situation have we been for Six days past, fortunately the wind luled and the river became calm about 3 oClock we loaded in great haste and set out passed the blustering Point below which is a sand beech. The waves became very high Evening fare & pleasent.

Gass This morning the weather appeared to settle and clear off, but the river remained still rough. About 1 o'clock when the water became more calm we loaded and set out from our disagreeable camp. Went about 3 miles into a bay and halted on a sand beach, in full view of the ocean, at this time more raging than pacific.

Ordway a wet morning. About 10 oClock AM cleared off   the after part of the day calm and pleasant.

SATURDAY, NOVEMBER 16

Sunrise			4 p.m.			River		
Temp	Weather	Wind	Temp	Weather	Wind	Rise/Fall	Feet	Inches
	f	*WSW*						

Clark   Cool the latter part of the last night this morning Clear and butifull. The Sea is fomeing and looks truly dismal to day, from the wind which blew hard to day from the SW. The evening proved Cloudy and I could not take any Luner observations. On the Lard side...a pinical of which is now covered with Snow or hail,

Gass   This was a clear morning and the wind pretty high. We could see the waves, like small mountains, rolling out in the ocean, and pretty [h]ard in the bay. We are now at the end of our voyage, which has been completely accomplished according to the intention of the expedition, the object of which was to discover a passage by the way of the Missouri and Columbia rivers to the Pacific Ocean; notwithstanding the difficulties, privations and dangers, which we had to encounter, endure and surmount. The day being clear we got our baggage dried

Ordway   a clear cool morning.

Whitehouse   A clear cool morning. We are now in plain view of the *Pacific Ocean*.   The waves rolling, & the surf roaring very loud. We are now of opinion that we cannot go any further with our Canoes, & think that we are at an end of our Voyage to the Pacific Ocean, and as soon as discoveries necessary are made, that we shall return a short distance up the River & provide our Selves with Winter Quarters.

SUNDAY, NOVEMBER 17

Sunrise			4 p.m.			River		
Temp	Weather	Wind	Temp	Weather	Wind	Rise/Fall	Feet	Inches
	c a f	*E*						

Clark   A fair cool morning wind from the East. The tides rises at this place 8 feet 6 inches and comes in with great wave brakeing on the Sand beech on which we lay with great fury

Gass   We had a fine pleasant clear morning.

Ordway   a clear morning.

Whitehouse   This morning we had clear pleasant weather.

MONDAY, NOVEMBER 18

	Sunrise			4 p.m.			River	
Temp	Weather	Wind	Temp	Weather	Wind	Rise/Fall	Feet	Inches
	f a c	SE						

Weather Diary   Cloudy …. (Lewis)   I proceed the Ocean. (Clark)

Clark   A little Cloudy this morning. here I found Capt. Lewis name on a tree, I also engraved my name, & by land the day of the month and year, as also Several of the men. Some rain in the after part of the night. Men appear much Satisfied with their trip beholding with estonishment the high waves dashing against the rocks & this emence Ocian.

Gass   The morning was cloudy.

Ordway   Cloudy. Set out in order to go down and see the passiffic ocean. —towards evening we arived at the Cape disapointment on the Sea Shore. Went over a bald hill where we had a handsom view of the ocean.

Whitehouse   We had a cloudy morning.

TUESDAY, NOVEMBER 19

	Sunrise			4 p.m.			River	
Temp	Weather	Wind	Temp	Weather	Wind	Rise/Fall	Feet	Inches
	c a r	SE						

Clark   I arose early this morning from under a Wet blanket caused by a Shower of rain which fell in the latter part of the the last night.   raind.   …it Comenced raining and continued moderately untill 11 oClock A M.

Gass   We had a cloudy, rainy morning

Ordway   cloudy   a light Sprinkling of rain the later part of last night

Whitehouse   A cloudy morning.

WEDNESDAY, NOVEMBER 20

	Sunrise			4 p.m.			River	
Temp	Weather	Wind	Temp	Weather	Wind	Rise/Fall	Feet	Inches
	f a r	SE						

Weather Diary   rained moderately from 6 AM. (Lewis)   On the 20th untill 1 PM the 21st after which it became Cloudy without rain (Clark)

Clark   Some rain last night. The Morning Cleared up fare and we proceeded on. The tide being out we walked home on the beech—

Gass   We had a fine clear morning.   This day continued clear and pleasant throughout.

Ordway   a fair morning.

Whitehouse   A clear pleasant morning.

THURSDAY, NOVEMBER 21

Sunrise			4 p.m.			River		
Temp	Weather	Wind	Temp	Weather	Wind	Rise/Fall	Feet	Inches
	c a r	SE						

Weather Diary   rained all last night untill 1 PM and Cleared away and was Cloudy without rain

Clark   A cloud morning. the Wind blew hard from the SE which with the addition of the flood tide raised verry high waves which broke with great violence against the shore throwing water into our camp the fore part of this day Cloudy   Morng. dark & Disagreeable, a Supriseing Climent. We have not had One cold day Since we passed below the last falls or great Shute & Some time before. The Climent is temperate, and the only change we have experienced is from fair weather to rainey windey weather— at 12 oClock it began to rain and continud all day moderately. Some wind from the SE   Waves too high for us to proceed on our homeward bound journey.

Gass   A cloudy morning. The wind blew so violent to day, and the waves ran so high, that we could not set out on our return, which it is our intention to do as soon as the weather and water will permit. The season being so far advanced, we wish to establish our winter quarters as soon as possible.   The night was very wet and disagreeable.

Ordway   a cloudy and a little rain.

Whitehouse   A cloudy morning, and a light sprinkling of rain fell. The swell in the River rain so high that it detain'd us, at our Camp from going up the River again, to look our for Winter Quarters, which our officers intended as soon as the Weather would permit, and the Season of the Year advancing made it absolutely necessary that it should be the case— The Season of the Year, is generally cold at this place, but at the present time it was very pleasant. The evening was rainey

FRIDAY, NOVEMBER 22

Sunrise			4 p.m.			River		
Temp	Weather	Wind	Temp	Weather	Wind	Rise/Fall	Feet	Inches
	r	SSE						

Weather Diary   rained all day   wind violent from the SE (Lewis) The wind violent from the SSE throwing water of the R over our Camp and rain continued all day. (Clark)

Clark   A moderate rain all the last night with wind, a little before Day light the wind which was from the SSE blew with Such Violence that we wer almost overwhelmed with water blown from the river, this Storm did not sease at day but blew with nearly equal violence throughout the whole day accompaned with rain. O! how horriable is the day—   waves brakeing with great violence against the Shore throwing the Water into our Camp &c. all wet and Confind to our Shelters. The Storm Continued all day with equal violence accompanied with rain.

Gass   This was a rainy and stormy morning.   The wind blew very hard from the south, and the river was rougher than it has been since we came here. At noon the tide was higher than common. The rain and wind continued all day violent.

Ordway   a hard Storm arose the later part of last night and continues raining and the wind high from the SW   the waves rolled so high and the tide raised much higher than common.   Dashed one of our canoes against the logs and was near Splitting it before we got it out. Damaged it and obledged us to move some of our Camps—

Whitehouse   A hard Storm arose in the course of last night accompanied with Rain, & it continued raining very hard & the wind High from the So West. This caused the tide of flood to rise much higher, than it commonly did at this place. The Swell ran also to an amazing height. We had also to move some of our Camps, the water being all round them & a rising. It continued raining hard all day—

SATURDAY, NOVEMBER 23

	Sunrise			4 p.m.			River	
Temp	Weather	Wind	Temp	Weather	Wind	Rise/Fall	Feet	Inches
	c a r	SW						

Weather Diary   rained all last night to day   Cloudy

Clark   A calm Cloudy morning, a moderate rain the greater part of the last night. Rained at intervales all day. I marked my name the day & year on a Alder tree.

Gass   The weather was somewhat cloudy but more calm. In the evening the weather cleared and we had a fine night.

Ordway   Still continues rainy and high wind

Whitehouse   We had a hard wind blowing the greater part of last night, & it rained powerfully. This morning it moderated, both with regard to Wind & Rain. The evening was pleasant

SUNDAY, NOVEMBER 24

	Sunrise			4 p.m.			River	
Temp	Weather	Wind	Temp	Weather	Wind	Rise/Fall	Feet	Inches
	f a r	W						

Weather Diary   rained moderately for a Short time this morning

Clark   A fare morning. This day proved to be fair which gave us an oppertunity of drying our wet articles. ....advantages of being near the Sea Coast one most Striking one occurs to me i'e, the Climate which must be from every appearance much milder than that above the 1ˢᵗ range of mountains (where the climate must be more Severe—). The Indians are Slightly Clothed and give an account of but little Snow, and the weather which we have experienced Since we arrived in the neighbourhood of the Sea Coast has been verry warm, and maney of the fiew days past disagreeably So. if this Should be the Case it will most Certainly be the best Situation of our naked party dressed as they are altogether in leather.

Gass   The morning was fine with some white frost. As this was a fine clear day, it was thought proper to remain here in order to make some observations, which the bad weather had before rendered impossible.   At the head of the bay the river is 3 miles and 660 yards wide. At night, the party were consulted by the Commanding Officers, as to the place most proper for winter quarters; and the most of them were of opinion, that it would be best, in the first place, to go over to the south side of the river, and ascertain whether good hunting ground could be found there.

Ordway   a clear pleasant morning. A white frost.   The Calumbian River at this place is three miles 660 yards wide.   Our officers conclude with the opinion of the party to cross the River and look out a place for winter quarters

Whitehouse   A white frost this morning, & the weather clear & pleasant. The river Columbia at this place is 3 miles from the Sea & 660 Yards wide. Our officers went out and took down Notes on several remarkable points &ca. which they could not before have done, on account of the badness of the weather. In the evening our Officers had the whole party assembled in order to consult which place would be the best, for us to take up Winter Quarter at. The greater part of our Men were of opinion; that it would be best, to cross the River...

Monday, November 25

	Sunrise			4 p.m.			River	
Temp	Weather	Wind	Temp	Weather	Wind	Rise/Fall	Feet	Inches
	c a r	ESE						

Weather Diary   Some Showers of rain last night

Clark   A fin day. The Swells too high to cross the river. The evening Cloudy the Winds of to day is generally ESE   Mt. St. Hilians Can be Seen from the mouth of this river.

Gass   The morning was pleasant, though cloudy, with a white frost.

Ordway   a clear pleasant morning. Attempt to cross the river but the waves so high that the canoes were near filling. So we turned back to Shore again.

Whitehouse   We had a clear pleasant morning. Up river 9 miles...attempted to cross it, but the Waves ran so high that we found it impracticable.

TUESDAY, NOVEMBER 26

	Sunrise			4 p.m.			River	
Temp	Weather	Wind	Temp	Weather	Wind	Rise/Fall	Feet	Inches
	r	*ENE*						

Weather Diary   rained all day with Some hard Showers. the wind not so violent as it has been for Sevral days past. (Lewis)   Some rain on the morning of the 23rd and night of the 24th instant. (Clark)

Clark   Cloudy and some rain this morning from 6 oClock. Wind from the ENE. We had rain all the day   all wet and disagreeable.   We found much difficuelty in precureing wood to burn, as it was raining hard, as it had been the greater part of the day.

Gass   The morning of this day was cloudy and wet.   The whole of the day was wet and unpleasant

Ordway   a Cloudy wet morning. The day rainy and cold.

Whitehouse   A cloudy wet morning, & we set out early. We continued on still down the River; the day being wet, cold and very disagreeable.

WEDNESDAY, NOVEMBER 27

	Sunrise			4 p.m.			River	
Temp	Weather	Wind	Temp	Weather	Wind	Rise/Fall	Feet	Inches
	r	*SW*						

Weather Diary   violent wind and hard all day. Campd. At Pt. William (Lewis) rained moderately all day   a hard wind from the SW which compelled us to lie by on the isthmus of point William on the South Side. (Clark)

Clark   Rain all the last night and this morning it Continues moderately— The Swells became high and rained so hard we Concluded to halt and dry our Selves, Soon after our landing the wind rose from the East and blew hard accompanied with rain, this rain obliged us to unload & draw up our Canoes. The water at our Camp Salt that above the isthmus fresh and fine—

Gass   a wet morning; coasted round, and turned a sharp cape [Tongue Point] about a mile; when we found swells running so high that we had to halt. Had a very wet night.

Ordway   rained all last night. We could perceive a considerable of current in the River. The waves ran So high that obledged us to halt at an old fishery.   Hard rain.

Whitehouse   A rainey wet morning & cold. The Rain continued hard all this day—

THURSDAY, NOVEMBER 28

	Sunrise			4 p.m.			River	
Temp	Weather	Wind	Temp	Weather	Wind	Rise/Fall	Feet	Inches
	r	SW & NW						

Weather Diary   a tremendious Storm from the NW in the after part of the day. rained all last night and to daye. (Lewis)   The wind which was from the SW Shifted in the after part of the day to the NW and blew a Storm which was tremendious. rained all the last night and to day without intermission. (Clark)

Clark   Wind Shifted about to the SW and blew hard accompanied with hard rain. Rained all the last night. Wind to high to go either back or forward. This is our present Situation,! truly disagreeable. About 12 oClock the wind Shifted about to the NW and blew with great violence for the remainder of the day   at maney times it blew for 15 ro 20 minits with Such violence that I expected every moment to See trees taken up by the roots. Many were blown down. Those Squals were Succeeded by rain, !O how Tremendious is the day. This dredfull wind and rain Continued with intervales of fair weather all the latter part of the night. O! How disagreeable is our Situation dureing this dreadfull weather.

Gass   We had a wet windy morning.   It rained all day; an we had here no fresh water, but what was taken out of the canoes as the rain fell.

Ordway   a hard Storm. The wind high from the N. West. Hard rain all day—

Whitehouse   We had a very heavy Storm during the whole of last night, & the wind blowing hard from the Westward this morning. It rained the greater part of this day. The Wind rose from the North West & became a perfect storm

FRIDAY, NOVEMBER 29

	Sunrise			4 p.m.			River	
Temp	Weather	Wind	Temp	Weather	Wind	Rise/Fall	Feet	Inches
	r	SW						

Weather Diary   rained all last night hard, and to day moderately   I decend with 5 men in a canoe to examine the Country.

Lewis   The wind being so high the party were unable to proceed with the perogues. It rained upon us by showers all day.   Encamped at an old Indian hunting lodge which afforded us a tolerable shelter from rain, which continued by intervales throughout the night—

Clark   Blew hard and rained the greater part of the last night and this morning much more moderate. The waves Still high and rain Continues. The Swells and waves being too high for us to proceed down in our large Canoes, in Safty— The winds are from Such points that we cannot form our Camp So as to provent the Smoke which is emencely disagreeable, and painfull to the eyes.

Gass   The weather continues cloudy and wet. There were some showers of rain and hail during the day.

Ordway   Showery and Some hail in the course of the day.

Whitehouse   It rained very hard all last night, & continued showery this morning. The weather continued showery & some hail fell during this day—

SATURDAY, NOVEMBER 30

	Sunrise			4 p.m.			River	
Temp	Weather	Wind	Temp	Weather	Wind	Rise/Fall	Feet	Inches
	f a r & h	SW						

Weather Diary   rained and hailed with short intervales throughout the last night, Some thunder and lightninge.

Lewis   cloudy morning. It rained but little on us today tho' it was cloudy generally— Wind from the NE.

Clark   Some rain and hail with intervals of fair weather for the Space of one or two hours at a time dureing the night untill 9 oClock this morning, at which time it Cleared up fair and the Sun Shown.

Gass   This was a fair day. The whole of the day was fair, pleasant and warm for the season.

Ordway   the after part of the day clear.

Whitehouse   We had several hard showers of rain, & some hail fell during last night, and this morning after day light it cleared off.

Note   *The expedition searches for a location to establish winter camp.*

SUNDAY, DECEMBER 1

	Sunrise			4 p.m.			River	
Temp	Weather	Wind	Temp	Weather	Wind	Rise/Fall	Feet	Inches
	c a r	E						

Weather Diary   rained last night and Some this morning.

Lewis   Cloudy morning wind from the SE.

Clark   Cloudy windey morning wind from the East. The Wind rose so high that I could not proceed. Began to rain hard at Sun Set and Continud. The Sea which is imedeately in front roars like a repeeted roling thunder. The emence Seas and waves which breake on the rocks & Coasts to the SW & NW roars like an emence fall at a distance, and this roaring has continued ever Since our arrival in the neighbourhood of the Sea Coast which has been 24 days Since we arrived in Sight of the Great Western; (for I cannot Say Pacific) Ocian as I have not Seen one pacific day Since my arrival in its vicinity, and its waters are forming and [perpetually] breake with emenc waves on the Sands and rockey Coasts, tempestous and horiable.

Gass   The whole of this day was cloudy.

Ordway   a cloudy morning.

Whitehouse   A dark cloudy morning. We had in the course of this day a little rain, & all anxiously waiting for the arrival of Captain Lewis.

MONDAY, DECEMBER 2

	Sunrise			4 p.m.			River	
Temp	Weather	Wind	Temp	Weather	Wind	Rise/Fall	Feet	Inches
	c a r	SW						

Weather Diary   rained all the last night and untill meridian   cloudy the remained of the day.

Clark   Cloudy with some little rain this morning. The evening....fair moon Shineing night—

Gass   The day was agin cloudy and wet. In the evening the weather became clear, and we had a fine night.

Ordway   a Cloudy wet morning.

Whitehouse   A cloudy wet morning.

TUESDAY, DECEMBER 3

	Sunrise			4 p.m.			River	
Temp	Weather	Wind	Temp	Weather	Wind	Rise/Fall	Feet	Inches
	f a r	E						

Weather Diary   rained all the last night & to day untill meridian and became fair &c. (Lewis)   fair from 12 to 2 PM rained all the last night & this morning. Rained the night of the 1st and morning of the 2 and Cloudy the remainder of the day.   rained at intervales the night of the 2d instant with constant hard and Sometimes violent winds. (Clark)

Clark   a fair windey morning wind from the East. Wind Continues to blow. I marked my name on a large pine tree imediately on the isthmus William Clark December 3rd 1805. By Land from the U. States in 1804 & 1805—   Some rain this afternoon and evening.

Gass   The morning was foggy.   The greater part of the day was fair, but in the evening it clouded over and rained again.

Ordway   Cloudy

Whitehouse   This morning cloudy.

WEDNESDAY, DECEMBER 4

	Sunrise			4 p.m.			River	
Temp	Weather	Wind	Temp	Weather	Wind	Rise/Fall	Feet	Inches
	r	*SE*						

Weather Diary   rained all day

Clark   Some little rain all the last night, this morning the rain and wind increased from the SE. a Spring tide to day rose 2 feet higher than Common flood tides and high water at 11 oClock. Hard wind from the SE this afternoon Hard wind from the South this evening. Rained moderately all day and the waves too high for me to proceed in Safty to the bay as I intended.

Gass   We had a cloudy rainy morning. The river was so rough we could not set with the canoes.   The rain continued all day.

Ordway   a rainy wet morning. Continued Storming & high wind all day—

Whitehouse   A rainey wet morning. The day continued Rainey, the Wind blew hard & the weather was stormy.

THURSDAY, DECEMBER 5

	Sunrise			4 p.m.			River	
Temp	Weather	Wind	Temp	Weather	Wind	Rise/Fall	Feet	Inches
	r	*SW*						

Weather Diary   rained all last night and today   I return to Capt Clark (Lewis)   rained yesterday, last night, and moderately to day; all day wind violent in the after part of the day. .... (Clark)

Clark   Some hard showers of rain last night, this morning Cloudy and drisley, in the bay the Shower appear harder. High water to day at 12 oClock   this tide is 2 inches higher than that of yesterday. All our Stores again wet by the hard Showers of last night. Capt. Lewis's long delay below has been the cause of no little uneasiness on my part for him, 1000 conjectures has crouded into my mind respecting his probably Situation & Safty— rained hard.   The repeeted rains and hard winds which blows from the SW renders it impossible for me to move with loaded Canoes along an unknown Coast we are all wet & disagreeable;   Rain continued all the after pt. of the day accompanied with hard wind from the SW which provents our moveing from this Camp.

Gass Again we had a wet stormy day, so the men were unable to proceed with the canoes. There is more wet weather on this coast, than I ever new in any other place; during a month, we had three fair days' and there is no prospect of a change.

Ordway rainy dissagreeable weather.

Whitehouse We had hard rain & stormy weather; which was very disagreable. It continued raining the whole of this day—

FRIDAY, DECEMBER 6

Sunrise			4 p.m.			River		
Temp	Weather	Wind	Temp	Weather	Wind	Rise/Fall	Feet	Inches
	r	SW						

Weather Diary rained last night and all day to day wind no violent in the after part of the day fair in the eving. (Lewis) rained all last night and to day untill 6 oClock at which time it Clear'd away and became far. the winds also Seased to blow violent. (Clark)

Clark The wind blew hard all the last night with moderate rain, the waves verry high, This morning the wind which is Still from the SW increased and rain continued all day, about Dusk the wind Shifted to the North and it Cleared up and became fair weather. The high tide of today at 12 oClock is 13 inches higher than yesterday.

Gass We had another wet morning At noon it rained very hard, and the tide flowed so high, that in some part of our camp the water was a foot deep; we had therefore to remove to higher ground. In the afternoon it still continued it still to rain hard.

Ordway about noon the Storm arose and the tide raised about 2 feet higher than common so that the water came in to our Camp So that we moved our Camps to higher ground. The Storm cont. all day.

Whitehouse A rainey disagreeable morning, & the Wind continued high. About 1 oClock PM it blew a storm, and the tide rose about 2 feet perpendicular higher, than it had been, since we are at this place, & over flowed some of our Camps, which obliged us to move them to higher ground, than they were first at; the Storm still continued, & the Rain extinguish'd our fires, & made it exceedingly disagreeable to us. Towards evening the Weather cleared up, & it became a little more pleasant,

SECTION 6 Fort Clatsop

December 7, 1805 to March 22, 1806

"Wet and disagreeable" was the motto of the Corps of Discovery during the winter of 1805/06 at Fort Clatsop, near present-day Astoria, Oregon. Using dead reckoning, Clark estimated the mileage of their winding journey from St. Louis to the mouth of the Columbia River at 4,162 miles. He was within 40 miles of the actual distance (Duncan and Burns 1997, 159; Ambrose 1998, 175).

Endless days of mist, drizzle, rain, thunder, and lightning punctuated by three weeks of snow and freezing temperatures resulted in a dreary stay along the Oregon coast. The party occupied their time hunting and boiling ocean water to obtain much-needed salt. The captains spent the hours translating journal entries, and visiting neighboring Indian nations, and Clark led a small contingent to the ocean to view a whale.

Anxiously wanting to return home with news of their exploits, the Corps set out from Fort Clatsop on March 23, 1806 and proceeded up the flood-swollen Columbia River.

As in the previous section, each day's entry begins with the data from the Weather Diary's Observation Tables (when available), followed by the corresponding remarks (when available), and finally excerpts that pertain to weather and climate from the expedition members' daily narrative journals. No data for rivers was recorded while the party was camped at Fort Clatsop.

Weather data reported here are from Coues (1893, 3: 1276, 1291–94), Moulton (1990, 6: 258–end and 1991, 7: 48), and Thwaites (1904f, 6: 202–211).

Note   *The expedition establishes Fort Clatsop on December 7.*

SATURDAY, DECEMBER 7

	Sunrise			4 p.m.			River	
Temp	Weather	Wind	Temp	Weather	Wind	Rise/Fall	Feet	Inches
	f a r	NE						

Weather Diary   rained from 10 to 12 and at 2 PM   leave Pt. William (Lewis)   last night fair day leave Point William a hard wind from the NW and a Shower of rain at 2 PM. (Clark)

Clark   some rain from 10 to 12 last night, this morning fair. We proceeded on against the tide the waves verry high. This day fair except about 12 oClock the wind became hard from the NE and some rain which lasted 2 hours and cleared away.

Gass   About 12 last night the rain ceased and we had a fine clear morning. The swells being too high here to land we went two miles further

Ordway   the morning clear. The waves ran verry high. The hunters...no meat....the distance so great and the weather so bad that they brought no meat. The waves

325

roled verry high. The River [Lewis and Clark River] is about 100 yds wide at this place but the tide water extends further up.

Whitehouse   This morning clear & cold. The wind rose, & the wind caused the Waves to rise also. The Waves ran so high, that we could not land. Wet set off, the Waves running verry high—

SUNDAY, DECEMBER 8

	Sunrise			4 p.m.			River	
Temp	Weather	Wind	Temp	Weather	Wind	Rise/Fall	Feet	Inches
	_c a r_*	NE						

Weather Diary   Cloudy after a moderate rain last night.

Clark   a Cloudy morning. Some rain this evening. We made a Camp of the Elk Skin to keep off the rain which Continued to fall

Gass   We had a fine fair morning, with some white frost. In the evening, it began to rain again. The country towards the south is mountainous at some distance off, and there is some snow on the mountains.

Ordway   one canoe taken away from the landing by the tide last night.   A hard white frost this morning and cold.

Whitehouse   We had a hard white frost & cold, & windy morning. The latter part of the day was cold & cloudy, & in the Evening we had a little Rain & high Wind from the North East—

MONDAY, DECEMBER 9

	Sunrise			4 p.m.			River	
Temp	Weather	Wind	Temp	Weather	Wind	Rise/Fall	Feet	Inches
	_c r_	NE						

Weather Diary   cloudy and rained moderately untill 3 PM.

Clark   rained all the last night   we are all wet. In the evening it began to rain with a tremendous storm and Continud accompanied with a Violent wind from the SW until 10 oClock PM—

Gass   The morning was cloudy and wet. It continues cloudy and wet all day.

Ordway   rained the greater part of last night   rained hard all day   found the canoe which the tide took off the other night.

Whitehouse   We had rain the greater part of last night, & it continued raining this morning.

---

*According to Moulton (1990, 6: 261), Clark's Journal Codex I lists this weather data as "c."

TUESDAY, DECEMBER 10

	Sunrise			4 p.m.			River	
Temp	Weather	Wind	Temp	Weather	Wind	Rise/Fall	Feet	Inches
	r	NE						

Weather Diary   a violent wind last night 6 to 9 PM. river fast with rain. rained all day. (Lewis)   rained all day and the air cool I return from the Ocean a violent wind last night from the SW rained the greater part of the night of the 8th and all day the 9th int. (Clark)

Clark   a Cloudey rainey morning. I proceeded on to my Camp thro a heavy Cold rain. The day was Cloudy I could not See distinctly— rained nearly all day

Gass   We had another wet cloudy morning; and all hands were employed at work notwithstanding the rain.

Ordway   rained hard the most of the day

Whitehouse   It rain'd the most part of this day.

WEDNESDAY, DECEMBER 11

	Sunrise			4 p.m.			River	
Temp	Weather	Wind	Temp	Weather	Wind	Rise/Fall	Feet	Inches
	r	SW						

Weather Diary   rained moderately all last night and to day

Clark   rained all the last night moderately.   The rain continued moderately all day.

Gass   This day was so cloudy and wet.

Ordway   rained the greater part of the day.

Whitehouse   A wet morning & the party continued cutting logs... It continued raining the greater part of this day—

THURSDAY, DECEMBER 12

	Sunrise			4 p.m.			River	
Temp	Weather	Wind	Temp	Weather	Wind	Rise/Fall	Feet	Inches
	r	SW						

Weather Diary   rained moderately all last night and to day

Clark   Some moderate showers of rain at intervals all last night and to day.
Gass   This morning was cloudy without rain Some rain fell in the evening.
Whitehouse   It was cloudy the whole of this day.

FRIDAY, DECEMBER 13

	Sunrise			4 p.m.			River	
Temp	Weather	Wind	Temp	Weather	Wind	Rise/Fall	Feet	Inches
	r	SW						

Weather Diary   rained moderately all last night and to day

Clark   Some showers of light rain last night, and to day several verry hard Showers.

Gass   We had a cloudy, but fine morning. The day continued cloudy and some rain fell
in the evening.

Ordway   cloudy & rain

Whitehouse   We had rain & Cloudy weather, during the whole of this day.

SATURDAY, DECEMBER 14

	Sunrise			4 p.m.			River	
Temp	Weather	Wind	Temp	Weather	Wind	Rise/Fall	Feet	Inches
	r	SW						

Weather Diary   rained moderately all last night and to day

Clark   a cloudy day & rained moderately. All our last supply of Elk has Spoiled in the
repeeted rains which has been fallen ever Since our arrival at this place, and for a
long time before, Scerce one man in Camp can bost of being one day dry Since we
landed at this point

Gass   In the course of the day a good deal of rain fell; the weather here still continues
warm, and there has been no freezing, except a little white frost.

Ordway   continues wet and rainy.

Whitehouse   This day we had moderate Rain.

SUNDAY, DECEMBER 15

	Sunrise			4 p.m.			River	
Temp	Weather	Wind	Temp	Weather	Wind	Rise/Fall	Feet	Inches
	c a r	SW						

Weather Diary   rained all last night and untill 8 AM to day after which it was Cloudy all day. (Lewis)   rained at Short intervales from the 10th instant untill 8 AM to day I with 16 men Set out after meat. (Clark)

Clark   Cloudy all Day. Some rain in the evening. Serjt. Ordway, Colter, Colins, Whitehouse & McNeal Staid out all night without fire and in the rain—

Gass   The morning was cloudy. Some light showers fell during the day.

Whitehouse   We had cloudy weather. We were obliged to s[t]ay out during the Night. It rained all that night & the wind blew very cold & being without fire, we suffered considerably both from the Rain & wind.

MONDAY, DECEMBER 16

	Sunrise			4 p.m.		River		
Temp	Weather	Wind	Temp	Weather	Wind	Rise/Fall	Feet	Inches
	r	SW						

Weather Diary   rained all the last night. air Cold wind violent from the SW accompanied with rain.—

Clark   Rained all the last night.   We covered our selves as well as we could with elk skins & set up the greater part of the night, all wet, I lay in water verry cold, a most dreadfull night the rain continues, with Tremendious gusts of wind. The winds violent from the SE. With some risque proceeded on thro high waves in the river, a tempestious disagreeable day. Trees falling in every direction, whorl winds, with gusts of rain Hail & Thunder, this kind of weather lasted all day, Certainly one of the worst days that ever was!

Gass   This was a wet morning with high wind. The men who were out…had a very bad night, as the weather was stormy and a great deal of rain fell. Notwithstanding this, a serjeant and four men, who had got lost, lay out all night without fire. The whole of the day was stormy and wet.

Ordway   rained hard all last night and cold   we Suffered with wet & cold all last night. Hard rain and high wind.

Whitehouse   It rained very hard during this day. We had hard Rain & some hail in the afternoon also.

TUESDAY, DECEMBER 17

	Sunrise			4 p.m.		River		
Temp	Weather	Wind	Temp	Weather	Wind	Rise/Fall	Feet	Inches
	f a r & h	SW						

Weather Diary   rained all last night and to day untill 9 AM when we had a Shower of hail for an hour and Cleared off.

Clark   Some rain last night and a continuation of it this morning. The fore part of this day rained hailed and blew hard, the after part of the day fair & Cool. The mountain which lies SE about 10 miles distant is covered with Snow on its top which is ruged and uneavin.

Gass   This was another cloudy day, with some light showers of rain and hail.

Ordway   a little Snow and hail fell last night and continues this morning.

Whitehouse   We had during last night some Snow & hail & it continued the same this morning. A little snow remained on the pine trees the whole of this day.

WEDNESDAY, DECEMBER 18

	Sunrise			4 p.m.			River	
Temp	Weather	Wind	Temp	Weather	Wind	Rise/Fall	Feet	Inches
	c a r s h	SE						

Weather Diary   rained Snowed and hailed at intervales all the last night and to day untill meridian.

Clark   rained and Snowed alternetly all the last night, and Spurts [Clark later rewrites as "gusts"] of Snow and Hail Continued untill 12 oClock, which has chilled the air, Cold disagreeable Dreadfull day, the wind hard and unsettled. At 12 the Hail and Snow Seased, and the after part of the day was Cloudy with Some rain.

Gass   Snow fell last night about an inch deep, and the morning was stormy. In the middle of the day the weather became clear, and we had a fine afternoon.

Ordway   cloudy and rain. A little hail and frozen rain & cold—

Whitehouse   This day was cloudy with some Rain. The day grew very cold, & some hail fell.

THURSDAY, DECEMBER 19

	Sunrise			4 p.m.			River	
Temp	Weather	Wind	Temp	Weather	Wind	Rise/Fall	Feet	Inches
	h r & c	SW						

Weather Diary   rained (and hailed) last night and Several Showers of Hail and rain to day. the air Cool.

Clark   Some rain with intervales of fair weather last night, this morning Clear & the wind from the SW. The after part of the Day Cloudy with Hail and rain.

Gass   This was a fine clear cool morning; and we expected to have some fair pleasant weather, but at noon it became cloudy again and began to rain.

Ordway   rained all last night, and continues hard this morng.

Whitehouse   It rained hard all last night, & continued the same this morning.

FRIDAY, DECEMBER 20

	Sunrise			4 p.m.			River	
Temp	Weather	Wind	Temp	Weather	Wind	Rise/Fall	Feet	Inches
	f a r h	*SW*						

Weather Diary   Some rain and hail last Night   the rain Contd. untill 10 AM

Clark   Some rain and hail last night and this morning it rained hard untill 10 oClock. The after part of the day Cloudy with Several Showers of rain—

Gass   The morning was cloudy and wet— about 10 o'clock the weather became clear; but before night it rained as fast as before.

Ordway   cloudy and rain. About 10 oClock cleared off, but rained again before evening—

Whitehouse   A Cloudy wet morning, & continued so the whole of this day. We continued on building our huts, notwithstanding the badness of the weather—

SATURDAY, DECEMBER 21

	Sunrise			4 p.m.			River	
Temp	Weather	Wind	Temp	Weather	Wind	Rise/Fall	Feet	Inches
	r	*SW*						

Weather Diary   rained last night and to day

Clark   rained as useal all the last night and contd. moderately all day to day without any intermition.

Gass   had occasional rain and high winds but the weather continued warm

Ordway   Still continues raining by we Still kept at work finishing our huts to make ourselves comfortable &C—

Whitehouse   A cloudy wet dy as usual, but rather warm.

SUNDAY, DECEMBER 22

	Sunrise			4 p.m.			River	
Temp	Weather	Wind	Temp	Weather	Wind	Rise/Fall	Feet	Inches
	r	*SW*						

Weather Diary   rained last night and to day

Clark rained Continued all the last night and to day without much intermition. We discover that part of our last Supply of meat is Spoiling from the [warmth] of the weather not withstanding a constant Smoke kept under it day and night.

Gass had occasional rain and high winds but the weather continued warm

Ordway high wind all last night. The weather rainy warm & wet.

Whitehouse Cloudy & wet weather, the Air Warm & Wind blowing from the Southwest

MONDAY, DECEMBER 23

	Sunrise			4 p.m.			River	
Temp	Weather	Wind	Temp	Weather	Wind	Rise/Fall	Feet	Inches
	r h & l	SW						

Weather Diary rained all last night and moderately to day with Several Showers of Hail accompanied with hard Claps of Thunder and Sharp Lightning. (Lewis) rained 21 & 22 all day & night (Clark)

Clark rained without intermition all the last night and to day with Thunder and some Hail the morning and evening with rain this day.

Gass had occasional rain and high winds but the weather continued warm

Ordway nothing extraordinary hapened more than common this day—

Whitehouse Cloudy & wet weather, the Air Warm & Wind blowing from the Southwest

TUESDAY, DECEMBER 24

	Sunrise			4 p.m.			River	
Temp	Weather	Wind	Temp	Weather	Wind	Rise/Fall	Feet	Inches
	r	SW						

Weather Diary rained at intervales last night and to day.

Clark Some hard rain at Different times last night, and moderately this morning and the rest of the day without intermission. A hard rain in the evening.

Gass had occasional rain and high winds but the weather continued warm

Ordway hard rain as usal.

Whitehouse Cloudy & wet weather, the Air Warm & Wind blowing from the Southwest

WEDNESDAY, DECEMBER 25

	Sunrise			4 p.m.			River	
Temp	Weather	Wind	Temp	Weather	Wind	Rise/Fall	Feet	Inches
	c r	SW						

Weather Diary   rained at intervales last night and to day.

Clark   Some rain at different times last night and Showers of hail with intervales of fair Starr light. w[a]rm Day. The day proved Showerey wet and disagreeable.

Gass   Was another cloudy wet day. had occasional rain and high winds but the weather continued warm

Ordway   rainy & wet. Disagreeable weather.   We all moved in to our new Fort.   We expect this to be the last winter that we will have to pass in this way—

Whitehouse   We had hard rain & Cloud weather as usual.

THURSDAY, DECEMBER 26

	Sunrise			4 p.m.			River	
Temp	Weather	Wind	Temp	Weather	Wind	Rise/Fall	Feet	Inches
	rat&l	SW						

Weather Diary   raind with violent wind all last night and to day with Hard Claps of thunder & Sharp Lightning.

Clark   rained and blew hard with great violence SE last night, some hard claps of Thunder, The rain continued as usial all day and wind blew hard from the SE.

Gass   cloudy with rain

Ordway   we found that our huts Smoaked by the high winds and hard Storms hard rain continues as usal—

Whitehouse   We had Stormy weather the whole of this day. It rain'd most part of this day—

FRIDAY, DECEMBER 27

	Sunrise			4 p.m.			River	
Temp	Weather	Wind	Temp	Weather	Wind	Rise/Fall	Feet	Inches
	r	SW						

Weather Diary   rained moderately last night and to day

Clark   rained last night as usial and the greater part of this day. w[a]rm weather. Elk meet nearly Spoiled; & this accident of Spoiled meet, is owing to w[a]rmth & the repeeted rains, which cause the meet to tante before we Can get it from the woods. Musquestors troublesom

Gass   cloudy with rain

Ordway   hard rain all day—

Whitehouse   It continued raining hard during the whole of this day. ...a large fish was drove by the Wind & waves on the shore near to where their lodges were...

SATURDAY, DECEMBER 28

Sunrise			4 p.m.			River		
Temp	Weather	Wind	Temp	Weather	Wind	Rise/Fall	Feet	Inches
	r	SE						

Weather Diary   rained moderately last night and to day

Clark   rained as usial the greater part of the last night, and this morning rained and the wind blew hard from the SE. This day is verry w[a]rm, and rained all day without intermition.

Gass   cloudy with rain

Ordway   Capt. Lewis...three men got ready to go with a canoe to See the whail as we expect it is, but the wind and Storm arose So high that they could not go.

Whitehouse   This morning it rained & the wind was so high, that it prevented us from going to see the Whale.

SUNDAY, DECEMBER 29

Sunrise			4 p.m.			River		
Temp	Weather	Wind	Temp	Weather	Wind	Rise/Fall	Feet	Inches
	c a r	SE						

Weather Diary   rained moderately last night and to day   untill 7 AM after Cloudy the remained of the day   wind hard from the SE (Lewis)   rained moderately without much intermittion from the 26th untill 7 AM this morning hard wind from the SE. (Clark)

Clark   rained last night as usial, this morning cloudy without rain, a hard wind from the SE. The wind has proved too high as yet for him to Set out in Safty. I have the

Satisfaction to Say that we had but little rain in the Course of this day, only not as much as would wet a person. But hard wind and Cloudy all day but verry light rain.

Gass   This was a cloudy morning, but a fair day succeeded

Ordway   a fair day.

Whitehouse   This day was fine clear pleasant weather, the first fair day we had for a long time past.

MONDAY, DECEMBER 30

	Sunrise			4 p.m.			River	
Temp	Weather	Wind	Temp	Weather	Wind	Rise/Fall	Feet	Inches
	far	SE						

Weather Diary   Hard wind & rain last night.   to day tolerably fair.

Clark   Hard wind and Some rain last night. This morning fair and the Sun Shown for a Short time. This day proved the fairest and best we have had Since our arrival at this place, only 3 Showers of rain this whole day, Wind the fore part of the day. Cloudy nearly all day, in the evening the wind luled and the fore part of the night fair and clear.

Gass   Heavy shower of rain fell last night, but the morning was fair, and we had some sunshine, which happens very seldom; light showers of rain fell during the day.

Ordway   a fair morning and a little Sun shine which is verry uncommon at this place.

Whitehouse   We had several showers of Rain during last night, and this morning was fair; and the Sun shone a little which was very uncommon to us—

TUESDAY, DECEMBER 31

	Sunrise			4 p.m.			River	
Temp	Weather	Wind	Temp	Weather	Wind	Rise/Fall	Feet	Inches
	r	SW						

Weather Diary   rained last night and moderately all day to day.

Clark   A fair night. Last night was Cloudy and Some rain, this day proved cloudy and Some Showers of rain to day. The fore part of this night fair and clear.

Gass   Another cloudy morning

Ordway   a cloudy morning. We built a box for the centinel to Stand in out of the rain

Whitehouse   A cloudy morning.

Wednesday, January 1

	Sunrise			4 p.m.			River	
Temp	Weather	Wind	Temp	Weather	Wind	Rise/Fall	Feet	Inches
	c a r	*SW*		*r a c*	*S*			

Weather Diary   sun visible for a few minutes about 11 AM. the changes of the weather are exceedingly suddon. sometimes tho' seldom the sun is visible for a few moments the next it hails & rains, then ceases, and remains cloudy the wind blows and it again rains; the wind blows by squalls most generally and is almost invariably from SW   these visicitudes of the weather happen tow three or more times half a day. Snake seen 25th Decembr.

Clark   This morning proved cloudy with moderate rain, after a pleasant w[a]rm night during which there fell but little rain— Some fiew Showers of rain in the Course of this day. Cloudy all the day.

Gass   The year commenced with a wet day; but the weather still continues warm; and the ticks, flies and other insects are in abundance, which appears to us very extraordinary at this season of the year, in a latitude so far north.

Ordway   a pleasant morning.

Whitehouse   The Morning was pleasant. The Winters here are not very Cold, & the ground has not as yet been cover'd with Snow this Winter.

Thursday, January 2

	Sunrise			4 p.m.			River	
Temp	Weather	Wind	Temp	Weather	Wind	Rise/Fall	Feet	Inches
	c a r	*SW*		*r*	*SW*			

Clark   A cloudy rainey morning after a wet night. The day proved Cloudy and wet.

Gass   This was a cloudy wet day

Ordway   rained the greater part of last night and continues this morning.

Whitehouse   It rained the greater part of last night, and continued to rain hard this morning. We had hard Showers of rain during this whole day.

Friday, January 3

	Sunrise			4 p.m.			River	
Temp	Weather	Wind	Temp	Weather	Wind	Rise/Fall	Feet	Inches
	c a r h t & l	SW		*c a r h f*	SW			

Weather Diary the sun visible for a few minutes only. The thunder and lightning of the last evening was violent. a Singular occurrence for the time of year. the loss of my Thermometer I most sincerely regret. I am confident that the climate here is much warmer than in the same parallel of Latitude on the Atlantic Ocean tho' how many degrees is now out of my power to determine. Since our arrival in this neighbourhood on the 7th of November, we have experienced one slight white frost only which happened on the morning of the 16th of that month. we have yet seen no ice, and the weather so warm that we are obliged cure our meat with smoke and fire to save it. we lost two parsels by depending on the air to preserve it, tho' it was cut in very thin slices and sufficiently exposed to the air.

Clark The Sun rose fair this morning for the first time for Six weeks past, the Clouds Soon obscure it from our view, and a Shower of rain Suckceeded— last night we had Sharp lightening a hard thunder Suckceeded with heavy Showers of hail, and rain, which Continud with intervales of fair moon Shine dureing the night.

Gass The weather is still cloudy and wet. Rain still continued

Ordway hard Thunder hail and rain the greater part of last night.

Whitehouse We had hard thunder, hail & Rain the greater part of last night, & light showers of rain this morning.

Saturday, January 4

	Sunrise			4 p.m.			River	
Temp	Weather	Wind	Temp	Weather	Wind	Rise/Fall	Feet	Inches
	c a r & h	SW		*r a f & r*	SE			

Weather Diary the sun visible about 2hours

Gass The morning was wet.

Ordway Small Showers of rain and hail as usal.

Whitehouse We had small showers of rain & some hail this morning. The Rain continued the greater part of this day—

SUNDAY, JANUARY 5

Sunrise			4 p.m.			River		
Temp	Weather	Wind	Temp	Weather	Wind	Rise/Fall	Feet	Inches
	r	SE		r	SE			

Gass   This was a very wet day. I, however, notwithstanding the cold, stript and swam to the raft, brought it over and then crossed on it in safety. The rain and wind continued so violent that we agreed to stay at these camps all night.

Ordway   a wet rainy morning.

Whitehouse   A wet rainey morning.

MONDAY, JANUARY 6

Sunrise			4 p.m.			River		
Temp	Weather	Wind	Temp	Weather	Wind	Rise/Fall	Feet	Inches
	c a r	SE		f	E			

Weather Diary   the sun shown about 5 hours this evening & it continued fare during the night.

Lewis   The humidity of the air has been so excessively great.

Clark   all last night rained without intermition, & the morning. Soon after I arrived in the Bay the wind Sprung up from the NW and blew So hard and raised waves so high that we were obliged to put into a Small Creek Short of the village. The evening a butifull Clear moon Shiney night, and the 1st fair night which we have had for 2 months

Gass   We had a fair morning and the weather cleared up, after two months of rain, except 4 days.

Ordway   about 9 oClock AM cleared off pleasant and warm.

Whitehouse   About 7 oClock the Weather cleared off, & became warm & pleasant which continued during the whole of this day—

TUESDAY, JANUARY 7

Sunrise			4 p.m.			River		
Temp	Weather	Wind	Temp	Weather	Wind	Rise/Fall	Feet	Inches
	f	NE		c a r*	SE			

---

*According to Moulton (1990, 6: 261), Clark's Journal Codex I lists this weather data as "c a f."

Weather Diary   it clouded up just about sunset, but shortly after became fare.

Lewis   this is the first day during which we have had no rain since we arrived at this place. Nothing extraordinary happened today—

Clark   Some frost this morning. I hesitated a moment & view this emence mountain the top of which (apd) was obscured in the clouds.

Gass   Another fine day.

Ordway   clear and pleasant. Contn. Clear all day which is a very uncommon thing at this place.

Whitehouse   We had a clear pleasant night, & still continues so this morning; which is rare to be met with at this place at this Season of the Year.

WEDNESDAY, JANUARY 8

	Sunrise			4 p.m.			River	
Temp	Weather	Wind	Temp	Weather	Wind	Rise/Fall	Feet	Inches
	f	NE		c a f	SE			

Weather Diary   lost my PM obstn. for Equal Altitudes.

Lewis   In the consequence of the clouds this evening I lost my PM observation for Equal Altitudes, and from the same cause have not been able to take a single observation since we have been at this place.

Clark   The last night proved fair and Cold wind hard from the SE. a fine morning wind hard from the SE. The high tide obliged me to delay untill late before the tide put out. A fair night wind blew from the SE.

Gass   Another fine day.

Ordway   a clear warm morning–

Whitehouse   A fine warm morning.

THURSDAY, JANUARY 9

	Sunrise			4 p.m.			River	
Temp	Weather	Wind	Temp	Weather	Wind	Rise/Fall	Feet	Inches
	f	Sw		c a f	SW			

Weather Diary   began to rain at 10 PM and continued all night.

Clark   a fine morning wind NE. Day Clouded up. Day proved fine. Rained the greater part of the night.

Gass   fair and pleasant. during the night some rain fell.

Ordway   rained the greater part of lat night but cleared off pleasant this morning, and continues warm

Whitehouse   It rained the greater part of last night   The Weather cleared off this morning & became warm & pleasant.

FRIDAY, JANUARY 10

	Sunrise			4 p.m.		River		
Temp	Weather	Wind	Temp	Weather	Wind	Rise/Fall	Feet	Inches
	f a r	*Sw*		*c a f*	SW			

Weather Diary   Various flies and insects now alive and in motion.

Clark   arived at the fort, wet and Cold at 9 oClock PM. I arrived at the Canoes about Sunset, the tides was Comeing in.

Gass   The morning was fine

Ordway   a clear pleasant day.

Whitehouse   A Clear pleasant day. …several small Indian Villages…which lived on the Whales that were thrown ashore by the Waves, in tempestuous Weather.

SATURDAY, JANUARY 11

	Sunrise			4 p.m.		River		
Temp	Weather	Wind	Temp	Weather	Wind	Rise/Fall	Feet	Inches
	c	SW		*c a r*	SW			

Lewis & Clark   this morning the Sergt. Of the guard reported the absence of our Indian Canoe, on enquiry we found that those who came in it last evening had been negligent in securing her and the tide in the course of the night had taken her off.

Gass   Pleasant

Whitehouse   And this morning, we had pleasant weather. We had rain towards night—

SUNDAY, JANUARY 12

	Sunrise			4 p.m.		River		
Temp	Weather	Wind	Temp	Weather	Wind	Rise/Fall	Feet	Inches
	f a c	NW		*c*	NW			

Weather Diary   cool this morning but no ice nor frost   at miday sand flies and insects in motion   the wind from any quarter off the land or along the NW Coast causes the air to become much cooler. every species of waterfowl common to this country at any season of the year still continues with us.

Ordway   a fair morning.
Whitehouse   A Clear pleasant Morning.

Monday, January 13

	Sunrise			4 p.m.			River	
Temp	Weather	Wind	Temp	Weather	Wind	Rise/Fall	Feet	Inches
	r	SW		r	SW			

Lewis   The bay in which this trade is carryed on is spacious and commodious, and perfectly secure from all except the S and SE winds, these however are the most prevalent and strong winds in the Winter season.

Clark   The Bay in which this trade is Carried on is Spacious and Commodious, and perfectly Secure from all except the S and SE winds and those blow but Seldom the most prevalent & Strong winds are from the SW and NW in the winter season.

Gass   The weather changed and we had a cloudy wet day
Ordway   rained hard all last night and continues this morning.
Whitehouse   It rained during the whole of last night and continues Raining this morning.

Tuesday, January 14

	Sunrise			4 p.m.			River	
Temp	Weather	Wind	Temp	Weather	Wind	Rise/Fall	Feet	Inches
	f a r	NE*		c a r	S			

Weather Diary   weather perfectly temperate   I never experienced a winter so warm as the present has been.

Lewis & Clark   This morning the Sergt. of the Guard reported the absence of one of the large peroques, it had broken the chord by which it was attached and the tide had taken it off....found her. We now directed three of the perogues to be drawn up out of reach of the tide.

Gass   The morning was pleasant
Ordway   the tide water took away one of our canoes but we Soon found it again
Whitehouse   A fine pleasant morning.

---

*According to Moulton (1990, 6: 261), Clark's Journal Codex I lists this wind direction as "NW."

WEDNESDAY, JANUARY 15

	Sunrise			4 p.m.			River	
Temp	Weather	Wind	Temp	Weather	Wind	Rise/Fall	Feet	Inches
	r a c & r	SE		*r a r*	S			

Weather Diary   Saw several insects, weather warm, we could do very well without fire, I am satsifyed that the murcury would stand at 55 a 0.

Lewis   we had determined to send out two hunting parties today but it rained so incessantly that we posponed it.

Clark   rained hard all day.

Gass   wet throughout

Ordway   rained hard the greater part of the day.

Whitehouse   It rained hard, & we had stormey weather

THURSDAY, JANUARY 16

	Sunrise			4 p.m.			River	
Temp	Weather	Wind	Temp	Weather	Wind	Rise/Fall	Feet	Inches
	r a r	SW		*r a r*	SW			

Weather Diary   wind hard this morning   rained incessently all night.

Lewis   ....every one appears content with his situation and his fare. It is true that we could even travel now on our return as far as the timbered country reaches, or to the falls of the river; but further it would be madness for us to attempt to preceede untill April, as the indians inform us that the snows lye knee deep in the plains of Columbia during the winter, and in the these plains we could scarecely get as much fuel as would cook our provision as we descended the river; and even were we happyly over the plains and again in the woody country at the foot of the Rocky Mountains we could not possibly pass that immence barrier of mountains on which the snows ly in winter to the debth in many places of 20 feet; in short the Indians inform us that they are impracticable untill about the 1st of June, at which time even there is an abundance of snow but a scanty subsistence may be obtained for the horses. We should not therefore forward ourselves on our homeward journey be reaching the rocky mountains. Early than the 1st of June, which we can easily effect by setting out from hence on the 1st of April.

Gass   wet throughout

Ordway   the rain & Storm high wind continues as usal.

Whitehouse   It rained hard, & we had stormey weather

Friday, January 17

Sunrise			4 p.m.			River		
Temp	Weather	Wind	Temp	Weather	Wind	Rise/Fall	Feet	Inches
	c a r	SW		c	SW			

Weather Diary   rained incessently all night, insect in motion

Gass   some cloudy, but about 10 o'clock they disapeared and we had a fine day

Whitehouse   It continued stormey all last night, and this morning Wet & rainey.

Saturday, January 18

Sunrise			4 p.m.			River		
Temp	Weather	Wind	Temp	Weather	Wind	Rise/Fall	Feet	Inches
	r a r	SW		c a r	SW			

Weather Diary   rained very hard last night

Gass   Last night was very dark; and early in it rain came on and continued all night. This day is also wet.

Ordway   hard rain all last night, and continues as usal.

Whitehouse   It rained hard all last night, & still continued the same this morning. It continued Raining during the whole of this day.

Sunday, January 19

Sunrise			4 p.m.			River		
Temp	Weather	Wind	Temp	Weather	Wind	Rise/Fall	Feet	Inches
	c a r	S		c a r	SW			

Weather Diary   rained the greater part of last night.

Gass   morning was fair with flying clouds; but in the evening it began to rain again.

Ordway   moderate Showers of rain.

Whitehouse   This morning we had moderate showers of rain

MONDAY, JANUARY 20

Sunrise			4 p.m.			River		
Temp	Weather	Wind	Temp	Weather	Wind	Rise/Fall	Feet	Inches
	r a r	*SW*		*r a r*	*SW*			

Weather Diary   rained greater part of night   wind hard

Gass   It rained hard all day. The evening was so wet and stormy.

Ordway   rainy and wet.

Whitehouse   Wet & rainey weather during the whole of this day. Nothing material occured worth mentioning.

TUESDAY, JANUARY 21

Sunrise			4 p.m.			River		
Temp	Weather	Wind	Temp	Weather	Wind	Rise/Fall	Feet	Inches
	c a r	*SW*		*c a r*	*SW*			

Weather Diary   wind hard this morning   contued all day

Gass   rain

Ordway   Cloudy and rain.

Whitehouse   A Cold cloudy day with Rain.

WEDNESDAY, JANUARY 22

Sunrise			4 p.m.			River		
Temp	Weather	Wind	Temp	Weather	Wind	Rise/Fall	Feet	Inches
	r a r	*SW*		*c a r*	*SW*			

Weather Diary   wind violent last night & this morning

Clark   Some rain this day at intervales—

Gass   rain. I saw some amazingly large trees of the fir kind; they are from 12 to 15 feet in diameter.

Ordway   a hard Storm of rain and verry high wind. We had a disagreeable time of it.

Whitehouse   A hard storm of Wind & Rain. A very disagreeable time of it.

THURSDAY, JANUARY 23

	Sunrise			4 p.m.			River	
Temp	Weather	Wind	Temp	Weather	Wind	Rise/Fall	Feet	Inches
	c a r h t & l*	*SW*		*c a f*	*SW*			

Weather Diary the sun shown about 2 h in the fore noon   when the sun is said to shine ore the weather fair it is to be understood that it bearly casts a shaddow, and that the atmosphere is haizy of a milkey white colour.

Gass We had a fine clear cool morning   the day continued pleasant until about 4 o'clock in the afternoon, when the weather became cloudy, and it began to rain.

Ordway a little Thunder and hail in the course of last night   high wind &C—

Whitehouse We had during last night thunder & some hail Showers. It rained & we had high wind during this day—

FRIDAY, JANUARY 24

	Sunrise			4 p.m.			River	
Temp	Weather	Wind	Temp	Weather	Wind	Rise/Fall	Feet	Inches
	c a r & s	*SE*		*c a r h & s*	*E*			

Weather Diary this morning the snow covered the ground and was cooler than any wether we have had, but no ice

Clark The nativs of this neighbourhood ware no further Covering than a light roabe, their feet legs & every other part exposed to the frost Snow & ice &c.

Gass At daylight some snow fell, and there were several snow showers during the day. The Indians were barefooted notwithstanding the snow on the ground; and the evening was so bad we permitted them to stay in the fort all night.

Ordway a light Snow fell the later part of last night. Several Showers of rain and hail this morning.

Whitehouse Last night we had a light snow, which hardly made the ground white, & some showers of rain & hail fell during this day.

---

*According to Moulton (1990, 6: 261), Clark's Journal Codex I lists this weather data as "c a r t & l."

SATURDAY, JANUARY 25

	Sunrise			4 p.m.			River	
Temp	Weather	Wind	Temp	Weather	Wind	Rise/Fall	Feet	Inches
	h a r h *& s*	NE		*c a r h* *& s*	NE			

Weather Diary   the ground covered with snow this morning $1/2$ inch deep   ice on the water in the canoes $1/4$ of an inch thick. it is now preceptably [colder] than it has been this winter.—

Gass   The morning was cloudy and some showers of snow fell in the course of the day; and in the night it fe[l]l to the depth of 8 inches.

Ordway   froze a little last night, and a little more Snow fell intermixet with hail. Continues Squawlly this morning.

Whitehouse   We had snow during last night & it continued snowing lightly this morning. The ground had froze a little.

SUNDAY, JANUARY 26

	Sunrise			4 p.m.			River	
Temp	Weather	Wind	Temp	Weather	Wind	Rise/Fall	Feet	Inches
	c a h & s	NE		*c a s*	NE			

Weather Diary   at 4 PM last evening the snow was one Inch deep at sunrise this morning 4 $1/2$ [Clark has 4 $3/4$ written down] inches deep   icesickles of 18 Inches in length hanging to the eves of the houses. coulder than it has been the snow this evening is 4 $3/4$ inches deep, the icesickles of 18 inches in length continued suspended from the eves of the houses during the day. it now appears something like winter for the first time this season.

Lewis & Clark   (evergreen huckleberry) this shrub retains [its] virdure very perfectly during the winter and is a beautifull shrub.

Gass   some light showers during the day, but the evening the weather cleared up and it began to free[ze] hard. This is the first freezing weather of any consequence we have had during the winter.

Ordway   considerable of Snow fell in the course of last night and continues this morning, and cold freezing weather   the Snow is this evening about 5 Inches deep on a level—

Whitehouse   During last night we had considerable Snow & it continued snowing this morning. The weather was cold & freezing & the Snow lay on the ground during this day 5 Inches deep on a level   It continued Snowing 'till the Evening.

Monday, January 27

	Sunrise			4 p.m.			River	
Temp	Weather	Wind	Temp	Weather	Wind	Rise/Fall	Feet	Inches
	f a s	*NE*		*f*	*NE*			

Weather Diary   the sun shone more bright this morning than it has done since our arrival at this place. the snow since 4 PM yesterday has increased to the debth of 6 Inches, and this morning is perceptibly the [coldest] that we have had. I suspect the Murcury would stand at about 20^0 above naught' the breath is perceptible in our room by the fire.

Lewis & Clark   a county almost inaccessible from the fallen timber, brush and sink-holes, which were now disgused by the snow.

Gass   This was a clear cold frosty morning, and the snow about 9 inches deep. Where the sun shone on it during the day, a considerable quantity melted; but these places were few as the whole face of the country near this is closely covered with fire timber.

Ordway   froze hard last night   a clear cold morning

Whitehouse   It froze hard during last night, & this morning was clear & cold.

Tuesday, January 28

	Sunrise			4 p.m.			River	
Temp	Weather	Wind	Temp	Weather	Wind	Rise/Fall	Feet	Inches
	f	*NE*		*f*	*NE*			

Weather Diary   last night exposed a vessel of water to the air with a view to discover the debth to which it would friez in the course of the night, but unfortunately the vessel was only 2 inches deep and it freized the whole thickness; how much more it might have frozen had the vessel been deeper is therefore out of my power to decide. it is the [coldest] night that we have had, and I suppose the murcury this morning would have stood as low as 15^0 above 0.—

Lewis & Clark   the badness of the weather and the difficulty of the road had caused their delay. The Elk had been killed just before the snow fell which had covered them and so altered the apparent face of the county that the hunters could not find the Elk

Gass   A clear cold morning, and the weather continued cold all day. About half of our men were employed in bringing home meat; and it was found a very cold uncomfortable business.

Ordway   a clear cold morning, and freezing hard.

Whitehouse   A Clear cold morning, & freezing weather. I got during this day my feet severely frost bit—

WEDNESDAY, JANUARY 29

Sunrise			4 p.m.			River		
Temp	Weather	Wind	Temp	Weather	Wind	Rise/Fall	Feet	Inches
	f	NE		*f*	NE			

Weather Diary not so [cold], water in a vessel exposed to the [air] during the night freized
³/₈ths of an inch only.

Lewis & Clark *Sac a commis* - this shrub is an evergreen, the leaves retain their virdure
most perfectly through the winter even in the most rigid climate as on lake
Winnipic. The frost appears to take no effect on it.

Gass We had a cold clear morning; and the day continued clear throughout.

Ordway froze hard last night   a clear cold morning. We do nothing except git wood for
our fires &C.

Whitehouse It froze very hard during last night, & this morning was clear cold weather.

THURSDAY, JANUARY 30

Sunrise			4 p.m.			River		
Temp	Weather	Wind	Temp	Weather	Wind	Rise/Fall	Feet	Inches
	s a s	N		*c a s*	W			

Weather Diary the weather by no means as [cold] as it has been   snow feell about
one inch deep

Lewis & Clark the Clatsops - they never wear leggins or mockersons which the mildness
of this climate I presume has rendered in a great measure unnecessary

Gass the weather was cloudy and not so cold as the day before; and some snow fell.

Ordway Cloudy and cold. Some fine Snow fell this morning. The evening clear and cold—

Whitehouse This day Cold & Cloudy, & some Snow fell, in the fore part of this day. In the
Evening we had clear cold weather.

FRIDAY, JANUARY 31

Sunrise			4 p.m.			River		
Temp	Weather	Wind	Temp	Weather	Wind	Rise/Fall	Feet	Inches
	f a c	NE		*f*	NE			

Weather Diary this morning is plesant, the night was clear and cold. nothwithstanding the
[cold] weather the Swan white Brant geese & ducks still continue with us;

the sandhill crain also continues.— the brown or speckled brant are mostly gone some few are still to be seen   the Cormorant loon and a variety of other waterfowls still remain. The Winds from the Land brings us [cold] and clear weather while those obliquely along either coast or off the Oceans bring us warm damp cloudy and rain weather. the hardest winds are always from the SW.

The blue crested Corvus bird has already began to build [its] nest. their nests are formed of small sticks; usually in a pine tree.—

Great numbers of Ravens, and a Small black Crow are continually about us. The pale yellwo Streiked and dove coloured robin is about, also the little brown ren or fly-catsch which is a little larger than the humming bird.

Lewis & Clark   Sent a party of eight men up the river this morning to renew their surch for Elk and also to hunt; they proceded but a few miles before they found the river so obstructed with ice that they were obliged to return.

Gass   This was a clear cold morning.

Ordway   a clear cold freezeing morning. Six men set out a hunting   took a canoe   found Ice in the River so that they turned back.

Whitehouse   A Clear cold morning with frost. Sergt. Gass & party returned, & informed us that the River was froze across a short distance up it & that the could not proceed—

SATURDAY, FEBRUARY 1

Sunrise			4 p.m.			River		
Temp	Weather	Wind	Temp	Weather	Wind	Rise/Fall	Feet	Inches
	f	*NE*		*f*	*NE*			

Weather Diary   the weather by no means as [cold] as it was tho' it freized last night

Lewis & Clark   This morning 5 went to find the Elk which had been killed some day since, and which could not be found in consequence of the snow.

Gass   We had a fine clear cold morning.

Ordway   a clear cold morning.

Whitehouse   A clear cold morning.;

SUNDAY, FEBRUARY 2

Sunrise			4 p.m.			River		
Temp	Weather	Wind	Temp	Weather	Wind	Rise/Fall	Feet	Inches
	f	*NE*		*c a s*	*SW*			

Weather Diary   the bald Eagle still remains.

Gass   The morning was pleasant and the weather more moderate. About the middle of the day it began to thaw and in the evening rain.

Ordway   the weather moderate. In the afternoon cloudy & a little Snow— but not any worth menting.

Whitehouse   We had a clear morning, & the day was moderate. In the evening it was cloudy & a little Snow fell—

MONDAY, FEBRUARY 3

	Sunrise			4 p.m.			River	
Temp	Weather	Wind	Temp	Weather	Wind	Rise/Fall	Feet	Inches
	c a s & r	NW		*c a f*	NE			

Weather Diary   the snow fell about half an inch, but the rain which succeded soon melted it at 9 AM the sun shone. the rain which feel in the latter part of the night freized and formed a slight incrustation on the snow which fell some days past, and also on the boughs of the trees &c. yesterday it continued fair until 11 AM when the wind vered about to SW and the horizon was immediately overcast with clouds, which uniformly takes place when the wind is from that point.

Lewis & Clark   the winds was so high that they were unable to set out untill a little before sunset, when they departed; at 10 PM they return excessively [cold] and informed us that they could not make land on this side of the bay nor get into the creek in consequence of the tide being out and much lower than usual.

Gass   Some light showers of rain fell in the course of the night; and this day is still somewhat wet and cloudy.

Ordway   a little frozen rain. Six men set out with a canoe after the meat, but the wind So high that obledged them to return

Whitehouse   We had a little frost, & the weather has moderated since Yesterday. Six men went out...but soon returned the wind being too high for them to proceed. In the evening they attempted it again, but the tide was so low that they could not get near the Shore

TUESDAY, FEBRUARY 4

	Sunrise			4 p.m.			River	
Temp	Weather	Wind	Temp	Weather	Wind	Rise/Fall	Feet	Inches
	f	NE		*f*	NE			

Weather Diary   the last night clear and [cold]   the Netul frozen over in several places. all the waterfowls before innumerated still continue with us. the bird which resembles the robbin have now visited us in small numbers   saw two of them yesterday about the fort; they are gentle.

Gass This was a fine clear morning. This day continued throughout clear and pleasant.
Ordway a clear pleasant morning. Tide high—
Whitehouse A Clear pleasant morning. We had a very high tide this day.

WEDNESDAY, FEBRUARY 5

	Sunrise			4 p.m.			River	
Temp	Weather	Wind	Temp	Weather	Wind	Rise/Fall	Feet	Inches
	f	NE		f	NE			

Lewis & Clark I sent sergt. Gass and party of men over; the tide being in, they took advantage of a little creek...recovered our Indian Canoe, so long lost and much lamented.

Gass clear cool day
Ordway a beautiful pleasant morning.
Whitehouse We had a beautiful pleasant cool morning.

THURSDAY, FEBRUARY 6

	Sunrise			4 p.m.			River	
Temp	Weather	Wind	Temp	Weather	Wind	Rise/Fall	Feet	Inches
	f	NE		c	SW*			

Weather Diary very cold last night think it reather the coldest night that we have had. cloudy at 9 AM

Gass We had a cool fair morning. 10 of us had to camp out, with the assistance of the elk skins and our blankets, we lodged pretty comfortable, though the snow was 4 or 5 inches deep.
Whitehouse This morning we had pleasant weather.

FRIDAY, FEBRUARY 7

	Sunrise			4 p.m.			River	
Temp	Weather	Wind	Temp	Weather	Wind	Rise/Fall	Feet	Inches
	c	SW		c	SW			

*According to Moulton (1990, 6: 365), Clark's Journal Voorhis No. 2 lists this wind direction as "NW."

Weather Diary   continued cloudy all night   a little snow at 10 AM.

Gass   The morning was fair. It rained hard and we had a disagreeable night.

Ordway   hard rain &C—

Whitehouse   The weather continued pleasant. A short time after dark we had a hard Rain—

SATURDAY, FEBRUARY 8

Sunrise			4 p.m.			River		
Temp	Weather	Wind	Temp	Weather	Wind	Rise/Fall	Feet	Inches
	c a s r & h	*SW*		*c a f r h & s*	*SW*			

Weather Diary   it was principally rain which fell since 4 PM yesterday, it has caused the snow to disappear   the rain of the last night has melted down the snow wich has continued to cover ground since the 24th of January; the feeling of the air and other appearances seem to indicate, that the rigor of the winter have passed; it is so warm that we are apprehensive that our meat will spoil, we therefore cut it in small peices and hang it seperately on sticks. Saw a number of insects flying about. the small brown flycatch continues with us. this is the smallest of all the American birds except the humming bird.

Gass   About noon there were showers of rain and hail

Ordway   rained verry hard all last night   we had several showers of hail this evening.

Whitehouse   It rained very hard the greater part of last night; the Men that went after the meat had a very disagreeable time of it. We had several small showers of rain & hail in the Evening.

SUNDAY, FEBRUARY 9

Sunrise			4 p.m.			River		
Temp	Weather	Wind	Temp	Weather	Wind	Rise/Fall	Feet	Inches
	c a r & h	*SW*		*c a r & h*	*SW*			

Weather Diary   principally rain which has fallen

Gass   We had a fine morning; but in the course of the day we had sometimes sunshine, and sometimes showers of rain.

Ordway   Several Showers of hail in course of the day.

Whitehouse   We had small showers of rain during this day

MONDAY, FEBRUARY 10

	Sunrise			4 p.m.			River	
Temp	Weather	Wind	Temp	Weather	Wind	Rise/Fall	Feet	Inches
	c a r h & s	*N*		*c a f & c*	*SW*			

Weather Diary Snow covered the ground this morning disappeared before evening. sun shown 2 hours

Gass A light snow fell last night, the morning was pleasant

Ordway a fair morning. A little Snow fell last night.

Whitehouse We had some Snow fell during last night and this morning the weather was clear & pleasant.

TUESDAY, FEBRUARY 11

	Sunrise			4 p.m.			River	
Temp	Weather	Wind	Temp	Weather	Wind	Rise/Fall	Feet	Inches
	c a f & c	*SW*		*r a f & r*	*SW*			

Gass This was a fine morning. We went out to hunt and remained out until the 17th during which time there was a great deal of heavy rain, and the weather changeable and disagreeable— During one of the most disagreeable nights, myself and another lay out in our shirts and overalls, with only one elk skin to defend us from a violent night's rain. Our shirts and overalls being all of leather made it the more disagreeable

Ordway the after part of the day rainy.

Whitehouse We had a fine clear day. The latter part of the day was rainey—

WEDNESDAY, FEBRUARY 12

	Sunrise			4 p.m.			River	
Temp	Weather	Wind	Temp	Weather	Wind	Rise/Fall	Feet	Inches
	r a r & c	*SW*		*r a c & r*	*SW*			

Weather Diary it rained the greater part of last night.

Ordway continues raining.

Whitehouse This day was rainey & wet.

THURSDAY, FEBRUARY 13

	Sunrise			4 p.m.			River	
Temp	Weather	Wind	Temp	Weather	Wind	Rise/Fall	Feet	Inches
	c a r	SW		c a r	SW			

Weather Diary   Wind very hard last evening and all night
Ordway   cloudy and rain.
Whitehouse   It rained the greater part of last night, and this morning was cloudy.

FRIDAY, FEBRUARY 14

	Sunrise			4 p.m.			River	
Temp	Weather	Wind	Temp	Weather	Wind	Rise/Fall	Feet	Inches
	c a f & s	SW		r a r f & r	SW			

Weather Diary   very small quantity of snow fell last night   not enough to cover the ground   somewhat colder this morning. the sun shown only a few moments.

Lewis & Clark   I completed a map of the Countrey through which we have been passing from the Mississippi at the Mouth of Missouri to this place. We now discover that we have found the most practicable and navigable passage across the Continent of North America. We discovered that there were no salmon in the Flathead river.

Ordway   the morning warm and Showers of rain through the course of this day—

Whitehouse   This morning was warm, & we had showers of rain during the whole of this day—

SATURDAY, FEBRUARY 15

	Sunrise			4 p.m.			River	
Temp	Weather	Wind	Temp	Weather	Wind	Rise/Fall	Feet	Inches
	c a r & f	S		c a r & f	SW			

Weather Diary   fair most of last night   hard frost this morning. the ground white with it. The robbin returned and were singing which reminded me of spring. some other small birds passed on their flight from the South, but were so high that we would not distinguish of what kind they were. the robbin had left this place before our arrival in November.

Lewis & Clark   Bratton informed that the cause of Sergt. Pryor's delay was attributeable to the winds which had been so violent for several days as to render it impossible to get a canoe up the creek. The SW winds are frequently very violent on the coast when we are but little sensible of them at Fort Clatsop. [Eastern Columbia Valley] no rain scarcely ever falls in these plains and the grass is short and but thin.

Ordway   a fair day.

Whitehouse   A clear morning.

SUNDAY, FEBRUARY 16

	Sunrise			4 p.m.			River	
Temp	Weather	Wind	Temp	Weather	Wind	Rise/Fall	Feet	Inches
	r a s & f	SW		r a f & r	SW			

Weather Diary   but a small quantity of snow   nearly all disolved by morning with the succeeding rain. at 11 AM it became fair and the insects were flying about. at $1/2$ after 12 O'Ck it again clouded up and began to rain.

Ordway   hard rain in the course of last night.

Whitehouse   We had hard rain during last night

MONDAY, FEBRUARY 17

	Sunrise			4 p.m.			River	
Temp	Weather	Wind	Temp	Weather	Wind	Rise/Fall	Feet	Inches
	c a r h & s	SW		r a f h s & r	SW			

Weather Diary   the hail and snow covered the ground this morning

Gass   The day was stormy

Ordway   a little Snow fell this afternoon.

Whitehouse   during last night some Snow fell. This Morning the weather clear.

TUESDAY, FEBRUARY 18

Sunrise			4 p.m.			River		
Temp	Weather	Wind	Temp	Weather	Wind	Rise/Fall	Feet	Inches
	c a r & h	*SW*		*r a r & h*	*SW*			

Weather Diary   wind violent greater part of the day and all night

Lewis & Clark   Sergt. Gass returned and reported that the waves ran so high in the bay that he could not pass to the entrance of the creek.

Gass   The morning of this day war cloudy. The weather was so stormy, we could not get round the bay, and we all returned to the fort.

Ordway   we found the wind so high at the bay that we ha[d] to return to the Fort. Had Several Squawls of wind & rain.

Whitehouse   This morning clear & pleasant weather. The party that were going to the Salt Camp on arriving at the bay, found the wind blowing so hard, that they returned to the fort—   We had several squalls of wind attended with rain in the course of this day—

WEDNESDAY, FEBRUARY 19

Sunrise			4 p.m.			River		
Temp	Weather	Wind	Temp	Weather	Wind	Rise/Fall	Feet	Inches
	r a r	*SW*		*r a r*	*SW*			

Weather Diary   wind violent all day.

Gass   The day was very wet and stormy

Ordway   a hard Storm of wind and rain. The frozen rain beat in our faces verry hard. Sand flew & waves rold.

Whitehouse   We had a hard Storm of Rain & high Wind, blowing from the SW. We proceeded on about half way, when the Storm was so high in the Priari, & on the Coast, that we could not proceed without suffering by the Sand blowing in our faces— and the Rain that fell froze & cut our faces likewise. We crossed a Creek, which took us middle deep, which benumbed & Chilled the party very much. We came to an Old deserted Indian hut, in which we made a fire. We staid at this place all night in expectation of the weather being better by morning—

THURSDAY, FEBRUARY 20

	Sunrise			4 p.m.			River	
Temp	Weather	Wind	Temp	Weather	Wind	Rise/Fall	Feet	Inches
	c a r	SW		c a r	SW			

Weather Diary   wind violent all night and the greater part of the day.

Gass   This was a cloudy morning. The Indians hats are made of cedar bark and silk grass...they keep the rain out. But little rain fell to day

Ordway   the wind continued verry high from the SW   we Set out eairly and proced. on along the coast faceing the wind the Sand cut our faces waided a creek rapid curret   the waves roles verry high and white froth flying &C. [at the Salt Works, near Seaside, Oregon]

Whitehouse   The wind still continued very high, blowing from the SW. The wind fell a little.

FRIDAY, FEBRUARY 21

	Sunrise			4 p.m.			River	
Temp	Weather	Wind	Temp	Weather	Wind	Rise/Fall	Feet	Inches
	r a c & r	SW		r a c & r	SW			

Weather Diary   the wind continues high this morning & untill evening.

Lewis & Clark   Drewyer and Collins went in pursuit of some Elk, the tracks of which Collins had discovered yesterday; but it rained so hard that they could not pursue them by their tracks and returned unsuccessfull.

Gass   had a very unpleasant day, as it rained hard during the whole of it

Ordway   when we got half way Set in to Storming & rained verry hard & the wind blew so high that we could not cross the creek in a canoe and waided across and got to the fort about half past 12 oClock   the day verry disagreeable and Stormey &C.

Whitehouse   A Cloudy morning. The party from the Salt works...had come about half way, when it set in to raining very hard, and the wind blew so hard, that they could not cross the Creek in the Canoe. This party had to wade this Creek. It continued raining very hard which occasioned that party to hurry on & they walked very fast till the arrived at the fort, which was half past 12 oClock AM.

SATURDAY, FEBRUARY 22

	Sunrise			4 p.m.		River		
Temp	Weather	Wind	Temp	Weather	Wind	Rise/Fall	Feet	Inches
	f a r	NE		c a f	NE			

Weather Diary   the wind scarcely perceptable

Gass   This was a fine clear day.

Ordway   a fair morning.

Whitehouse   We had a pleasant morning, but cool.

SUNDAY, FEBRUARY 23

	Sunrise			4 p.m.		River		
Temp	Weather	Wind	Temp	Weather	Wind	Rise/Fall	Feet	Inches
	f	SW		c a f	SW			

Weather Diary   heavy frost this morning. at eleven AM it clouded up and continued so all day.

Gass   clear and pleasant

Ordway   a fair morning.

Whitehouse   A pleasant morning.

MONDAY, FEBRUARY 24

	Sunrise			4 p.m.		River		
Temp	Weather	Wind	Temp	Weather	Wind	Rise/Fall	Feet	Inches
	c a f & c	SW		r a c & r	S			

Weather Diary   the wind became hard this evening. much warmer this morning than usual. the aquatic and other birds heretofore enumerated continue with us still. the Sturgeon and a small fish like the Anchovey begin to run. they are taken in the Columbia about 40 mils. above us. the anchovey is exquisitely fine.—

Lewis & Clark   we purchased all the articles which these people brought us; we suffered these people to remain all night as it rained, the wind blew most violently and they had their women and children with them.

Gass   the morning was cloudy and at 10 o'clock it began to rain hard. The rain continued with high stormy wind; and we suffered the Indians to remain in the fort all night.

Ordway   Cloudy

Whitehouse   This morning we had Cloudy weather.

TUESDAY, FEBRUARY 25

	Sunrise			4 p.m.			River	
Temp	Weather	Wind	Temp	Weather	Wind	Rise/Fall	Feet	Inches
	r a r	*S*		*r a r*	*S*			

Weather Diary   the wind violent all night and this morning   continued untill late in the evening when it ceased.

Lewis & Clark   It continued to rain and blow so violently that there were no movement of the party today. I am mortifyed at not having it in my power to make more celestial observations since we have been at Fort Clatsop, but such has been the state of the weather that I have found it utterly impracticable—

Gass   The rain continued and the weather was story. About 10 o'clock the natives went away, though it continued to rain very fast.

Ordway   a hard Storm of wind and rain. I feel a little better. The Storm contnd' thro the course of the day—

Whitehouse   This morning a hard Storm of wind arose accompanied with Rain. The storm continued during the whole of this day—

WEDNESDAY, FEBRUARY 26

	Sunrise			4 p.m.			River	
Temp	Weather	Wind	Temp	Weather	Wind	Rise/Fall	Feet	Inches
	f a r	*NE*		*c a f & r*	*S*			

Weather Diary   at 9 AM it clouded up again

Gass   had a fair morning
Ordway   the morning fair.
Whitehouse   A pleasant morning & Clear weather.

THURSDAY, FEBRUARY 27

	Sunrise			4 p.m.			River	
Temp	Weather	Wind	Temp	Weather	Wind	Rise/Fall	Feet	Inches
	c a r	*SW*		*r a r*	*SW*			

Gass   cloudy wet day
Ordway   a rainy wet morning.
Whitehouse   We had a rainey wet morning.

Friday, February 28

	Sunrise			4 p.m.			River	
Temp	Weather	Wind	Temp	Weather	Wind	Rise/Fall	Feet	Inches
	r a r	SW		*c a c & f*	SW			

Weather Diary   it rained constantly during the last night. the sun shown about 9 AM partially a few minutes   saw a variety of insects in motion this morning some small bugs as well as flies. A brown fly with long legs about half the size of the common house fly was the most common. this has been the first insect that appeared   it is generally about the sinks or filth of any kind. The yellow and brown flycatch has returned. It is a very small bird with a tail as long proportiably as a Sparrow.

Gass   This was a foggy morning, and the afternoon cloudy. The greater part of this day was fair and pleasant.

Ordway   rained very hard the greater part of last night.

Whitehouse   It rained the greater part of last night & this day proved wet & Rainey.

Saturday, March 1

	Sunrise			4 p.m.			River	
Temp	Weather	Wind	Temp	Weather	Wind	Rise/Fall	Feet	Inches
	f a r & c	SW		*r a c & r*	SW			

Weather Diary   The clouds interfered in such manner that no observations could be made this morning.— a great part of this day was so warm that fire was unnecessary, notwithstanding [its] being cloudy and raining.

Gass   had a cloudy wet morning.

Ordway   a fair morning. The day Showery and wet.

Whitehouse   A pleasant morning. The afternoon proved Showery & wet.

Sunday, March 2

	Sunrise			4 p.m.			River	
Temp	Weather	Wind	Temp	Weather	Wind	Rise/Fall	Feet	Inches
	r a c & r	S		*r a c & r*	S			

Gass   This day was also wet.

Ordway   a rainy morning.

Whitehouse   This morning rainey & Wet.

MONDAY, MARCH 3

	Sunrise			4 p.m.			River	
Temp	Weather	Wind	Temp	Weather	Wind	Rise/Fall	Feet	Inches
	c a r	S		*c a r*	S			

Weather Diary   rained and the wind blew hard all night. Air perfectly temperate.

Gass   It rained all this day

Ordway   hard rain all last night. A rainy wet day.

Whitehouse   We had hard rain all last night, & this morning it sill continued the same, & lasted during the whole of this day.

TUESDAY, MARCH 4

	Sunrise			4 p.m.			River	
Temp	Weather	Wind	Temp	Weather	Wind	Rise/Fall	Feet	Inches
	r a c & r	S		*r a r*	S			

Weather Diary   rained constantly most of the night. Saw a Snail, this morning, they are very large.

Gass   It rained all this day

Ordway   rained hard all last night and continues all this day.

Whitehouse   It rained hard all last night, & continued the same during the whole of this day.

WEDNESDAY, MARCH 5

	Sunrise			4 p.m.			River	
Temp	Weather	Wind	Temp	Weather	Wind	Rise/Fall	Feet	Inches
	c a r	NE		*c a r*	S			

Weather Diary   the air is considerably colder this morng but nothing like freizing.—

Clark   a high mountain [Saddle Mountain] is Situated S 60⁰ W about 18 miles from Fort Clatsop on which there has been Snow Since Nov.

Gass   About 12 o'clock last night, the rain ceased, and we had a fine morning.
Ordway   a fair morning.
Whitehouse   a pleasant morning.

THURSDAY, MARCH 6

	Sunrise			4 p.m.			River	
Temp	Weather	Wind	Temp	Weather	Wind	Rise/Fall	Feet	Inches
	f a r	SE		*c a f*	SE			

Weather Diary   altho' it is stated to be fair this morning the sun is so dim that no observations can be made   Saw a spider this morning, tho' the air [is] perceptably colder than it has been since the 1ˢᵗ inst.— at 9 AM it clouded up and continued so the ballance of the day. Even the Easterly winds which have heretofore given us the only fair weather which we have enjoyed seem now to have lost their influence in this rispect.—

Gass   This day continued fair throughout.
Ordway   a fair morning
Whitehouse   We had a pleasant morning

FRIDAY, MARCH 7

	Sunrise			4 p.m.			River	
Temp	Weather	Wind	Temp	Weather	Wind	Rise/Fall	Feet	Inches
	r a r & h	SE		*r a f r h c & f*	SE			

Weather Diary   Sudden changes & frequent, during the day, scarcly any two hours of the same discription. the Elk now being to shed their horns. a bird of a scarlet colour as large as a common pheasant with a long tail has returned, one of them was seen today near the fort by Capt. Clark's black man, I could not obtain a view of it myself.

Lewis & Clark   The wind was so high that Comowol did not leave us untill late this evening.

Gass   This was a wet morning, and some shower fell occasionally during the day.

Ordway   a little hail last night and Showers of hail and rain this morning.

Whitehouse   A Rainey wet morning

SATURDAY, MARCH 8

	Sunrise			4 p.m.			River	
Temp	Weather	Wind	Temp	Weather	Wind	Rise/Fall	Feet	Inches
	h & r ah & s	S		r ar & h	SE			

Weather Diary   the ground covered with hail and snow this morning, air cool but not freizing.—

Gass   Some snow fell last night, and the morning was stormy and disagreeable.

Ordway   we had Showers of hail and rain last night and continues this morning   the day cold and Showery—

Whitehouse   We had showers of hail during last night; and we have Showers of Rain & hail this morning.

SUNDAY, MARCH 9

	Sunrise			4 p.m.			River	
Temp	Weather	Wind	Temp	Weather	Wind	Rise/Fall	Feet	Inches
	s & h ars & h	SW		rah & r	SW			

Weather Diary   Snow and hail 1 inch deep this morning   air Still cold more so than yesterday but not freizing

Lewis & Clark   Sergt. Pryor and the fishing party not yet arrived, suppose they are detained by the winds.

Gass   There were some light showers of snow this afternoon, but during the greater part of it, the sun shone clear and warm.

Ordway   a little snow & hail this morning and cold.

Whitehouse   This morning we had Snow & hail

MONDAY, MARCH 10

	Sunrise			4 p.m.			River	
Temp	Weather	Wind	Temp	Weather	Wind	Rise/Fall	Feet	Inches
	s&rahr &s	*SW*		*farh &s*	*SW*			

Weather Diary   Snow nearly disappeared by this morning. the air considerably warmer.

Lewis & Clark   About 1 PM it became fair and we sent out two parties of hunters. It blew hard all day.

Gass   changeable weather with snow showers

Ordway   Showers of hail and a little Snow intermixed   High winds &C.

Whitehouse   We had Showers or rain, with Snow and hail. The wind blew hard this day.

TUESDAY, MARCH 11

	Sunrise			4 p.m.			River	
Temp	Weather	Wind	Temp	Weather	Wind	Rise/Fall	Feet	Inches
	farh&s	*SE*		*far&h*	*SE*			

Weather Diary   snow 1 inch deep this morning air cold but no ice. some insects seen in the evening in motion   I attemted to make an observation for Equal Altitudes but the PM obsevtn. was lost in consequence of clouds. it became cloudy at 1 AM and rained attended with some hail   at six PM it became fair and the wind changing to NE it continued fair during the night. the snow had all disappeared by 4 PM this evening.—

Lewis & Clark   Sergt. Pryor...the wind had prevented his going to the fishery on the opposite side of the river.

Gass   The weather was nearly the same as yesterday

Ordway   a little Snow fell last night.

Whitehouse   We had during last night some Snow & this morning we have fair Weather.

WEDNESDAY, MARCH 12

	Sunrise			4 p.m.			River	
Temp	Weather	Wind	Temp	Weather	Wind	Rise/Fall	Feet	Inches
	f a c	*NE*		*c a f*	*NE*			

Weather Diary   white frost this morning and ice on the pools of standing water.— it being fair in the morning I again attempted Equal Altitudes but it be came cloudy 3 PM and continued so during the day. (Lewis)   without any rain (Clark)

Lewis & Clark   The Whale...tho I believe it is much more frequently killed by running on the rocks of the Coast to the SSW in violent Storms, and thrown on different parts of the Coast by the winds and tide—

Gass   the morning was pleasant; but towards the evening the day became cloudy.

Ordway   a white frost. Clear and cold.

Whitehouse   A fair morning.

THURSDAY, MARCH 13

	Sunrise			4 p.m.			River	
Temp	Weather	Wind	Temp	Weather	Wind	Rise/Fall	Feet	Inches
	f a r	*NE*		*f*	*NE*			

Weather Diary   slight frost this morning. A little rain fell in the latter part of the night. Saw a number of insects in motion; among others saw for the fist time this spring and winter a downey black fly about the size of the common house fly. the plants begin to appear above the ground, among others the rush of which the natives eat the root. and the plant, the root of which resembles in flavor the sweet potato also eaten by the natives.

Lewis & Clark   took equal altitudes to day this being the only fair day for Sometime past.

Gass   the morning was fine.

Ordway   a fair cold morning. High winds.

Whitehouse   A Clear cold morning. I took two Men & a canoe...to purchase fish...the wind blew so hard that I was forced to return with them to the fort.

FRIDAY, MARCH 14

	Sunrise			4 p.m.			River	
Temp	Weather	Wind	Temp	Weather	Wind	Rise/Fall	Feet	Inches
	c a f	*NE*		*c*	*NE*			

Weather Diary   yesterday and last night were the most perfectly fair wether we have seen at this place

Lewis & Clark   The Indians tell us that the Salmon begin to run early in the next month.

Gass   We had a fine morning. While out to day I saw a number of musquitoes flying about.

Ordway   Cloudy

Whitehouse   This morning was Cloudy.

SATURDAY, MARCH 15

Sunrise			4 p.m.			River		
Temp	Weather	Wind	Temp	Weather	Wind	Rise/Fall	Feet	Inches
	c a c	NE		f	NE			

Weather Diary   the temperature of the air is perfectly pleasent without fire.— became fair at 8 AM.— the sorrel with an oval, obtuse and ternate leaf has now put forth [its] leaves. some of them have nearly obtained their growth already. The birds were singing very agreably this morning particularly the common robin.—

Gass   There was a fine plesant morning.
Ordway   a fair morning.
Whitehouse   A pleasant morning

SUNDAY, MARCH 16

Sunrise			4 p.m.			River		
Temp	Weather	Wind	Temp	Weather	Wind	Rise/Fall	Feet	Inches
	r a f & c	SW		c a f c r	SW			

Weather Diary   wind hard greater part of the day. The Anchovey has ceased to run; the white salmon trout have succeeded them. The weather so warm that the insects of various speceis are every day in motion.—

Lewis & Clark   Drewyer and party did not return…we suppose he was detained by the hard winds of today.
Gass   Last night it became cloudy and began to rain; and the rain has continued all day—
Ordway   a rainy wet morning. Rained the greater part of the day.
Whitehouse   A Rainey wet morning.

MONDAY, MARCH 17

Sunrise			4 p.m.			River		
Temp	Weather	Wind	Temp	Weather	Wind	Rise/Fall	Feet	Inches
	c a r	SW		rafhas & r*	SW			

---

*According to Moulton (1991, 7: 46), Clark's Journal Codex J lists this weather data as "r a f h s & r."

Weather Diary   rained all night. Air somewhat colder this morning. frequent and sudden changes [Clark uses "showers" in his version] in the course of the day.—

Lewis & Clark   we have had our perogues prepared for our departure, and shal set out as soon as the weather will permit. The weather is so precarious that we fear by waiting untill the first of April that we might be detained several days longer before we could get from this to the Cathlahmahs as it must be calm of we cannot accomplish that part of our rout.

Gass   it rained occasionally during the whole of the day.

Ordway   Showers of rain intermixed with Snow. Showery all day.

Whitehouse   A Cloudy day, and showery.

TUESDAY, MARCH 18

	Sunrise			4 p.m.		River		
Temp	Weather	Wind	Temp	Weather	Wind	Rise/Fall	Feet	Inches
	rac&r	*SW*		*rafr&h*	*SW*			

Weather Diary   frequent showers through the day

Lewis & Clark   the frequent showers rain in the course of the day prevented the Canoes drying Sufficient to pay them even with the assistance of fire—

Gass   The weather was much like that of yesterday, and some hail fell in the course of the day.

Ordway   a Showery morning of rain and hail.

Whitehouse   We had showers of rain, some hail & thunder this morning

WEDNESDAY, MARCH 19

	Sunrise			4 p.m.		River		
Temp	Weather	Wind	Temp	Weather	Wind	Rise/Fall	Feet	Inches
	r&hacr &h	*SW*		*rafr &h*	*SW*			

Weather Diary   frequent and suddon changes during the day wind not so hard as usual.

Lewis & Clark   It continued to rain and hail today in such a manner that nothing further could be done to the canoes. ...the dress of the man [Native American] consists of a small robe   ....a mat is sometimes temperarily thrown over the sholders to protect them from rain. They have no other article of cloathing whatever neither winter nor summer. It continued to rain so constantly today that Sergt. Pryor could not pitch his canoes—

Gass   the morning was stormy, some hard showers of hail fell and it continued cloudy through the day.

Ordway   hard Showers of rain intermixed with Snow and hail.

Whitehouse   We had a fair morning. In the afternoon we had showery disagreeable weather.

THURSDAY, MARCH 20

	Sunrise			4 p.m.			River	
Temp	Weather	Wind	Temp	Weather	Wind	Rise/Fall	Feet	Inches
	rar&h	*SW*		*r*	*SW*			

Weather Diary   rained all day without intermission.

Lewis & Clark   It continued to rain and blow so violently today that nothing could be done towards forwarding our departure. ...but the rain rendered our departure so uncertain that we declined this measure for the present. nothing remarkable happened during the day. We have yet Several days provisions on hand, which we hope will be Sufficient to Serve us dureing the time we are compell'd by the weather to remain at this place—

Gass   The whole of this day was wet and disagreeable. We intended to have set out today on our return, but the weather was too bad.

Ordway   rained hard the grater part of last night and continues this morning. So we are only waiting for good weather to Start.

Whitehouse   A Rainey wet day. We are now waiting for fair weather in Order to make a Start to the United States.

FRIDAY, MARCH 21

	Sunrise			4 p.m.			River	
Temp	Weather	Wind	Temp	Weather	Wind	Rise/Fall	Feet	Inches
	r a r	*SW*		*c a r*	*NE*			

Weather Diary   rained all night at 9 AM wind changed to NE and the rain ceased ["a short time"—Clark]. Cloudy the ballance of the day.

Gass   We had a cloudy wet morning.

Ordway   rained hard all last night, and continues this morning.

Whitehouse   It rained hard all last night & contined the same this morning.

SATURDAY, MARCH 22

	Sunrise			4 p.m.			River	
Temp	Weather	Wind	Temp	Weather	Wind	Rise/Fall	Feet	Inches
	r a r	*SW*		*rac&r*	*SW&NE*			

Weather Diary   rain continued without intermission greater part of the night. Air temperate. The leaves and petals of the flowers of the green Huckleburry have appeared. Some of the leaves have already obtained $1/4$ of their size.

Lewis & Clark   the air is perfectly temperate, but it continues to rain in such a manner that there is no possibility of geting our canoes completed in order to Set out on our homeward journey.

Gass   We had a cloudy wet morning.

Ordway   continues rainy.

Whitehouse   It continued raining. We are all getting in readiness so start which we expect if the weather permits will be tomorrow—

SECTION 7   Return to St. Louis

March 23 to September 23, 1806

The expedition left its winter quarters at Fort Clatsop on March 23, 1806 and headed up the swollen Columbia River. Clark toured up the Multnomah (Willamette) River, missed during their foggy, rainy descent, reaching a point near modern-day Portland, Oregon. The rapid early spring waters and strong Columbia River Gorge winds created difficulties for the corps. Near Dalles, Oregon, the expedition purchased horses and traveled by land to the confluence of the Columbia and Walla Walla rivers. Instead of returning up the Snake River, they followed an overland Indian trail along the Walla Walla River to the confluence of the Snake and Clearwater rivers.

The Expedition proceeded overland past present-day Lewiston/Clarkston and Canoe Camp into the heart of the Nez Perce nation. Finding deep snow in the Bitterroot Mountains, the Indian chiefs explained that the earliest attempt at the Lolo Trail should be made after the flood waters of the Clearwater reduced for at least five days. On June 10, they moved out of the Clearwater Valley and started up the Lolo Trail. Excessive snowpack from the numerous winter storms produced depths between 10 and 18 feet. The party retreated and made a second attempt on June 25. Safely over the Bitterroots by the end of June, the expedition split into two parties. Lewis led one group back to the Great Falls of the Missouri and then led a smaller party to scout the Marias River. Lewis's party rejoined the larger group below the falls near the end of July and proceeded down the Missouri to rendezvous with Clark at the confluence of the Yellowstone River. Clark took the other party back to the supplies left at the headwaters of the Missouri (Jefferson River) and then proceeded to the Yellowstone River and followed it back to the Missouri.

Rainy and thunderstorm-laden afternoons plagued the corps during the summer of 1806. Both parties reunited east of Williston, North Dakota, on August 12 and together reached the Mandan villages on August 14. Staying only a couple of days, the party said good-bye to interpreter Charbonneau, his wife Sacagawea, their son "little Pomp," and expedition member John Colter, and proceeded down the Missouri. Moving at times 60, 70, or 80 miles a day, the expedition passed Sioux City, Iowa, on September 4 and paid respects to the only expedition member to die, Sergeant Charles Floyd. Returning to the sultry midsection of America, the corps passed Kansas City on September 15. Just as it had bid them farewell in May of 1804, rain greeted their return to the confluence of the Missouri and Mississippi rivers. They moved on to St. Louis on their last day of travel, September 23, 1806.

The systematic entries for the Lewis and Clark Expedition daily narrative journals as well as those of the army sergeants and privates were made every day in 1806. However, not every journalist noted weather, water, or climate data each day. As in the previous section, each day's entry begins with the data from the Weather Diary's Observation Tables (when available), followed by the corresponding remarks (when available), and finally excerpts that pertain to weather and climate from the expedition members' daily narrative journals.

River rise and fall observations begin again during the trip up the Columbia River. Although not like previous recording episodes, the data usually is not for a 24-hour period unless they are encamped. In most cases, when the party stopped for the evening, it is surmised, following Lewis's habit from previous journal entries, that a mark was made and measured the next morning. Thus many observations were made for only an 8- to 10-hour-long period.

Different journals and notebooks were used during the expedition. Weather data reported here are from Coues (1893, 3: 1277–1281, 1293–1298), Moulton (1991, 7: 42–end and 1993,

8: 375), and Thwaites (1904f, 6, part 2: 208–229). For a more detailed explanation on the journals and entry practices consult Cutright (1976) and Moulton (1986).

Note   *The expedition leaves Fort Clatsop on the Pacific Ocean to begin its long journey back to St. Louis in the Louisiana Territory of the United States.*

SUNDAY, MARCH 23

	Sunrise			4 p.m.			River	
Temp	Weather	Wind	Temp	Weather	Wind	Rise/Fall	Feet	Inches
	r a r	SW		*f a c & r*	SW			

Weather Diary   it became fair at 12 Ock. and continued cloudy and fair by intervales without rain till night

Lewis   the wind is pretty high but it seems to be the common opinion that we can pass point William— at 1 PM we bid a final adieu to Fort Clatsop. The wind was not very hard—

Clark   This morning proved So raney and unceratain that we were undetermined for Some time whether we had best set out & risque the river which appeared to be riseing or not. the rained seased and it became fair about Meridian, at which time we loaded our canoes & at 1 P.M. left Fort Clatsop. not withstanding the repeated fall of rain which has fallen almost Constantly Since we passed the long narrows... indeed we have had only   days fair weather since that time. and proceeded on, thro' Meriwethers Bay, there was a Stiff brccse from the SW which raised Considerable Swells around Meriwethers point which was as much as our canoes could ride.

Gass   There was a cloudy wet morning—   The afternoon was fair.

Ordway   this morning provcd so rainy and uncertain that our officers were undetermined for Some time whether they had best Set out & risque the [wind] which appeared to be riseing or not. The rained Seased and it became fair about meridian at which time we loaded our canoes & at 1 PM left Fort Clatsop on our homeward bound journey. Notwithstanding the reputed fall of rain which has fallen continualy Since we passed the long narrows on the Novr last, indeed we have had only   days fair weather since that time. There was a stiff breeze from the SW which raised considerable Swells around Merewethers Point.

Whitehouse   It rained very hard, during the whole of last night. This morning it still continued raining, & the Weather appeared very uncertain. About 12 oClock AM it ceaased raining; & the weather became Clear & pleasant, & we loaded our Canoes, & got every thing in readiness to ascend the Columbia River. At 1 oClock we embarked...we went around a point of land called by our officers Merryweather point [Astoria, OR] when the wind rose & blew hard from the South West, & the waves ran very high.. We proceeded on, & passed another point of land called point William [Tongue Point].

Monday, March 24

	Sunrise			4 p.m.			River	
Temp	Weather	Wind	Temp	Weather	Wind	Rise/Fall	Feet	Inches
	rac&r	SW		*f a c*	NW & SW			

Weather Diary at 9 AM it became fair and continued fair all day and greater pt. of the night. The brown bryery shrub with a broad pinnate leaf has began to put forth [its] leaves. The pole-cat Colwort, is in blume. Saw the blue crested fisher. Birds are singing this morning. The black Alder is in blume.

Lewis the tide being out this morning we found some difficulty in passing through the day below the Cathlahmah village. The night was cold tho' wood was abundant

Gass After a bad night's rest, on account of the rain, 15 men went out to hunt. The morning was fair

Tuesday, March 25

	Sunrise			4 p.m.			River	
Temp	Weather	Wind	Temp	Weather	Wind	Rise/Fall	Feet	Inches
	c a f	SE		*rac&r*	SE			

Weather Diary cold this morning, but no ice nor frost. The Elder, Gooseberry, & honeysuckle are not putting fourth their leaves. The nettle and a variety of other plants are now springing up. The flower of the broad leafed thorn is nearly blown. Several small plants in blume.

Lewis The morning being disagreeably cold. Continued on the south coast of the river against the wind and strong current, out progress was of course but slow. I observe that the green bryer …retains [its] leaves all winter— the wind in the evening was very hard.

Clark Last night and this morning are cool wend hard a head and tide going out. The winds in the evening was verry hard, it was with some dificuelty that we Could find a Spot proper for an encampment.

Gass had a fair morning. Around 12 o'clock halted, the wind and tide being both against us. When the tide began to rise we went on again. Later in the afternoon the wind rose and blew very hard accompanied with rain, notwithstanding we proceeded on till night

Ordway the winds hard a head and tide against us so we delayed untill 1 oClock PM at which time we set out

Whitehouse the wind & tide being against us, We had to delay at our encampment, untill 1 oClock PM

WEDNESDAY, MARCH 26

Sunrise			4 p.m.			River		
Temp	Weather	Wind	Temp	Weather	Wind	Rise/Fall	Feet	Inches
	c a r	NW		*c a f & c*	SE			

Weather Diary   cold and rainy last night. Wind hard this morning fair at 9 AM   Cloudy at 1 PM [Clark writes "and cleared off at 1 PM" in his version]   The humming bird has appeared. Killed one of them and found it the same with those common to the United States.

Lewis & Clark   The wind blew so hard this morning that we delayed untill 8 AM.

Gass   After a disagreeable night's rain, and wind, we continued on our voyage.

Ordway   the wind ran high last night and the tide rose higher than common and came in under my blankets before I awoke and obledged me to move twise.

Whitehouse   The wind & tide rose very high during last night. The water raised so much that it obliged several of our party to move their Camps.

THURSDAY, MARCH 27

Sunrise			4 p.m.			River		
Temp	Weather	Wind	Temp	Weather	Wind	Rise/Fall	Feet	Inches
	r a c	SE		*r a c & r*	SE			

Weather Diary   blew hard about noon. Rained greater part of the day. The small or bank martin appeared today, saw one large flock of them. Waterfowl very scarce, a few Comorant, geese, and the redheaded fishing duck are all that are to be seen. The red flowering currant are in blume, this I take to be the same speceis I first saw in the <waters of the columbia>, Rocky Mountains; the fruit is a deep purple berry covered with a gummy substance and not agreeably flavoured. There is another speceis uncovered with gum which I first found on the waters of the Columbia about the 12th of August last.

Lewis   But as the weather would not permit us to dry our canoes in order to pitch them we declined their friendly invitation, and resumed our voyage at 12 Ock. The Coweliskee is 150 yards wide, is deep The night as well as the day proved cold wet and excessively disagreeable.

Clark   a rainey disagreeable night   rained the greater part of the night. Proceeded on...the wind rose and the rain became very hard Soon after we landed The night as well as the day proved Cold wet and excessively disagreeable.

Gass   There was a cloudy wet morning.

Ordway   rain commenced this morning and continued thro the day.

Whitehouse   This morning early it commenced raining, which continued during the whole of this day. We have still hard rain this evening.

FRIDAY, MARCH 28

	Sunrise			4 p.m.			River	
Temp	Weather	Wind	Temp	Weather	Wind	Rise/Fall	Feet	Inches
	c a r	N		faf&r	SW			

Weather Diary   rained by showers greater part of last night   frequent showers in the course of the day. This evening we saw many swan passing to the North as if on a long flight. vegitation is not by several days as forward here as at Fort Clatsop when we left that place. The river rising fast, the water is turbed; the tide only swells the water a little, it does not stop the current. It is now within 2 feet of [its] greatest hight. (Lewis)   which appears to increas as we assend. (Clark)

Lewis & Clark   at $1/2$ after ten AM it became fair. Pitched our canoes. We determined to remain this evening and dry our beding baggage &c. the weather being fair.

Gass   The morning was cloudy. It rained at intervals during the day. The Columbia river is now very high, which makes it more difficult to ascend.

Ordway   rained the greater part of last night. They day proved Squawlley   high winds &C.

Whitehouse   It continued raining the greater part of last night. The remainder of this day was Squally & the wind high.

SATURDAY, MARCH 29

	Sunrise			4 p.m.			River	
Temp	Weather	Wind	Temp	Weather	Wind	Rise/Fall	Feet	Inches
	car&f	S		c a r	SW			

Weather Diary   frequent showers through the night. Very cold this morning.—

Lewis   the river rising fast

Clark   the morning was very cold wind Sharp, and keen off the rainge of Mountains to the East Covered with snow. The river is now riseing very fast and retards our progress very much. Crossed the mouth of the Chah-wa-nahi-ooks River whish about 200 yards wide and a great portion of water into the columbia at this time it being high.

Gass   The morning was pleasant with some white frost

Whitehouse   We set out early this morning & proceeded on, the Columbia River being very high & the current running very swift.

Note   *Between March 30 and April 6, the party stays in the vicinity of present-day Portland, Oregon. Captain Clark leads a small party up the Multnomah River (present-day Willamette River).*

SUNDAY, MARCH 30

	Sunrise			4 p.m.		River		
Temp	Weather	Wind	Temp	Weather	Wind	Rise/Fall	Feet	Inches
	c	*S*		*f a c*	*SW*			

Weather Diary at 10 AM [Clark lists this as PM] it became fair and continued so weather moderately warm. Saw a leather winged bat the grass is about 16 Inches high in the river bottoms. The frogs are now abundant and are crying in the swamps and marshes.—

Lewis we had a view of mount St. helines and Mount Hood. The 1st is the most noble looking object of [its] kind in nature. Both of these mountains are perfectly covered with snow; at least the parts of them which are visible.

Clark we made 22 miles only to day the wind and a Strong current being against us all day, with rain. Discovered a high mountain SE Covered with Snow which we call Mt. Jefferson.

Gass The morning was fair with some dew. The river is very high, overflowing all its banks.

Ordway The river still riseing & is now So high that the tide has no effect to be perceived at this time   considerable of drift wood floating down the river. Saw mount rainey [Mt. St. Helens] and Mount Hood which is verry white with Snow &C.

Whitehouse The River still continuing rising, and is so high, that the tide has no Effect, as high up the River as where we now are. We saw a considerable quantity of drift wood floating down the River. We say this day Mount Rainey [probably Mt. St. Helens] & Mount hood; they appeared white & was covered with Snow —

MONDAY, MARCH 31

	Sunrise			4 p.m.		River		
Temp	Weather	Wind	Temp	Weather	Wind	Rise/Fall	Feet	Inches
	f	*SE*		*f*	*SE*			

Weather Diary The Summer or wood duck has returned. Butterflies and Several Species of insects appear. Musquitoes are troublesome this evening Encamp opposit quick San River The Summer Dick has returned   I saw Several to day in a small pond. This evening the Musquetors were verry troublesom this evening, it is the first time they have been so this Spring. The waterfowls are much plentyer about the enterance of quick Sand river than they were below. Observed a species of small wild onion growing among the moss of the rocks, they resemble the Shives of our gardins and grow remarkably close together forming a perfect tuft; they are quite as agreeably flavoured as the Shives.

Lewis the water is very clear. Saw a summer duck or wood duck….this is the same with those of our country and is the first I have seen since I entered the rocky mountains last summer—

Clark the Columbia is at present on a Stand.

Gass This was a beautiful clear morning

Ordway a clear pleasant morning. The wind rose from the Southward. The waves high.

Whitehouse A clear pleasant morning.

TUESDAY, APRIL 1

	Sunrise			4 p.m.		Columbia River		
Temp	Weather	Wind	Temp	Weather	Wind	Rise/Fall	Feet	Inches
	c a f	SE		c a f	SE	R		1

Weather Diary at 6 PM last evening it became cloudy. Cottonwood in blume. From the best opinion I could form of the State of the Columbia on the 1st of April it was about 9 feet higher than when we decended it in the beginning of November last. The rising and falling of the river as set down in the diary is that only which took place from sunseting to sunrise or thereabouts it being the time that we usually remain at our encampments.—

Lewis & Clark The last evening and this morning were so cloudy that I could neither obtain any Lunar observations nor equal altitudes— it was so cloudy at the time of this observation that cannot vouch for any great accuracy—

Gass We had a cloudy morning

Ordway we discovred yesterday the top of a high white Mountain some distance to the Southward our officers name it Mount Jefferson.

Whitehouse we saw a high mountain laying a great distance off to the Southward of us, which appeared to be covered with snow. Our Officers named this Mountain Jefferson Mountain.

WEDNESDAY, APRIL 2

	Sunrise			4 p.m.		Columbia River		
Temp	Weather	Wind	Temp	Weather	Wind	Rise/Fall	Feet	Inches
	c	SE		c a f	SE	F		$^1/_8$

Weather Diary heavy dew last night. Cloudy all night.

Lewis the night and morning being cloudy I was again disappointed in making the observations I wished

Clark   Multnomah had fallen 18 inches from [its] greatest annual height. from the enterance of this river, I can plainly See Mt. Jefferson which is high and covered with snow SE Mt. Hood East, Mt. St. Helians a high humped Mountain [Mount Adams] to the East of Mt. St. Helians. The current of the Multnomar is jentle as that of the Columbia glides Smoothly with an eavin surface, and appears to be Sufficiently deep for the largest Ship.

Ordway   the after part of the day clear and pleasant.

Whitehouse   The great River is called by the natives the Mult-no-mack...it is 500 yard wide at its mouth. The Indian guide that was with us, told us that it heads Near the head Waters of the California. The guide also mentioned...that the Tide water runs up it to a falls which is 40 feet high.*

THURSDAY, APRIL 3

	Sunrise			4 p.m.		Columbia River		
Temp	Weather	Wind	Temp	Weather	Wind	Rise/Fall	Feet	Inches
	c a r	SW		*c a r*	W	F		3 $^1/_2$

Weather Diary   a slight rain about day light this morng.

Clark   The water had fallen in the course of last night five inches [Multnomah River]. I set out and proceeded up a short distance and attempted a second time to fathom the river with my cord of 5 fathom but could find no bottom. the mist was So thick that I could See but a short distance up this river.

Gass   it rained so hard we could not dry the meat

Ordway   a foggey morning. Slight Showers of rain n the course of the day. Indians inform us of a verry large River & is 500 yd wide and is Supposed to head with the waters of the California [Willamette]

FRIDAY, APRIL 4

	Sunrise			4 p.m.		Columbia River		
Temp	Weather	Wind	Temp	Weather	Wind	Rise/Fall	Feet	Inches
	c a r	SW		*c a r*	SW	F		4 $^1/_2$

Weather Diary   the rains have been very slight.

Lewis & Clark   this evening being fair I observed time and distance

---

*This is the final entry of the known journal writings of Joseph Whitehouse.

Gass   a cloudy morning. Captain Clarke found a large river [Willamette River], 500 yards wide; and its source supposed to be near the head waters of some of the rivers, which fall into the gulph of California.

Ordway   the after part of the day pleasant.

SATURDAY, APRIL 5

	Sunrise			4 p.m.		Columbia River		
Temp	Weather	Wind	Temp	Weather	Wind	Rise/Fall	Feet	Inches
	c a r	SW		caf&c	SW	F		2 1/2

Weather Diary   rain but slight, air colder than usual this morning.—

Lewis & Clark   This morning was so cloudy that I could not obtain any lunar observations with Aquila as I wished. The weather has been so damp that there was no possibility of pounding the meat as I wished—   Observed Magnetic Azimuth and altitude of the sun with Circumferenter and Sextant. Immediately after this observation the sun was suddenly obscured by a cloud and prevented my taking Equal Alitudes. I therefore had recourse to two altitudes in the evening which I obtained as the sun happened to shine a few minutes together through the passing cloudys.

Gass   The weather was pleasant.

SUNDAY, APRIL 6

	Sunrise			4 p.m.		Columbia River		
Temp	Weather	Wind	Temp	Weather	Wind	Rise/Fall	Feet	Inches
	f a c	SW		f	SW	F		1

Weather Diary   this is the most perfectly fair day we have seen for a Some time musquetoes troublesome this evening   the cottonwood has put forth its leaves begin to assume a green appearance at a distance. The sweet willow has not yet generaly birst [its] budscales wile the leaves of the red and broad leafed willow are of some size; it appears to me to be the most backward in vegetating of all the willows. The narrow leafed willow is not found below tide water on this river.—

Lewis   from the appearance of a rock near which we were encamped on the 3rd of November last I could judge better of the rise of the water than I could at any point below. I think the flood of this spring has been about 12 feet higher than it was at that time; the river is here about 1 1/2 miles wide; [its] general width from the beacon rock which may be esteemed the head of tide water, to the marshey island is from

one to 2 miles tho' in many places it is still wider. It is only in the fall of the year
when the river is low that the tides are persceptable as high as the beacon rock.
Mount Jefferson bears SE this is a noble mountain…is a regular cone and is covered
with eternal snow.

Gass   We had a fine morning, with some fog

Ordway   a clear pleasant morning.

MONDAY, APRIL 7

	Sunrise			4 p.m.			Columbia River	
Temp	Weather	Wind	Temp	Weather	Wind	Rise/Fall	Feet	Inches
	f	*SW*		*f*	*SW*	R		*¹/₂*

Weather Diary   the air temperate, bird singing, the pizmire, flies, beetles, in motion.

Lewis & Clark   the day has been fair and weather extreemly pleasant.

Gass   This was a pleasant day, but cloudy.

Ordway   a fair morning. The musquetoes trouble us a little &C.

TUESDAY, APRIL 8

	Sunrise			4 p.m.			Columbia River	
Temp	Weather	Wind	Temp	Weather	Wind	Rise/Fall	Feet	Inches
	f	*E*		*f*	*E*	R		*1 ¹/₂*

Weather Diary   wind commenced at 5 AM [Clark lists this as PM] and continued to blow
most violently all day. Air temperate the male flowers of the cottonwood
are falling. The goosburry has cast the petals of [its] flowers, and [its] leaves
obtained their full size. The Elder which is remarkably large has began
to blume. Some of [its] flowerets have expanded their corollas. The
serviceburries, chokecherris, the growth which resembles the beach,
the small birch and grey willow have put fourth their leaves.—

Lewis   The wind blew so violently this morning that we were obliged to unlode our
perogues and canoes, soon after which they filled with water. The wind continued
without intermission to blow violently all day.

Clark   This morning about day light I heard a Considerable roreing like wind at a distance
and in the Course of a Short time wavs rose very high which appeared to come
across the river and in the course of an hour become so high that we were obliged
to unload the canoes,   at 7 oClock A.M. the winds Suelded and blew so hard and
raised the Wave So emensely high from the N.E. and tossed our canoes against the
Shore in such a manner as to render it necessary to haul them up on the bank,

finding from the appearance of the winds that it is probable that we may be detained all day. ....in the evening late an old man his Son & Grand Son and their wives &c. came down dureing the time the waves raged with great fury. ....the wind continued violently hard all day, and threw our canoes with such force against the shore that one of them split before we could get it out. The Wind Continued violently hard all day—

Gass This was a fine morning, but the wind blew so hard from the north-east, that it was impossible to go on; and about 8 o'clock the swells ran so high, that we had to unload our canoes, and haul some of them out of the water to prevent their being injured. Some of the men are complaining of rheumatick pain; which are to be expected from the wet and cold we suffered last winter, during which from the 4th of November 1805 to the 25th of March 1806, there were not more than twelve days in which it did not rain, and of these but six were clear.

Ordway a fair morning. The wind raised so high a head that in Stead of our setting our as we intended had to unload our canoes. The waves ran high and filled them with water &C. The River rises a little the wind continued high all day &C.

Note   *The expedition is at the farthest inland reaches of the lower Columbia River system's tidal water. From April 9–12, they begin their ascent toward the top of the Cascades of the Columbia (the "Great Shute" or the "Great Rapids of the Columbia"), just above the present-day Bonneville Dam.*

WEDNESDAY, APRIL 9

	Sunrise			4 p.m.			Columbia River	
Temp	Weather	Wind	Temp	Weather	Wind	Rise/Fall	Feet	Inches
	f	W		*f*	W			

Weather Diary   the wind lulled a little before day, and became high at 11 AM continued til dark   the vining honeysuckle has put forth shoots of several inches   the dogtoothed violet is in blume as is also both the speceis of the mountain holley, the strawburry, the bears claw, the cowslip, the violet, common striped; and the wild cress or tongue grass.

Lewis We passed several beautifull cascades which fell from a great hight over the stupendious rocks...the most remarkable of these cascades falls about 300 feet perpendicularly over a solid rock into a narrow bottom of the river on the south side [probably Multnomah Falls]. The evening being far spent and the wind high raining and very cold we thought best not to attempt the rapids this evening. The fir has been lately injured by a fire near this place

Clark the wind high and a rainey disagreeable evining. ...evening wet & disagreeable

Gass The morning was pleasant. In the afternoon the weather became cloudy and some rain fell.

Ordway a fair morning and calm   commenced raining hard & high winds from NW the River much higher at this time than it was last fall when we passd.

down. Some Spots of Snow is now on the tops of these Mountains Near the River.

THURSDAY, APRIL 10

	Sunrise			4 p.m.		Columbia River		
Temp	Weather	Wind	Temp	Weather	Wind	Rise/Fall	Feet	Inches
	c a r	W		c a r	SW	R		1

Weather Diary   some snow fell on the river hills last night. Morning cold, slight sowers through the day.

Lewis & Clark   in crossing the River which at this place is not more than 400 yards wide we fell down a great distance owing to the rapidity of the Current.

Gass   during the night some showers of rain fell.

Ordway   rained hard the grater part of last night. A cloudy & Showery morning. Soon came to bad rapids where we had to two one canoe up at a time.

FRIDAY, APRIL 11

	Sunrise			4 p.m.		Columbia River		
Temp	Weather	Wind	Temp	Weather	Wind	Rise/Fall	Feet	Inches
	r a r	W		c a r	SW	R		2

Weather Diary   cold raining night   the geese are yet in large flocks and do not yet appear to have mated. What I have heretofore termed the broad leafed ash is now in blume. The fringetree has cast the corolla and [its] leaves have nearly obtained their full size. The sac a commis is in blume.—

Lewis   As the tents and skins which covered both our men and baggage were wet with the rain which fell last evening, and as it continued still raining this morning we concluded to take our canoes first to the head of the rapids, hoping that by evening the rain would cease and afford us a fair afternoon to take our baggage over the portage. These rapids are much worse than they were fall when we passed them...the water appears to be [considerably] upwards of 20 feet higher than when we decended the river. ...I observe snow-shoes in all the lodges of the natives above the Columbean vally.

Clark   rained the greater part of the last night and continued to rain this morning, as the Skins and the Covering of both the mend and loading wcrc wet we determined to take the Canoes over first in hopes that by the evening the rain would Sease and afford us a fair afternoon. Vegitation is rapidly progressing.

Gass   We had a cloudy morning.

Ordway   rained the greater part of the last night and continues this morning.

SATURDAY, APRIL 12

Sunrise			4 p.m.			Columbia River		
Temp	Weather	Wind	Temp	Weather	Wind	Rise/Fall	Feet	Inches
	c a r	W		*r a c & r*	W	R		2

Weather Diary   cold. Snow on the mountains through which the river passes at the rapids. The duckinmallard which bread in this neighbourhood is now laying [its] eggs,— vegetation is rapidly progressing in the bottoms tho' the snow of yesterday and today reaches within a mile of the base of the mountains at the rapids of the Columbia.—

Lewis   It rained the greater part of last night and still continued to rain this morning. ...it contineud to rain by showers all day, as the evening was rainy cold and far advance and ourselves wet we determined to remain all night.

Clark   rained the greater part of the last night, and this morning untile 10 AM. The rain Continued at intervales all day. Mountains are high on each Side and Covered with Snow for about 1/3 of the way down.

Gass   This morning was wet. It rained at intervals all day; and upon the very high mountains on the south side of the river, snow fell and continued on the trees and rocks during the whole of the day.

Ordway   a rainy wet morning.

SUNDAY, APRIL 13

Sunrise			4 p.m.			Columbia River		
Temp	Weather	Wind	Temp	Weather	Wind	Rise/Fall	Feet	Inches
	r a c & r	W		*c a r & f*	W	R		2 1/2

Weather Diary   cold rainy night. Rained by showers through the day. Wind hard.

Lewis   departed...on south side of the river the wind being too high to pass over the enterance of Cruzatts river [Wind River]. Capt. C. informed me that the wind detained him several hours a little above Cruzatt's river.

Clark   ...the wind rose and raised the wavs to Such a hight that I could not proceed any further. Walked...the wind had lulled. At 1/2 passed 2 PM Set out and proceeded on

Gass   There was a cloudy morning. Passed Crusatte's River (Wind River), when the wind rose so high we could not go on, so we halted. In 3 hours...the wind fell and we went on

Ordway   the current swift. The wind rose So high that obledged us to halt at this bottom where we expected to find our hunters. The day proved fair   the wind cold and Snow laying low on the Mountains near the River.

MONDAY, APRIL 14

	Sunrise			4 p.m.			Columbia River		
Temp	Weather	Wind	Temp	Weather	Wind	Rise/Fall	Feet	Inches
	f	W		f	W	R		1

Weather Diary   wind arrose at 8 AM and contined hard all day. Service bury in blume.

Lewis & Clark   at 9 AM the wind arrose and continued hard all day but not so violent as to prevent our proceeding. The river from the rapids as high as the commencement of the narrows is from $1/2$ to $3/4$ of a mile in width, and possesses scarcely any current. The mountains through which the river passes nearly to the sepulchre rock, are high broken, rocky, partially covered with fir white cedar, and in many places exhibit very romantic seenes. Some handsome cascades are seen on either hand tumbling from the stupendious rocks of the mountains into the river. We find the trunks of maney large pines Standing erect as they grew, at present in 30 feet of water; at the lowest water of the river maney of those trees are in 10 feet water.

Gass   The morning was fine with some fog. The wind blew hard from the southwest and the weather was clear and cool, but there has been no frost lately, except on the tops of the hills.

Ordway   about noon the wind rose so high from the NW that we came too at a village. Soon &C Mount Hood appears near the River on the South Side which is covd. thick with Snow & very white   the wind high   we delayed about 2 hours and proceed. on   the wind continued aft and high So we run fast.

Note   *The expedition portages around the Long and Short Narrows of what is now known as The Dalles, in Oregon.*

TUESDAY, APRIL 15

	Sunrise			4 p.m.			Columbia River		
Temp	Weather	Wind	Temp	Weather	Wind	Rise/Fall	Feet	Inches
	f	W		f	W			

Weather Diary   wind blew tolerably hard today after 10 AM observed the Curloo and prarie lark.

Gass   The morning was fair.

Ordway   a clear pleasant morning.

WEDNESDAY, APRIL 16

	Sunrise			4 p.m.		Columbia River		
Temp	Weather	Wind	Temp	Weather	Wind	Rise/Fall	Feet	Inches
	f a c	SW		f	SW	F		2

Weather Diary   morning unusually warm. vegitation rapidly progressing.— at the rock fort camp saw the prarie lark, a speceis of the peawee, the blue crested fisher, the partycoloured corvus, and the black pheasant. A speceis of hiasinth native of this place blumed today, it was not in blume yesterday.

Lewis   a great portion of that extensive tract of country to the S and SW of the Columbia and [its] SE branch [Snake River], and between the same and the waters of the California must be watered by the Multnomah river— [Willamette River]

Gass   This was a pleasant day.

Ordway   a clear pleasant morning.

THURSDAY, APRIL 17

	Sunrise			4 p.m.		Columbia River		
Temp	Weather	Wind	Temp	Weather	Wind	Rise/Fall	Feet	Inches
	f	NE		c a f	SW	F		2

Weather Diary   weather warm; the sweet willow & white oak begin to put forth their leaves

Lewis   even at this place which merely on the border of the plains of Columbia the climate seems to have changed the air feels dryer and more pure. The earth is dry and seems as if there had been no rain for a week or ten days. The plain is covered with a rich virdure of grass and herbs from four to nine inches high and exhibits a beautifull seen particularly pleasing after having been so long imprisoned in mountains and those almost impenetrably thick forrests of the seacoast.

Clark   altho' the night was cold they could not rase as much wood as would make a fire.

Gass   This was a fine morning.

Ordway   a beautiful warm morning.

FRIDAY, APRIL 18

	Sunrise			4 p.m.		Columbia River		
Temp	Weather	Wind	Temp	Weather	Wind	Rise/Fall	Feet	Inches
	f a c	SW		f	SW	F		1

Weather Diary   rain but slight. Wind very hard all day—

Lewis   the long narrows are much more formidable than they were when we decended them last fall [the Dalles, Oregon]

Gass   We had fine weather, and all set out from this place, and proceeded with great difficulty and danger to the foot of the long narrows.

Ordway   a clear cool morning.

SATURDAY, APRIL 19

	Sunrise			4 p.m.			Columbia River		
Temp	Weather	Wind	Temp	Weather	Wind	Rise/Fall	Feet	Inches
	c a r	SW		c	SW	F		3

Weather Diary   raind, moderate showers, very cold   snow on the tops of the low hills

Clark   This morning early some rain. Several showers of rain in the after part of to day, and the SW wind very high. There was great joy with the nativs last in consequence of the arrival of the Salmon. They informed us that those fish would arive in great quantities in the Course of about 5 days.

Gass   The morning was cloudy   Some light showers of rain fell in the afternoon   In the evening the weather cleared up and we had a fine night.

Ordway   a clear cold morning. A little Snow fell on the hills last night.

SUNDAY, APRIL 20

	Sunrise			4 p.m.			Columbia River		
Temp	Weather	Wind	Temp	Weather	Wind	Rise/Fall	Feet	Inches
	f a r*	SW		c a r	SW	F		2 ¹/₂

Weather Diary   weather cold. Rain slight   snow on the hills adjacent— wind violent. Some frost this morning.

Lewis   some frost this morning.

Clark   This morning very Cold the western mountains Covered with Snow. At night when they [Native Americans] wish a light they burn dry Straw & Some fiew Small dry willows. Wind hard all day cold from NW.

Gass   This was a pleasant morning with some white frost.

Ordway   a clear cold morning.

---

*According to Moulton (1991, 7: 194), Clark's Journal Voorhis No. 3 lists this weather data as "c a r."

Note   *The expedition portages around the Great Falls of the Columbia, Celilo Falls, near Wishram, Oregon (along the present-day state line).*

MONDAY, APRIL 21

	Sunrise			4 p.m.		Columbia River		
Temp	Weather	Wind	Temp	Weather	Wind	Rise/Fall	Feet	Inches
	f	NE		*f*	E	F		2

Weather Diary   heavy frost this morning. Remarkably cold last night

Lewis   I ordered all the spare poles, paddles and the ballance of our canoe put on the fire as the morning was cold.

Clark   a fair Cold morning.

Gass   This was another pleasant morning with some white frost. About 3 in the afternoon we arrived at the great falls of the Columbia

Ordway   a clear cold morning.

TUESDAY, APRIL 22

	Sunrise			4 p.m.		Columbia River		
Temp	Weather	Wind	Temp	Weather	Wind	Rise/Fall	Feet	Inches
	f	NW		*f*	W	F		1

Weather Diary   night cold the day warm

Lewis   as one of our canoes was passing...we hailed them and ordered them to come over but the wind continued so high that they could not join us untill after sunset   ...the nights are cold and days warm.

Clark   I assended a high hill [possibly Haystack Butte, Washington] from which I could plainly See the range of Mountains which runs South from Mt. Hood as far as I could See. I also discovered the top of Mt. Jefferson which is Covered with Snow and is S 10° W. Mt. Hood is S 30° W. The range of mountains are Covered with Timber and also Mt. Hood to a sertain hite The range of mountains has Snow on them. Here we observed our 2 Canoes passing up on the opposit Side and the Wind too high for them to join us. The air I find extreemly Cold which blows Continularly from Mt. Hoods Snowey regions.

Gass   This was a pleasant morning and high wind. We proceeded on about 3 miles, when the wind became so violent, that we could not proceed any further, and halted an unloaded our canoes. Having remained here two hours....we proceeded on though the wind was high and river rough.

Ordway   a clear pleasant cold morning. The wind so high from the NW that the canoes being on the opposite Side of the river could not cross

WEDNESDAY, APRIL 23

	Sunrise			4 p.m.		Columbia River		
Temp	Weather	Wind	Temp	Weather	Wind	Rise/Fall	Feet	Inches
	f a c	E		*f*	NE	F		4

Lewis   the river is by no means as rapid as when we decended or at least not obstructed with those dangerous rapids   the water at present covers most of the rocks in the bed of the river.

Clark   the river is by no means as rapid as it was at the time we decended.

Gass   We had a cloudy morning.

Ordway   the day warm.

THURSDAY, APRIL 24

	Sunrise			4 p.m.		Columbia River		
Temp	Weather	Wind	Temp	Weather	Wind	Rise/Fall	Feet	Inches
	f	NW		*f*	NW	F		2

Lewis & Clark   the winds which set from Mount Hood or in a westerly direction are much more cold than those from the opposite quarter. There are now no dews in these plains, and from the appearance of the earth there appears to have been no rain for several weeks—

Gass   The weather was pleasant.

Ordway   a clear cool morning.

FRIDAY, APRIL 25

	Sunrise			4 p.m.		Columbia River		
Temp	Weather	Wind	Temp	Weather	Wind	Rise/Fall	Feet	Inches
	f	NE		*f*	NE	F		2

Gass   The morning was pleasant.

Ordway   a clear cool morning.

SATURDAY, APRIL 26

	Sunrise			4 p.m.		Columbia River		
Temp	Weather	Wind	Temp	Weather	Wind	Rise/Fall	Feet	Inches
	f a c	NW		*f*	NE	F		2 1/2

Weather Diary   the sweet willow has put forth its leaves. The last evening was cloudy it continued to threaten rain all night but without raining. The wind blew hard all night. The air cold as it is invariably when it sets from the westerly quarter.—

Lewis & Clark   we covered ourselves partially [Clark wrote "perfectly"] this evening from the rain by means of an old tent.

Gass   had a fine morning.

Ordway   the day warm. Saw considerable of Snow on the mountains to the South & S East.

Note   *The expedition reaches the confluence of the Walla Walla and Columbia rivers, having crossed the present-day Oregon–Washington state line. After resting a couple of days, they proceed on April 30 by land toward the Kooskooskee River (present-day Clearwater River).*

SUNDAY, APRIL 27

	Sunrise			4 p.m.		Columbia River		
Temp	Weather	Wind	Temp	Weather	Wind	Rise/Fall	Feet	Inches
	f a r	SE		*f*	NW	F		1 1/2

Weather Diary   had a shower of rain last night

Gass   The morning was cloudy with some light showers of rain. Some light showers of rain fell at intervals during the day.

Ordway   a little rain fell the latter part of last night.

MONDAY, APRIL 28

	Sunrise			4 p.m.		Columbia River		
Temp	Weather	Wind	Temp	Weather	Wind	Rise/Fall	Feet	Inches
	f a t	SW		*f*	NE	F		2

Lewis & Clark   I at length urged that there was no wind blowing and that the river was consequently in good order to pass our horses and goods

Gass   The morning was pleasant. From this place we can discover a range of mountains (Blue Mountains) covered with snow, in a southeast direction and about fifty miles distant. In the evening the weather was cloudy, and it thundered and threatened rain, a few drops of which fell.

Ordway   a clear pleasant morning.

TUESDAY, APRIL 29

	Sunrise			4 p.m.		Columbia River		
Temp	Weather	Wind	Temp	Weather	Wind	Rise/Fall	Feet	Inches
	f a c	*NW*		*f*	*NW*	*F*		*1*

Lewis & Clark   The Wallahwallah River discharges [its] self into the Columbia on [its] South Side 15 miles below the enterance of Lewis's River [Snake], or the S.E. branch. this is a handsome Stream about 4 $\frac{1}{2}$ feet deep and 50 yards wide. The water is Clear. the Indians inform us that it has [its] source in the range of Mountains in view of us to the E. and SE. Those mountains are Coverd with Snow at present tho' do not appear high. They [Native Americans] insisted on our dancing this evening but it rained a little the wind blew hard and the weather was cold, we therefore did not indulge them—

Gass   The day was fair.

WEDNESDAY, APRIL 30

	Sunrise			4 p.m.		Columbia River*		
Temp	Weather	Wind	Temp	Weather	Wind	Rise/Fall	Feet	Inches
	c a r	*NW*		*f a c*	*NW*	*F*		*2*

Weather Diary   rain slight.

Lewis & Clark   this stream…Wallah wallah…it is deep and has a bold current. There are many large banks of pure sand which appear to have been drifted up by the wind to the hight of 15 to 20 (20 or 30) feet.

Gass   This was a cloudy morning

Ordway   chilley and cold.

Note   After traveling overland for the day on May 1, the expedition camps near present-day Waitsburg, Washington.

---

*This is the last data collected on the Columbia River.

THURSDAY, MAY 1

	Sunrise			4 p.m.			River	
Temp	Weather	Wind	Temp	Weather	Wind	Rise/Fall	Feet	Inches
	c a r	SW		*c*	SW			

Weather Diary   had a pretty hard shower last night. Cold morning.— having left the river we could no longer observe [its] state; it is now declining tho' it has not been as high this season by five feet as it appears to have been the last spring. The indians inform us that it will rise higher in this month, which I presume is caused by the snows of the mountains.

Clark   Small portion of rain which fell last night Caused the road to be much furmer and better than yesterday. The morning Cloudy and Cool.

Gass   Some rain fell during the night, and the morning continues cloudy.

Ordway   Saw a timbred county a long distance to the SE & Mount of Snow.

FRIDAY, MAY 2

	Sunrise			4 p.m.			River	
Temp	Weather	Wind	Temp	Weather	Wind	Rise/Fall	Feet	Inches
	f a c	NE		*f*	SW			

Weather Diary   cold this morning, some dew.

Lewis & Clark   a branch falls in on the S side which runs south towards the SW mountains which appear to be about 25 miles distant low yet covered with snow.

Gass   We continued our journey up this branch, and saw to our right a range of high hills covered with timber and snow...encamped on the north fork, the creek having two forks....the south fork is the largest, and from its course is supposed to issue from those snow-topped hills on our right.

Ordway   a clear cold morning. High hills to our right covred with timber and partly covered with Snow. We crossed the branch in several places where it was 3 feet deep.

SATURDAY, MAY 3

	Sunrise			4 p.m.			River	
Temp	Weather	Wind	Temp	Weather	Wind	Rise/Fall	Feet	Inches
	c a h r & s	SW		*c a r h & s*	SW			

Weather Diary	rained last night and snows & hailed this morning. The air cold and wind hard. The mountains to our right seem to have experienced an increase of their snow last evening.
Lewis & Clark	here we encamped in small grove of cottonwood tree which in some measure broke the violence of the wind. it rained, hailed, snowed & blowed with Great Violence the greater portion of the day. it is fortunate for us that this storm was from the SW and of course on our backs. the air was very cold. The SW Mountains appear to become lower…they are Covered with timber and at this time Snow.
Gass	We had a wet uncomfortable morning. The wind was very high this forenoon, and rather cold for the season; with some rain. The wind continued to blow hard and some snow showers fell in the afternoon.
Ordway	a little rain the later part of last night, and continues Showery and cold a little hail & Snow intermixed. The wind blew verry high and cold Showers of hail & rain before noon considerable of Snow fell on the high hills Since yesterday. We had considerable of hail & verry high winds. The air is very cold—

Note   *The expedition arrives at the confluence of the Kooskooskee and Lewis's rivers (today known as the Clearwater and Snake rivers), at present-day Clarkston, Washington / Lewiston, Idaho. They cross the present-day state line from Washington into Idaho.*

SUNDAY, MAY 4

	Sunrise			4 p.m.			River	
Temp	Weather	Wind	Temp	Weather	Wind	Rise/Fall	Feet	Inches
	f a h	*SW*		*c a r & h*	*SW*			

Weather Diary	heavy white frost this morning ice 1/6 of an inch thick on standing water. (Lewis) thick on Standing water. (Clark)
Lewis & Clark	Collected our horses and set out early; the morning was cold and disagreeable. The lands though which we passed today are fertile consisting of a dark rich loam…the SW mountains are covered with snow at the present nearly to their bases. the evening was cold and disagreeable, and the natives crouded about our fire in great numbers insomuch that we could scarcely cook or keep ourselves warm.
Gass	We had a severe frost last night; and the morning was cold and clear.
Ordway	a hard frost & verry cold this morning.

Monday, May 5

Sunrise			4 p.m.			River		
Temp	Weather	Wind	Temp	Weather	Wind	Rise/Fall	Feet	Inches
	f	*SW*		*f*	*SW*			

Weather Diary   hard frost this morning   ice ⅛ of an inch thick on vessels of water

Gass   We had a fine morning.

Ordway   a white frost and verry cold this morning.

Tuesday, May 6

Sunrise			4 p.m.			River		
Temp	Weather	Wind	Temp	Weather	Wind	Rise/Fall	Feet	Inches
	rac&r	*NE*		*far*	*NE*			

Lewis & Clark   The Kooskooske river [Clearwater] may be safely navigated at present All the rocks of the sholes and rapids are perfectly covered; the current is strong, the water clear and cold. This river is riseing fast. The natives have a considerable salmon fishery up Colter's Creek [Potlach River]. Had a small shower of rain this evening—

Gass   There was a cloudy wet morning;

Ordway   a rainy wet morning.

Wednesday, May 7

Sunrise			4 p.m.			River		
Temp	Weather	Wind	Temp	Weather	Wind	Rise/Fall	Feet	Inches
	fac	*NE*		*f*	*SW*			

Weather Diary   the KoosKooski is rising   water cold and clear.

Lewis & Clark   The spurs of the rocky mountains which were in view from the high plain to day were perfectly covered with snow. The Indians inform us that the snow is yet so deep on the mountains that we shall not be able to pass them untill after the next full moon or about the first of June; others set the time at a more distant period. this unwelcom inteligence to men confirmed to a diet of horse-beff and roots, and who are as anxious as we are to return to the fat plains of the Missouri, and thence to our native homes. This evening was cold as usual—

Gass This was a fine morning

Ordway a fair morning. Saw the rockey mountains covered with Snow.

THURSDAY, MAY 8

	Sunrise			4 p.m.		River		
Temp	Weather	Wind	Temp	Weather	Wind	Rise/Fall	Feet	Inches
	f	SW		f	SW			

Gass The morning of this day was pleasant. Here some of the natives came to our camp, and informed us, that we could not cross the mountains for a moon and a half; as the snow was too deep, and no grass for our horses to subsist on.

Ordway a fair morning.

FRIDAY, MAY 9

	Sunrise			4 p.m.		River		
Temp	Weather	Wind	Temp	Weather	Wind	Rise/Fall	Feet	Inches
	f	SW		f a c	W			

Weather Diary Musquetors troublesom

Lewis the climate appears quite as mild as that of similar latitude on the Atlantic coast if not more so and it cannot be otherwise than healthy; it possesses a fine dry pure air. The grass and many plants are now upwards of knee high. I have no doubt but this tract of country if cultivated would produce in great abundance every article essentially necessary to the comfort and subsistence of civillized man. The situation of our camp was a disagreeable one in an open plain; the wind blew violently and was cold. At seven PM it began to rain and hail, at 9 it was succeeded by a heavy shower of snow which continued untill the next morning—

Clark the wind blew hard from the SW. accompanied with rain untill from 7 oClock untill 9 P.M. when it began to Snow and Continued all night.

Gass There was a cloudy morning

Ordway the evening cold rainy & windy—

SATURDAY, MAY 10

	Sunrise			4 p.m.			River	
Temp	Weather	Wind	Temp	Weather	Wind	Rise/Fall	Feet	Inches
	car&s	SW		f a s	SW			

Weather Diary   Snow was 8 inches deep this morning. It began to rain and hail about sunseting this evening which was shortly after succeeded by snow. It continued to fall without intermission untill 7 AM and lay 8 inches deep on the plain where we were. The air was very keen. A suddon transition this. Yesterday the face of the country had every appearance of summer. After none AM the sun shown but was frequently obscured by clouds which gave us light shower of snow. In the after part of the day the snow melted considerably but there was too great a portion to be disipated by the influence of one day's sun.

Lewis   This morning the snow continued falling $1/2$ after 6 AM when it ceased, the air keen and cold, the snow 8 inches deep on the plain. I was surprised to find on decending the hills of Commearp Cr. to find that there had been no snow in the bottoms of that stream. It seems that the snow melted in falling and decended here in rain while it snowed on the plains. The hills are about six hundred feet high about one fourth of which distance the snow had decended and still lay on the sides of the hills. The noise of their women pounding roots reminds me of a nail factory.

Clark   the air keen and cold the snow 8 inches deep on the plain. we collected our horses and set out for the village of the Chief with a flag, and proceeded on through an open plain. the road was slipry and the Snow Cloged and caused the horses to trip very frequently. the mud at heads of the streams which we passed deep and well supplied with the Car mash [camass].

Gass   At dark last night the weather became cloudy and it rained about an hour, when the rain turned to snow, and it continued snowing all night. In the morning the weather became clear. Where we are lying in the plains the snow is about five inches deep; and amidst snow and frost we have nothing whatever to eat. When we were about half way down the hill there was not a particle of snow nor the least appearance of it.

Ordway   the wind fell and the rain turned to Snow   Some time last night the Snow fell 6 Inches deep & continues chilly & cold this morning, & we had not any thing to eat. In the evening....the snow is gone in this bottom but lyes on the high plains & hills   We are now as near the mountains as we can git untill such times as the Snow is nearly gone off the mountains as we are too eairly to cross.

SUNDAY, MAY 11

	Sunrise			4 p.m.			River	
Temp	Weather	Wind	Temp	Weather	Wind	Rise/Fall	Feet	Inches
	f a r	SW		f a c	SW			

Weather Diary   the Crimson haw is not more forward now at this place than it was when we lay at rock fort camp in April.—

Lewis   after this council was over we amused ourselves with shewing them the power of magnetism, the spye glass, compass, watch, air-gun and sundry other articles equally novel and incomprehensible to them. About 3 PM Drewyer arrived with 2 deer which he had killed. He informed us that the snow still continued to cover the plain.

Clark   Some little rain last night.

Gass   This was a fine clear morning

Ordway   a fair morning.

MONDAY, MAY 12

	Sunrise			4 p.m.			River	
Temp	Weather	Wind	Temp	Weather	Wind	Rise/Fall	Feet	Inches
	f	E		*f*	SW			

Weather Diary   the natives inform us that the salmon have arrived at the entrance of the KoosKooskie in great numbers and that some were caught yesterday in Lewis's river opposite to us many miles above the entrance of that river. From this village of the broken arm Lewis's river is only about 10 miles distant to the SW.— the natives also inform us that the salmon appear [much] many days sooner in Lewis's river above the entrance of the Kooskoske than they do in that stream.

Lewis   [Chief said] that the snow was yet so deep in the mountain if we attempted to pass we would certainly perish, and advised us to remain untill after the next full moon when the said the snow would disappear and we could find grass for our horses—

Clark   a fine morning

Gass   We had another fine morning

Ordway   a clear pleasant morning.

TUESDAY, MAY 13

	Sunrise			4 p.m.			River	
Temp	Weather	Wind	Temp	Weather	Wind	Rise/Fall	Feet	Inches
	f	SW		*f*	SW			

Weather Diary   formed a camp on the Kooskooske

Clark   A fine morning I administered to the sick and gave directions. In the evening we tried the Speed of Several of our horses.

Gass   We had a fine morning with white frost.

Ordway   a clear frosty morning.

Note   *The expedition halts at Camp Chopunnish in the Kooskooskee (Clearwater) River valley and waits for the winter snows in the upper reaches of the Bitterroot Mountains to melt. They remain at this site, near present-day Kamiah, Idaho, from May 14 through June 10.*

WEDNESDAY, MAY 14

Sunrise			4 p.m.			River		
Temp	Weather	Wind	Temp	Weather	Wind	Rise/Fall	Feet	Inches
	f	SW		f	SW			

Lewis   The morning was fair. The river is 150 yds. wide at this place and extreemly rapid. Tho' it may be safely navigated at this season, as the water covers all the rocks which lie in [its] bed to a considerable debth. [near Kamiah, Idaho]

Clark   a fine day.

Gass   The morning was pleasant with some white frost.

Ordway   a clear frosty morning.

THURSDAY, MAY 15

Sunrise			4 p.m.			River		
Temp	Weather	Wind	Temp	Weather	Wind	Rise/Fall	Feet	Inches
	f	N		f a c	NW			

Weather Diary   the Kooskoske rising fast, the water is clear and cold.

Lewis   the party formed themselves very comfortable tents...made perfectly secure as well from the heat of the sun as from rain. About noon the sun shinces with intense heat in the bottoms of the river. The air on the top of the river hills, or high plains forms a distinct climate, the air is much colder, and vegitation is not as forward by at least 15 or perhaps 20 days. The rains which fall in the river bottoms are snows on the plain. At the distance of fifteen miles from the river on the Eastern border of this plain the Rocky Mountains commence and present us with winter it [its] utmost extreem. The snow is yet many feet deep even near the base of these mountains; here we have summer spring and winter within the short space of 15 to 20 miles—

Clark   some men sick...the Cause of those disorders we are unable to account for. Their diet and Sudin Change of Climate must contribute. the greater part of our Security from the rains &c. is the grass which is formed in a kind of ruff So as to turn the rain Completely and is much the best tents we have. as the days are w[a]rm &c. we have a bowry made to write under which we find not only comfortable but necessary, to keep off the intence heet of sun which has great effect in this low bottom. On the high plains off the river the climate is entirely different cool, Some Snow on the north hill Sides near the top and vegetation near 3 weeks later than in the river bottoms, and the rocky Mountains imedeately in view covered several say 4 to 5 feet deep with Snow. here I behold three different Climats within a fiew miles.

Gass   This was a fine morning, and some hunters went out early. The rest of the party were engaged in making places of shelter, to defend them from the stormy weather. Here we expect to remain a month before we can cross the mountains.

Ordway   a fair morning.

Note   *Kooskooskie (Clearwater) River observations did not begin until the expedition stopped in Nez Perce Country. As for previous observations of river rise and fall, they probably measured the river at the sunrise for a 24-hour reading. These observations continue until they leave for the Lolo Trail on June 10, 1806.*

FRIDAY, MAY 16

Sunrise			4 p.m.			Clearwater River		
Temp	Weather	Wind	Temp	Weather	Wind	Rise/Fall	Feet	Inches
	c	SE		c a r	SE	R		6

Weather Diary   last night was uncommonly warm   river rising fast. Say 9 inches

Clark   a cloudy morning with Some rain which continued untill Meridean at interavles, but very moderately.

Gass   The morning was cloudy and some rain fell; but in about two hours it cleared away and we had a fine day.

Ordway   a light rain in the fore part of the day. The after part pleasant.

SATURDAY, MAY 17

Sunrise			4 p.m.			Clearwater River		
Temp	Weather	Wind	Temp	Weather	Wind	Rise/Fall	Feet	Inches
	r a r	SE		c a r	SE	R		10 3/4

Weather Diary   rained hard the great part of the night wet the Chronometer by accedent. River rise 11 inches. The indians caught 3 salmon at their village on the Kooskooskee above our camp some miles. They say that these fish are now passing by us in great numbers but they cannot be caught as yet because those which first ascend the river do not keep near shore; they further inform us that in the course of a few days the fish run near the shore and then they take them with their skimming neitts in great numbers. Rained untill 12 Ock. by intervails.—

Lewis   It rained the greater part of the last night and this morning untill 8 Ock. The water passed through flimzy covering and wet our bed most perfectly in shot we lay in the water all the latter part of the night. Unfortunately my chronometer which for greater security I have woarn in my fob for ten days past, got wet last night; it seemed a little extraordinary that every part of me breechies which were under my head, should have escaped the moisture except the fob where the time peice was. I opened it and found it nearly filled with water which I carefully drained out exposed it to the air and wiped the works as well as I could with dry feathers after which I touched them with a little bears oil. Several parts of the iron and steel works were rusted a little which I wiped with all the care in my power. I set her to going and from her apparent motion hope she has sustained no material injury— it rained moderately the greater part of the day and snowed as usual on the plain. Sergt. Pryor informed me that it was shoe deep this morning when he came down. It is somewhat astonishing that the grass and a variety of herbatious plants which are now from a foot to 18 inches high on these plains sustain no injury from the snow or frost. I am pleased at finding the river rise so rapidly, it now doubt is attributeable to the melting snows of the mountains; that icy barrier which seperates me from my friends and Country, from all which makes life esteemable— patience, patience—

Clark   rained moderately all the last night and this morning untill, we are wet. The little river on which we are encamepd rise Sepriseingly fast. the rains of last night unfortunately wet the Crenomuter in the fob of Capt. L. breaches, which has never before been wet Since we Set out on this expedition. Her works were cautiously wiped and made dry by Capt. L. and I think she will receive no injury from this misfortune &c. rained moderately all day. at the same time Snowed on the mountains which is in the SE of us. The fiew w[a]rm days which we have had has melted the Snows in the Mountains and the river has rose considerably. that icy barrier which separates me from my friends and Country, from all which makes life estimable, is yet white with the Snow which is maney feet deep. I frequently Consult with the nativs on the subject of passing this tremendious barier which now presents themselves to our view for great extent. they all appear to agree as to the time those Mountains may be passed which is about the middle of June. Sergt. Pryor informs me that the snow on the high plains from the river was shoe deep this morning when he came down. At the distance of 18 miles from the river and on the Eastern border of the high Plain the Rocky Mountain commences and presents us with Winter here we have Summer, Spring and winter in the short space of twenty or thirty miles.

Gass   We had a cloudy wet morning and some light rain all day. Two hunters…said it snowed on the hills, when it rained at our camp in the valley.

Ordway   rained the greater part of last night and continues this morning.

SUNDAY, MAY 18

	Sunrise			4 p.m.			Clearwater River	
Temp	Weather	Wind	Temp	Weather	Wind	Rise/Fall	Feet	Inches
	c a r	SE		*c*	SE	R		2

Lewis   shortly after dark it began to rain and continued raining moderately all night. The air was extreemly cold and disagreeable and we lay in the water as the preceeding night—

Clark   Cloudy morning. The evening Cloudy, Soon after dark it began to rain and rained moderately all night—

Gass   The morning was cloudy, but without rain

Ordway   cloudy.

MONDAY, MAY 19

	Sunrise			4 p.m.		Clearwater River		
Temp	Weather	Wind	Temp	Weather	Wind	Rise/Fall	Feet	Inches
	r a r	SE		*c a r*	SE	F		4

Weather Diary   rained hard last night and untill 8 AM

Lewis   It continued to rain this morning untill 8 Ock. when it became fair.

Clark   Rained this morning untill 8 oClock when it Cleared off and became fair—

Gass   We had a cloudy wet morning. The day was fair during the whole of the afternoon.

Ordway   a light rain. About noon cleared off pleasant & warm.

TUESDAY, MAY 20

	Sunrise			4 p.m.		Clearwater River		
Temp	Weather	Wind	Temp	Weather	Wind	Rise/Fall	Feet	Inches
	r a r	NW		*c a r*	SE	R		2

Weather Diary   rained violently the great part of the night. Air raw and cold. A nest of the large blue or sand hill crain was found by one of our hunters. The young were in the act of leaving the shell. The young of the partycoloured corvus begin to fly.—

Lewis & Clark   it rained the greater part of last night and continued this morning untill noon when it cleared away about an hour and then rained at intervals untill 4 in the evening. Our covering is so indifferent that Capt. C. and myself lay in the water the greater part of last night. The hunters wounded a bear and deer...they were unable to pursue them and the snow which fell in the course of the night and this morning had covered the blood and rendered all further pursuit impracticable. Cruzatte...informed us that it was snowing on the plain while it was raining at our camp in the river bottom. Cloudy &c.

Gass   We again had a very wet morning. It continued raining till about noon, when we had fair weather with some sunshine. The hunters said it also snowed on the hills today, while it rained out our camp. In the evening there were some light showers.

Ordway   rained all last night and continues this morning, but Snows on the hills. Rained the greater part of the day—

WEDNESDAY, MAY 21

	Sunrise			4 p.m.		Clearwater River		
Temp	Weather	Wind	Temp	Weather	Wind	Rise/Fall	Feet	Inches
	c a r	SE		*f a c*	SE	F		1

Weather Diary   the air is remarkably dry and pure   it has much the feeling and appearance of the air in the plains of the Missouri.

Lewis   It rained a few hours this morning. As our tent was not sufficient to shelter us from the rain we had a lodge constructed…it is perfectly secure against the rain sun and wind and affords us much the most comfortable shelter we have had since we left Fort Clatsop.

Clark   Rained this morning. as our tent is not Sufficient to keep off the rain we are Compelled to have Some other resort for a Security from the repeeted Showers which fall. we have a small half circular place made and covered with grass which makes a very secure shelter for us to sleep under. …which is but a Scanty dependance for roots to take us over those Great snowey Barriers [Rocky Mountains] which is and will be the Cause of our Detention in this neighbourhood probably untill the 10 or 15 of June. they are at this time Covered with snow. The plains of the high Country above us is also covered with Snow.

Gass   There was a cloudy morning. At 10 o'clock the weather became clear, and in the evening was cold.

Ordway   continues rainy & wet.

THURSDAY, MAY 22

	Sunrise			4 p.m.		Clearwater River		
Temp	Weather	Wind	Temp	Weather	Wind	Rise/Fall	Feet	Inches
	f	SE		*f*	SE	F		2

Weather Diary   air colder this morning than usual   White frost tho' no ice. Since our arrival in this neighbourhood on the 7th inst. all the rains noted in the diary of the weather were snows on the plain and in some instances it snowed on the plains when only a small mist was perseptable in the bottoms at our camps. [The high plains are about 800 feet higher than the small bottoms on the river and creeks.]

Lewis   A fine morning.

Clark   a fine day we expose all our baggage to the Sun to air and dry   …this day proved to be fine fair which afforded us an oppertunety of drying our baggage which had got a little wet.

Gass   We had a fine clear morning with some white frost.

Ordway   a clear cold frosty morning.

FRIDAY, MAY 23

	Sunrise			4 p.m.		Clearwater River		
Temp	Weather	Wind	Temp	Weather	Wind	Rise/Fall	Feet	Inches
	f	NW		*f*	NW & SE	F		1 ½

Weather Diary   the air is cold in the morning but warm through the day. Some dew each morning.

Clark   a fair morning. The hunters…inform us that the high lands are very cold with snow which has fallen for every day or night for Several past.

Gass   We again had a fine morning

Ordway   clear & pleasant.

SATURDAY, MAY 24

	Sunrise			4 p.m.		Clearwater River		
Temp	Weather	Wind	Temp	Weather	Wind	Rise/Fall	Feet	Inches
	f	SE		*f*	NW	F		1

Weather Diary   air remarkably pleasant all day.

Lewis   this day has proved warmer than any of the proceeding since we have arrived here—

Clark   a fine morning. This day proved to be very w[a]rm.

Gass   This was another fine morning.

Ordway   a clear pleasant warm day.

SUNDAY, MAY 25

	Sunrise			4 p.m.		Clearwater River		
Temp	Weather	Wind	Temp	Weather	Wind	Rise/Fall	Feet	Inches
	car&t	NW		f	NW	R		·9 1/2

Weather Diary   rained moderately the greater part of last night and untill a little before sunrise. Thunder

Lewis & Clark   It rained the greater part of last night and continued untill 6 AM our grass tent is impervious to the rain.

Clark   rained moderately the greater part of last night and this morning untill 6 A.M.

Gass   There was a cloudy morning, and some light showers of rain fell. The weather became clear and we had a fine evening.

Ordway   a Thunder Shower eairly this evening.

MONDAY, MAY 26

	Sunrise			4 p.m.		Clearwater River		
Temp	Weather	Wind	Temp	Weather	Wind	Rise/Fall	Feet	Inches
	f a r	SE		f	NW	R		6

Weather Diary   the sun shone warm today, but the air was kept cool by the NW breezes

Lewis   Had frequent showers in the course of the last night. One of our men saw a salmon in the river today. The river still rising fast and snows of the mountains visibly diminish.

Clark   Some Small Showers of rain last night, and continued Cloudy this morning untill 7 AM when it Cleared away and became fair and w[a]rm. the [river rising] very fast and Snow appear to melt on the Mountains.

Gass   This day was fine and pleasant

Ordway   clear & pleasant. The river riseing   Our hunters returned the creek being so high they did not go to where was any hunting

TUESDAY, MAY 27

	Sunrise			4 p.m.		Clearwater River		
Temp	Weather	Wind	Temp	Weather	Wind	Rise/Fall	Feet	Inches
	c	SE		rafrtl	SE	R		6 1/2

Weather Diary   the dove is cooing which is the signal as the indians inform us of the approach of the salmon. The snow has disappeared on the high plains and seems to be diminishing fast on the spurs and lower region of the Rocky Mountains.

Clark   A cloudy morning. Up Collin's Creek...that Stream Still continue So high that they could not pass it—

Gass   The morning was fair and pleasant. In the afternoon some rain fell

Ordway   had a hard Thunder Shower

WEDNESDAY, MAY 28

	Sunrise			4 p.m.		Clearwater River		
Temp	Weather	Wind	Temp	Weather	Wind	Rise/Fall	Feet	Inches
	c a r t l	SE		*c a f r t & l*	SE	R		11

Weather Diary   had several heavy thunder showers in course of last evening and night. The river from sunrise yesterday to sun rise this morning raised 1 ft 10 Incs.— drift wood runing in considerable quantities and current incredibly swift tho' smooth.—

Clark   This Country would from an extensive Settlement; the Climate appears quit as mild as that of a Similar latitude on the Atlantic Coast; & it cannot be otherwise than healthy; it possesses a fine dry pure air. The Chiefs...would let us know before we left them. that the Snow was yet so deep in the Mountains that if we attempted to pass, we would Certainly perish, and advised us to remain untill after the next full Moon when the Snow would disappear on the South hill sides and we would find grass for our horses—

Gass   There was a cloudy foggy morning

Ordway   Some spots of Snow & falling timber. Had a hard Thunder Shower.

THURSDAY, MAY 29

	Sunrise			4 p.m.		Clearwater River		
Temp	Weather	Wind	Temp	Weather	Wind	Rise/Fall	Feet	Inches
	car & t	SE		*c a r*	NW	R	1	5

Weather Diary   frequent and heavy showers attended by distant thunder through the night. The river raised 6 inches in the course of yesterday and 1 foot 5 I. in the course of the last night. It is now as high as there are any marks of [its] having been in the spring 1805.—   a t 10 AM it arrived at [its] greatest hight having raised 1 1/2 inches from sunrise to that time. In the ballance of the day it fell 7 inches. The native inform us that it will take one more rise before it begins finally to subside for the season and then the passage of the mountains will be practicable.—

Lewis & Clark   We would have repeated the sweat today hd not been cloudy and frequently raining—

Gass   The morning was cloudy and wet, and the river is rising very fast; which gives us hope that the snow is leaving the mountains. At 10 o'clock the river ceased rising and the weather became clear.

Ordway   rained the greater part of last night. A rainy morning. We…descended the worst hills we ever saw a road made down. [They went to where the Salmon and Snake rivers meet.]

FRIDAY, MAY 30

Sunrise			4 p.m.			Clearwater River		
Temp	Weather	Wind	Temp	Weather	Wind	Rise/Fall	Feet	Inches
	c a r	SE		*f*	SE	F		6

Weather Diary   rain slight last night. The river continud to fall untill 4 AM having fallen 3 [inches] by that time since sunrise. It now was at a stand untill dark having which it began again to rise.

Lewis & Clark   very strong current. I sent Sergt. Pryor and a party over with the indian canoe in order to raise and secure ours but the depth of the water and the strength of the current baffled every effort.

Gass   The morning was fine, with a little fog. The river is so high that the trees stand some distance in the water.

SATURDAY, MAY 31

Sunrise			4 p.m.			Clearwater River		
Temp	Weather	Wind	Temp	Weather	Wind	Rise/Fall	Feet	Inches
	c a f	SE		*f*	SE	R	1	1

Weather Diary   within 3 Inches of its greatest hight on the 29th inst. and fell a little after which it rose again. The river rose 13 inches last night and continues to rise fast. From sunset on the 31st of May untill sun rise on th 1st of June it rose Eighteen inches and is now as high as any marks of [its] having been for several years past. A heavy thunder cloud passed around us last evening about sunset. Some rain fell in the fore part of the night only.

Gass   We had a fine clear morning with a heavy dew. In the evening the weather became cloudy, and we had some rain with sharp thunder and lightning.

Ordway   the Toomonamah river [Salmon] which is about 150 yards wide here. They [Native Americans] took us over a verry bad hill down on to the river

SUNDAY, JUNE 1

	Sunrise			4 p.m.		Clearwater River		
Temp	Weather	Wind	Temp	Weather	Wind	Rise/Fall	Feet	Inches
	f a r t & l	SE		*f a c*	NW	R	1	6

Weather Diary   about dark last evening had a slight rain from a heavy thunder cloud which passed to the E & NE of us.

Gass   We had a fine morning after some light showers of rain during the night.— Since last evening the river rose eighteen inches.

MONDAY, JUNE 2

	Sunrise			4 p.m.		Clearwater River		
Temp	Weather	Wind	Temp	Weather	Wind	Rise/Fall	Feet	Inches
	c a c	NW		*f a c*	SE	R		8

Weather Diary   have slept comfortably for several nights under one blankett only. The river from sunrise untill 10 AM yesterday raised 1 $1/2$ inches; from that time until dark fell 4 $1/2$, and in the course of last night raised again 8 inches as stated in the diary. The Indians inform us that the present rise of the river is the greatest which it annually takes, and that when the water now subsides to about the hight it was when we arrived here the mountains will be passable. I have not doubt but that the melting of the mountain snows in the beginning of June is what causes the annual inundation of the lower part of the Missouri form the 1st of the Middle of July.—

Lewis   in order to prepare in the most ample manner in our power to meet that wretched portion of our journy, the Rocky Mountain, where hungar and cold in their most rigorous forms assail the waried traveller; not any of us have yet forgotten our sufferings in those mountains in September last, and I think it probably we never shall. Sergt. Ordway & party returned from The East fork of Lewis's river they discribe as one continued rapid about 150 yds wide its banks are in most places solid and perpendicular rocks, with rise to a great hight; [its] hills are mountains high. On the tops of some of those hills which they passed, the snow had not entirely disappeared, and the grass was just springing up.

Clark McNeal and York were sent on a tradeing voyage over the river this morning. having exhosted all our Merchindize we were obliged to have recourse to every Subterfuge in order to prepare in the most ample manner in our power to meet that wretched portion of our journey, the Rocky Mountains, where hungar and Cold in their most rigorous form assail the waired traveller; not any of us have yet forgotten our sufferings in those mountains in September last, I think it probable we never shall. Both forks above the junction of Lewis's river appear to enter a high Mountainious Country.

Gass The morning was cloudy.

Ordway a fair morning. Found the river [Clearwater] very high indeed.

TUESDAY, JUNE 3

	Sunrise			4 p.m.		Clearwater River		
Temp	Weather	Wind	Temp	Weather	Wind	Rise/Fall	Feet	Inches
	caf&c	SE		*c a f*	SE	R		6

Weather Diary The weather has been much warmer for five days past then previously, particularly the mornings and nights.—

Lewis & Clark I begin to lose all hope of any dependance on the Salmon as this river will not fall sufficiently to take them before we shall leave it, and as yet I see now appearance of their runing near the shores as the indians inform us they would in the course of a few days. I find that all the salmon which they procure themselves they obtain on Lewis's river [Snake River].

Gass This was a cloudy morning with a few drops of rain; and there were some light showers during the forenoon at intervals. The river rises in the night, and falls in the day time; which is occasioned by the snow melting by the heat of the sun on the mountains, which are too distant for the snow water to reach this place until after night.

Ordway clouded up and Sprinkled a little rain.

WEDNESDAY, JUNE 4

	Sunrise			4 p.m.		Clearwater River		
Temp	Weather	Wind	Temp	Weather	Wind	Rise/Fall	Feet	Inches
	c a r	SE		*f a c*	NW	R		1 1/2

Weather Diary rained greater part of last night but fell in no great quantity—— yesterday the water was at [its] greatest hight at noon, between which and dark it fell 15 inches and in the course of the night raised 1 1/2 inches as stated in the

Weather Diary   river fell 9 In. yesterday (Lewis)   & 3 ½ last night (Clark)

Lewis   the river has been falling for several days and is now lower by near six feet than it has been; this we view as a strong evidence that the great body of snow has left the mountains, though I do not conceive that we are as yet loosing any time as the roads is in many parts extreemly steep rocky and must be dangerous if wet and slippry; a few days will dry the roads and will also improve the grass—

Clark   The flat head river is Still falling fast and nearly as low as it was at the time we arrived at this place. This fall of water is what the nativs have informed us was a proper token for us. When the river fell the Snows would be Sufficiently melted for us to Cross the Mountains. The greater length of time we delayed after that time, the higher the grass would grow on the Mountains—

Gass   This was a fine plesant day

Note   *The expedition leaves the Kooskooskee River, moves up onto the Weippe Prairie (Idaho), and starts its journey along the Lolo Trail over the Bitterroot Mountains.*

TUESDAY, JUNE 10

	Sunrise			4 p.m.		Clearwater River*		
Temp	Weather	Wind	Temp	Weather	Wind	Rise/Fall	Feet	Inches
	f	SE		f	NW	F		1

Weather Diary   do fell 5 ½ in. couse of yesterday having left the river today I could not longer keep [its] state; it appears to be falling fast and will probably in the course of a few days be as low as when we first arrived there. It is now about 6 feet lower than it has been.

Clark   we intend to delay a fiew days....by which time we Calculate that the Snows will have melted more off the mountains and the grass raised to a sufficient hight for our horses to live.

Ordway   clear & pleasant.

WEDNESDAY, JUNE 11

	Sunrise			4 p.m.		River		
Temp	Weather	Wind	Temp	Weather	Wind	Rise/Fall	Feet	Inches
	f	SE		f	NW			

Weather Diary   at the quawmash flats

---

*The last data collected on the Clearwater River.

Gass   We had a fine morning with some white frost.
Ordway   clear and pleasant.

THURSDAY, JUNE 12

	Sunrise			4 p.m.		River		
Temp	Weather	Wind	Temp	Weather	Wind	Rise/Fall	Feet	Inches
	f a r t l	SE		*f*	NW			

Weather Diary   slight sprinkle of rain in the forepart of the night.—

Lewis & Clark   the days are now very warm and the Musquetoes our old companions have become very troublesome.

Gass   We had a fine lovely morning with a heavy dew.

Ordway   a clear pleasant morning.

FRIDAY, JUNE 13

	Sunrise			4 p.m.		River		
Temp	Weather	Wind	Temp	Weather	Wind	Rise/Fall	Feet	Inches
	c	SE		*c a f*	NW			

Weather Diary   the days for several past have been warm, the Musquetoes troublesome

Lewis & Clark   we directed the meat to be cut thin and exposed to dry in the sun.
Gass   There was a fine morning. In the evening the weather became cloudy.
Ordway   a fair morning.

SATURDAY, JUNE 14

	Sunrise			4 p.m.		River		
Temp	Weather	Wind	Temp	Weather	Wind	Rise/Fall	Feet	Inches
	f	SE		*f*	NW			

Lewis   ....we have now been detained near five weeks in consequence of the snows; a serious loss of time at this delightfull season for traveling. I am still apprehensive that the snow and the want of food for our horses will prove a serious imbarrassment to us as at least four days journey of our rout in these mountains lies over hights and along a ledge of mountains never intirely destitute of snow. every body seems anxious to be in motion, convinced that we have not now any time to delay if the calculation is to reach the United States this season; this I am detirmined to accomplish if within the compass of human power.

Clark we expect to Set out early, and Shall proceed with as much expedition as possible over those Snowy tremendious mountains which has detained us near five weeks in this neighbourhood waiting for the Snows to melt Sufficent for us to pass over them. And even now I Shudder with the expectation with great dificueltes in passing those Mountains, form the debth of Snow and the want of grass Sufficient to Subsist our horses as about 4 days we Shall be on the top of the Mountain which we have every reason to believe is Covered with Snow the greater part of the year.

Gass We had a cloudy morning.

SUNDAY, JUNE 15

	Sunrise			4 p.m.			River	
Temp	Weather	Wind	Temp	Weather	Wind	Rise/Fall	Feet	Inches
	c	*NW*		*raf&r*	*NW*			

Weather Diary   it began to rain at 7 AM and contined by showers untill 5 PM

Lewis it rained very hard in the morning and after collecting our horses we waited for it to abait, but as it had every appearance of a settled rain we set our at 10 AM. The rains have rendered the road very slippery insomuch that it is with much difficulty our horses can get on

Clark Some hard Showers of rain detained us untill AM at which time we took our final departure from the quawmash fields and proceeded with much dificuelty owing to the Situation of the road which was very Sliprey   They frequently Sliped down both assending and decending those hills. the rain Seased and Sun Shown out. we passed through bad fallen timber and a high Mountain this evening. From the top of this Mountain I had an extensive view of the Rocky Mountains...several high pts. to the N & NW Covered with Snow. A remarkable high rugd mountain in the forks of Lewis's river nearly South and covered with Snow.

Gass This was a cloudy wet morning with some thunder. We had rain at intervals during the forenoon, but the afternoon was clear.

Ordway Soon set in to raining hard. About noon we had Thunder and hard Showers of rain.

MONDAY, JUNE 16

	Sunrise			4 p.m.			River	
Temp	Weather	Wind	Temp	Weather	Wind	Rise/Fall	Feet	Inches
	f a c	*SE*		*c a f*	*SE*			

Weather Diary   on the tops of the hills the dog tooth violet is just in bloom grass about 2 inches high small Huckkleberry just puting forth [its] leaves &c.

Lewis   the snow has increased in quantity so much that the greater part of our rout this evening was over the snow which has become sufficiently firm to bear our horshes, otherwise it would have been impossible for us to proceed as it lay in immence masses in some places 8 or ten feet deep. We found much difficulty in pursuing the road as it was so frequently covered with snow. The air is pleasant in the course of the day but becomes very cold before morning notwithstanding the shortness of the nights. Hungry Creek is but small at this place but is deep and runs a perfect torrent; the water is perfectly transparent and as cold as ice.

Clark   and proceed on through most intolerable bad fallen timber over a high Mountain on which great quantity of Snow is yet lying premisquissly through the thick wood, and in maney places the banks of snow is 4 feet deep. We nooned it or dined on a Small Creek in a small open valley where we found Some grass for our horses to eate, altho Serounded by Snow. the Snow has increased in quantity So much that the great part of our rout this evening was over the Snow which has become Sufficiently firm to bear our horses, otherwise it would have been impossible fur us to proceed as it lay in emince masses in Some places 8 to ten feet deep. We found much dificulty in finding the road, as it was so frequently covered with snow.

Gass   We had a pleasant morning,   The hills…great many banks of snow, some of them four or five feet deep. These banks are so closely packed and condensed, that they carry our horses, and are all in a thawing state. Had some rain at dinner. In the afternoon we found the snow banks more numerous; extensive and deep; in some of them the snow as much as eight feet deep.

Ordway   the morning fair. Towards noon we passed over high banks of Snow which bore up our horses, some places 5 or 6 feet deep light Showers of rain this afternoon   the Snow is more Common and much deeper. The bushes are all bent flat down by the deep Snow lying on them. The Snow must fall in these hallars in the winter 15 to 20 feet deep and perhaps the Snow drifts in and fills the hollars full.

Note   *On June 17, the expedition retreats (the first time doing so in the entire journey) due to snowpack.*

TUESDAY, JUNE 17

Sunrise			4 p.m.			River		
Temp	Weather	Wind	Temp	Weather	Wind	Rise/Fall	Feet	Inches
	c a r	E		*caf&r*	SE			

Weather Diary   rained slightly a little after sunset   Air cool. Rained frm 1 to 3 PM (Lewis) assend a mtn. Snow 15 feet deep on top. (Clark)

Lewis   this hill or reather mountain we ascended about 3 miles when we found ourselves invelloped in snow from 12 to 15 feet deep even on the south sides of the hills with the fairest exposure to the sun; here was winter with all [its] rigors; the air was cold, my hands and feet were benumbed. We could not hope for any food for our horses not even underwood itself as the whole was covered many feet deep with snow. The snow boar our horses very well and the travelling was therefor infinitely better that the obstruction of rocks and fallen timber which we met with in our passage over last fall when the snow lay on this part of the ridge in detached spots only. Under these circumstances we conceived it madnes in this stage of the expedition to proceed without a guide. This is the first time since we have been on this long tour that we have even been compelled to retreat or make a retrograde march. It rained on us most of this evening—

Clark   ....this mountain we ascended 3 miles when we found ourselves invelloped in snow from 8 to 12 feet deep even on the South Side of the mountain. I was in front and could only prosue the derection of the road by the trees which had been peeled by the nativs for the iner bark of which they scraped and eate. I with great difficulty prosued the direction of th road one mile further to the top of the mountain where I found the snow from 12 to 15 feet deep, but fiew trees with the fairest exposure to the Sun; here was Winter with all [its] rigors; the air was cold my hands and feet were benumed. We therefore come to the resolution to return...and again to proceed as soon as we could precure Such a guide, knowing from the appearance of the snows that if we remained untill it had disolved Sufficiently for us to follow the road that we Should not be enabled to return to the United States within this Season. we began our retragrade march at 1 P.M. haveing remain'd about three hours on this Snowey mountain. we returned by the rout we had advanced to hungary Creek, which we assended about 2 miles and encamped. ...the party were a good deel dejected, tho' not as much so as I had apprehended they would have been. It rained on us the most of this evening. On the top of the Mountain the Weather was very fluctiating and uncertain snowed cloudy & fair in a few minets.

Gass   There was a cloudy morning, but without rain. When we got about half way up the mountains, the ground was entirely covered with snow, three feet deep; and as we ascended it still became deeper, until we arrived at the top, where it was twelve or fifteen feet deep;   After remaining about two hours, we concluded it would be most adviseable to go back to some place where there was food for our horses. At this time it began to rain; and we proceeded down to Hungry creek again. The grass and plants here are just putting out, and the shrubs budding. It rained hard during the afternoon.

Ordway   we set out as usal the morning chilley and cloudy. When we got about half way up it the ground was covred with Snow 3 or 4 feet deep   as we ascended higher it go deeper untill we got to the top of the mountain where it was 12 to 15 feet in general even on the South Side where the Sun has open view but is So Settled So that it bears up our horses. Set in to hailling & raining at this time verry cold and disagreeable. So we turned back much against our expectations when we started.

WEDNESDAY, JUNE 18

	Sunrise			4 p.m.			River	
Temp	Weather	Wind	Temp	Weather	Wind	Rise/Fall	Feet	Inches
	c a r	E		*car&h*	SW			

Weather Diary   obliged to return

Gass The morning was cloudy and several showers of rain fell during the day. Where we had dinner on the 16th had a gust of rain, hail, thunder and lightning, which lasted an hour, when the weather cleared and we had a fine afternoon. We found the musquitoes very troublesome on the creek, notwithstanding the snow is at so short a distance up the mountains.

Ordway cloudy. About noon....at which time came up a hard Shower of hail and rain and hard Thunder, which lasted about an hour and cleared off. The musquetoes verry troublesome at this place.

THURSDAY, JUNE 19

Sunrise			4 p.m.			River		
Temp	Weather	Wind	Temp	Weather	Wind	Rise/Fall	Feet	Inches
	f a c	*SE*		*f*	*NW*			

Weather Diary returned to quawmash flats.

Lewis these trout are of a red kind they remain all winter in the upper parts of the rivers and creeks and are generally poor at this season.

Clark The SW Sides of the hills is fallen timber and burnt woods, the NE Sides of the hills is thickly timbered with lofty pine, and thick under growth.

Gass This was a fine morning

Ordway a fair morning.

FRIDAY, JUNE 20

Sunrise			4 p.m.			River		
Temp	Weather	Wind	Temp	Weather	Wind	Rise/Fall	Feet	Inches
	f	*SE*		*f*	*NW*			

Lewis & Clark we have determined to wrisk a passage on the following plan immediately, because should we wait much longer or untill the snow desolves in such mannger as to enable us to follow the road we cannot hope to reach the United States this winter. The travelling in the mountains on the snow at present is very good, the snow bears the horses perfictly; it is a firm couse snow without a crust. Although the snow may be stated on an average at 10 feet deep yet around the bodies of the trees it has desolved generally more than in other parts not being generally more than one or two feet deep immediately at the roots of the trees. The reason why the snow is comparitively so shallow about the roots of the trees I presume proceeds as well from the snow in falling being thrown off their bodies by their thick and spreading branches as from the reflection of the sun against the trees and the warmth which they in some measure acquire from the earth which is never frozen underneath these masses of snow.

Gass   There was a fine morning.

Ordway   a fair morning.

SATURDAY, JUNE 21

	Sunrise			4 p.m.			River	
Temp	Weather	Wind	Temp	Weather	Wind	Rise/Fall	Feet	Inches
	f	*SE*		*f*	*NW*			

Gass   We had again a fine morning

Ordway   a fair morning.

SUNDAY, JUNE 22

	Sunrise			4 p.m.			River	
Temp	Weather	Wind	Temp	Weather	Wind	Rise/Fall	Feet	Inches
	f	*NW*		*f*	*NW*			

Weather Diary   hard frost this morning tho' no ice. Strawberries ripe at the Quawmash flats, they are but small and not abundant.—

Lewis & Clark   the last evening was cool but the day was remarkably pleasent with a fine breize from the NW.

Gass   We had a pleasant day

Ordway   clear and pleasant

MONDAY, JUNE 23

	Sunrise			4 p.m.			River	
Temp	Weather	Wind	Temp	Weather	Wind	Rise/Fall	Feet	Inches
	f	*NW*		*f*	*NW*			

Weather Diary   hard frost this morning ice one eighth of an inch thick on standing water

Gass   We had again a fine morning

Ordway   a clear pleasant morning.

TUESDAY, JUNE 24

Sunrise			4 p.m.			River		
Temp	Weather	Wind	Temp	Weather	Wind	Rise/Fall	Feet	Inches
	f	NW		f a c*	NW			

Weather Diary   Set out a 2nd time from quawmash flats

Clark   we had fine grass for our horses this evening—

Gass   There was a cloudy morning. The day keeps cloudy, and the musquitoes are very troublesome.

Note   *On June 25, the expedition starts again along the Lolo Trail over the Bitterroot Mountains, hoping the snowpack has melted enough by this time to allow easier passage.*

WEDNESDAY, JUNE 25

Sunrise			4 p.m.			River		
Temp	Weather	Wind	Temp	Weather	Wind	Rise/Fall	Feet	Inches
	c a r	SE		c a r	NW			

Weather Diary   rained a little last night, some showers in the evening.

Lewis & Clark   last evening the indians entertained us with setting the fir trees on fire. they have a great number of dry limbs near their bodies which when Set on fire creates a very sudden and emmence blaize from bottom to top of those tall trees. They are a beautifull object in this situation at night. this exhibition reminded me of a display of fireworks. the natives told us that their object in Setting those trees on fire was to bring fair weather for our journey—

Gass   There was a light shower of rain this morning. A considerable quantity of rain had fallen during the afternoon.

Ordway   a little rain last night. We find the Snow has melted considerable Since we passd. The after part of the day Showery and wet.

THURSDAY, JUNE 26

Sunrise			4 p.m.			River		
Temp	Weather	Wind	Temp	Weather	Wind	Rise/Fall	Feet	Inches
	c a r	SE		f	SE			

---

*According to Moulton (1993, 8: 73), Clark's Journal Codex M lists this weather data as "f."

Weather Diary   Slight rain in the fore part of the last evening.— (Lewis) in the snowey region. (Clark)

Lewis   the snow has subsided near four feet since the 17th inst. We now measured it accurately and found from a mark which we had made on a tree when we were last here on the 17th that it was then 10 feet 10 inches which appeared to be about the common debth though it is deeper still in some places. It is now generally about 7 feet. Accordingly we set out with our guides who lead us over and along the steep sides of tremendious mountains entirely covered with snow except about the roots of the trees where the snow had sometimes melted and exposed a few square feet of the earth. Encamped…here we found an abundance of fine grass for our horses. This situation was the side of an untimbered mountain with a fair southern aspect where the snows from appearance had been desolved about 10 days.

Clark   …and assended to the summit of the mountain where we deposited our baggage on the 17th inst. found everything safe and as we had left them. the Snow which was 10 feet 10 inches deep on the top of the mountain, had sunk to 7 feet tho' perfectly hard and firm. we made Some fire Cooked dinner and dined, while our horses stood on snow 7 feet deep at least. Encamped…the grass was young and tender of course and had much the appearance of the Green Swoard.

Gass   We had a foggy morning; procceded on early; and found the banks of snow much decreased; at noon we arrived at the place where we had left our baggage and stores. The snow here had sunk twenty inches. We measured the depth of the snow here and found it ten feet ten inches. We proceeded over some very steep tops of the mountains and deep snow; but the snow was not so deep in the drafts between them. Some heavy showers of rain had fallen in the afternoon.

Ordway   we find the Snow has settled a little more than 2 feet Since we left this the other day. Proceeded on thro. Snow deep. In the evening we Came to the Side of a mountain where the Snow is melted away and a little young grass &C.

FRIDAY, JUNE 27

	Sunrise			4 p.m.			River	
Temp	Weather	Wind	Temp	Weather	Wind	Rise/Fall	Feet	Inches
	*far & t**	*SE*		*f*	*SE*			

Weather Diary   Thunder shower last evening   some rain a little before dark last evening

Lewis & Clark   we halted by the request of the Indians a few minutes and smoked the pipe. From this place we had an extensive view of these stupendous mountains principally covered with snow like that on which one unacquainted with them it would have seemed impossible ever to have escaped;   we arrived at a situation very similar to our encampment of the last evening tho' the ridge was somewhat higher and the snow had not been so long desolved of course there was but little grass. I doubt much whether we who had once passed them could find our way to Travellers rest in their present situation

---

*According to Moulton (1993, 8: 73), Clark's Journal Codex M lists this weather data as "f a r."

Gass We had a cloudy morning…proceeding over some of the steepest mountains I ever passed. The snow is so deep that we cannot wind along the sides of these steps, but must slide straight down. …about 5 o'clock in the evening, when we stopped at the side of a hill where the snow was off, and where there was a little grass. The day was pleasant throughout; but it appeared to me somewhat extraordinary, to be traveling over snow six or eight feet deep in the latter end of June. The most of us, however, had saved our socks, as we expected to find snow on these mountains.

Ordway a fair morning. We took an eairly breakfast and proceeded on verry fast over the high banks of Snow. The day warm and Snow melts fast—

SATURDAY, JUNE 28

	Sunrise			4 p.m.			River		
Temp	Weather	Wind	Temp	Weather	Wind	Rise/Fall	Feet	Inches
	f	SE		*f*	SE			

Weather Diary nights are cool in these mountains but no frost

Lewis & Clark the whole of the rout of this day was over deep snows. We find the traveling on the snow not worse than without it as the easy passage it gives us over rocks and fallen timber fully compensates for the inconvenience of sliping, certain it is that we travel considerably faster on the snow than without it. the snow sinks from 2 to 3 inches with a horse, is coarse and firm and seems to be formed of the larger and more dense particles of the Snow; the Surface of the Snow is reather harder in the morning than after the sun shines on it a fiew hours, but it is not in that situation so dense as to prevent the horses from obtaining good foothold.

Clark much fallen timber caused in the first instance by fire and more recently by a storm from the SW.*

Gass The morning was pleasant. On the south side of this ridge there is summer with grass and other herbage in aboundance; and on the north side, winter with snow six or eight feet deep.

Ordway a fair clear cool morning. The snow continues as yesterday. Had a bad Shower of hail and some thunder in the evening

Note *The expedition crosses the Lolo Pass, at the present-day Idaho–Montana state line.*

---

*From Clark's courses and estimated distances list, along a high ridge 4 $1/2$ miles, which turns off to the right and leads to the fishery at the entrance of Colt Creek (Moulton 1993, 8: 73).

SUNDAY, JUNE 29

	Sunrise			4 p.m.		River		
Temp	Weather	Wind	Temp	Weather	Wind	Rise/Fall	Feet	Inches
	f	*SE*		*farh & t*	*SE*			

Weather Diary night cold hard frost this morning. The quawmash and strawberries are just begining to blume at the flatts on the head of the Kooskooske. The sun flower also just beginning to blume, which is 2 months later that those on the Sides of the Western Mountains [Cascades] near the falls [Celilo] of the Columbia.

Lewis & Clark when we decended from this ridge we bid adieu to the snow. The Kooskooske at this place is about 30 yds. wide and runs with great volocity. the principal spring is about the temperature of the Warmest baths used at the Hot Springs in Virginia. in this bath which had been prepared by the Indians by stopping the river with Stone and mud, I bathed and remained in 10 minits it was with dificuelty I could remain this long and it caused a profuse sweat. two other bold Springs adjacent to this are much warmer, their heat being so great as to make the hand of a person Smart extreemly when immerced. We think the temperature of those Springs about the Same as that of the hotest of the hot Springs of Virgina.

Gass There was a foggy morning. Proceeded early…at which time there was a shower of rain, with hail, thunder and lightning, that lasted about an hour. At 10 o'clock we left the snow, and in the evening we arrived at the warm springs [Lolo Hot Springs].

Ordway a fair morning. The fog rose up thick from the hollars had a Shower of Hail and Thunder. Towards evening we arived at the hot Stream where we camped. A number of the party as well as myself bathed in these hot Springs, but the water so hot [has been measured at 111 °F], that it makes the Skin Smart when I first entered it. I drank Some of the water also—

Note *The expedition arrives at Traveler Rest, near present-day Lolo, Montana.*

MONDAY, JUNE 30

	Sunrise			4 p.m.		River		
Temp	Weather	Wind	Temp	Weather	Wind	Rise/Fall	Feet	Inches
	f	*SE*		*f*	*NW*			

Weather Diary   We are here Situated on Clark's river in a Vally between two high mountains of Snow. Night cold hard frost this morning.

Clark   Descended the mountain to Travellers rest leaving those tremendious mountain behind us— in passing of which we have experienced Cold and hunger of which I shall ever remember…from the 14th to the 19th of Septr. 1805 we marched through snow, which fell on us on the night of the 14th and nearly all the day of the 15 in addition to the cold rendered the air cool and the way difficuelt. On the 16th we met with banks of Snow and in the hollars and maney of the hill Sides of Snow was from 3 to 4 feet deep and Scercely any grass vegitation just commencing where the snow had melted— on the 17th at meridian, the Snow became So deep in every direction from 6 to 8 feet deep we could not prosue the road…on the 27th & 28th also passing over Snow 6 to 8 feet deep all the way   on 29th passed over but little Snow— but saw great masses of it lying in different directions.*

Gass   We continued our march early and had a fine morning.

Ordway   a clear morning. Musquetoes verry troublesome here—

Note   *Through July and early August, Lewis and Clark split the expedition and proceed on separate trails (to meet again on the Missouri River in August). Lewis takes his party, which includes Sergeant Gass, to the White Bear Island Camp near the Great Falls of the Missouri. There Lewis takes a smaller contingent to scout the Marias River. Clark takes the rest of the party up the Bitterroot Valley towards Camp Fortunate to pick the cache of goods left the last year. The Weather Diary data and remarks and the daily narrative journal entries made by members of the group are organized below by party under each day they are separated.*

TUESDAY, JULY 1

Lewis's Party

Sunrise			4 p.m.			River		
Temp	Weather	Wind	Temp	Weather	Wind	Rise/Fall	Feet	Inches
	c a f	NW		f	NW			

Weather Diary   a speceis of wild clover with a small leaf just in blume.

Gass   We had a fine morning

Clark's Party

Sunrise			4 p.m.			River		
Temp	Weather	Wind	Temp	Weather	Wind	Rise/Fall	Feet	Inches
	c a f	NW		f	NW			

---

*At the end of this narrative journal remark, Clark summarizes their experiences traveling across the mountains to the Pacific Ocean and back to Travellers Rest.

Weather Diary   a Species of wild [clover] in blume
Ordway   cloudy.

WEDNESDAY, JULY 2

Lewis's Party

Sunrise			4 p.m.			River		
Temp	Weather	Wind	Temp	Weather	Wind	Rise/Fall	Feet	Inches
	f	SE		f	SE			

Lewis   the tops of the high mountains on either side of this river are covered with snow.
Gass   We continued here during this day, which was fine

Clark's Party

Sunrise			4 p.m.			River		
Temp	Weather	Wind	Temp	Weather	Wind	Rise/Fall	Feet	Inches
	f	SE		f	NW			

Weather Diary   Musquetors very troublsom
Ordway   a clear pleasant morning.

Note   *Lewis and his party travel up the Blackfoot River through Lewis and Clark Pass, down the Sun River and into the Great Falls of the Missouri. Clark and his party venture back up the Bitterroot Valley, over the Continental Divide into the Big Hole Valley and to the cache at Camp Fortunate, south of Dillon, Montana.*

THURSDAY, JULY 3

Lewis's Party

Sunrise			4 p.m.			River		
Temp	Weather	Wind	Temp	Weather	Wind	Rise/Fall	Feet	Inches
	f	SE		f	NW			

Weather Diary   the turtle dove lays [its] eggs on the ground in these plains and is now seting, it has two eggs only and they are white.

Lewis we saddled our horses and set out I took leave of my worthy friend and companion Capt. Clark and the party that accompanied him. I could not avoid feeling much concern on this occasion although I hoped this seperation was only momentary. Passed the east branch of Clark's River...from 90 to 120 yds. wide. the musquetoes were so excessively troublesome this evening that we were obliged to kindle large fires for our horses these insects torture them in such manner untill they placed themselves in the smoke of the fires that I realy thought they would become frantic. About an hour after dark the air become so [cold] that the musquetoes disappeared.

Gass We had again a fine morning

Clark's Party

	Sunrise			4 p.m.			River	
Temp	Weather	Wind	Temp	Weather	Wind	Rise/Fall	Feet	Inches
	f	SE		*f*	SW			

Weather Diary Cap L. & my Self part at Travellers rest.

Clark those Creeks take their rise in the mountains to the West which mountains is at this time Covered with Snow for about ⅕ of the way from their tops downwards. Some Snow is also to be Seen on the high points and hollows of Mountains to the East of us.

Ordway I am with Capt. Clark up the [Bitterroot River]. We kept up the west Side as it is too high at this time to cross.

FRIDAY, JULY 4

Lewis's Party

	Sunrise			4 p.m.			River	
Temp	Weather	Wind	Temp	Weather	Wind	Rise/Fall	Feet	Inches
	f	SE		*f*	NW			

Lewis the evening was fine, air pleasent and no musquetoes.

Gass We had a beautiful morning

Clark's Party

	Sunrise			4 p.m.			River	
Temp	Weather	Wind	Temp	Weather	Wind	Rise/Fall	Feet	Inches
	f	SW		*f*	SW			

Weather Diary   a w[a]rm day. I saw a Speces of Honeysuckle with a redish brown flower in blume

Clark   the last Creek or river which we pass'd was So deep and the water So rapid that Several of the horses were Sweped down Some Distance. The water was So Strong, altho' the debth was not much above the horses belly, the water passed over the backs and loads of the horses. Those Creeks are emensely rapid has great decnt.

Ordway   a fair morning. Dined and proceeded on without finding the road. As we cannot ford the river yet.

SATURDAY, JULY 5

Lewis's Party

Sunrise			4 p.m.			River		
Temp	Weather	Wind	Temp	Weather	Wind	Rise/Fall	Feet	Inches
	f	NE		f	SW			

Weather Diary   a great number of pigeons breeding in this part of the mountains musquetoes not so troblesom as near Clark's river. some ear flies of the common kind and a few large horse flies.

Gass   We had another beautiful morning....crossed a river about 35 yards wide, which flows in with a rapid current from some snow topped mountains on the north.

Clark's Party

Sunrise			4 p.m.			River		
Temp	Weather	Wind	Temp	Weather	Wind	Rise/Fall	Feet	Inches
	f	NE		f	SW			

Weather Diary   Cool night. Some dew this morning the nights are Cool. the musquetors are troublesome untill at little after dark when the air become Cool and Musquetoes disappear.

Clark   I crossed the river which heads in a high peecked mountain Covered with Snow NE of the Valley at about 20 miles [near West Pintlar Peak Montana]

Ordway   a fair M. Set out to cross the right fork of the river which we found nearly Swimming.

Note   *Clark's party passes over the Continental Divide at present-day Gibbons Pass, Montana.*

SUNDAY, JULY 6

Lewis's Party

Sunrise			4 p.m.			River		
Temp	Weather	Wind	Temp	Weather	Wind	Rise/Fall	Feet	Inches
	f	*NE*		*f*	*SW*			

Weather Diary   the last night cold with a very heavy dew

Gass   We had a fine clear morning with some white frost

Clark's Party

Sunrise			4 p.m.			River		
Temp	Weather	Wind	Temp	Weather	Wind	Rise/Fall	Feet	Inches
	f	*SW*		*c a r t & l*	*SW*			

Weather Diary   cold night with frost. I slept cold under 2 blankets on head of Clark's river. I arived in an open plain in the middle of which a violent Wind from the N.W. accompanied with hard rain which lasted from 4 untill half past 5 P.M. quawmash in those plains at the head of wisdom River is just beginning to blume and the grass is about 6 inches high.

Clark   Some frost this morning. The last night was so cold that I could not Sleep. I observe great quantities of quawmash just beginning to blume. the Snow appears to lying in considerable masses on the mountain from which we decended on the 4th of Septr. Last. [Saddle Mountain, Ravalli County, Montana] Crossd. a large Creek [Ruby Creek] from the right which heads in a Snow Mountain. we had not proceeded more than 2 Miles in the last Creek, before a violent Storm of wind accompand. with hard rain from the SW imediately from off the Snow Mountains this rain was Cold and lasted 1 1/2 hours. I discovd. the rain wind as it approached and halted and formd. a solid column to protect our Selves from the Violency of the gust.

Ordway   a fair morning. Our Intrepters wife [Sacajawea] tells us that She knows the country & that this branch is the head waters of jeffersons river &C. Late in the afternoon we came to a large extensive plain [Big Hole Valley]   Came up a hard Thunder Shower of hail and hard wind. We halted a short time in the midst of it then proceed.

Note   *Lewis's party crosses over the Continental Divide at present-day Lewis and Clark Pass, Montana.*

MONDAY, JULY 7

Lewis's Party

	Sunrise			4 p.m.			River	
Temp	Weather	Wind	Temp	Weather	Wind	Rise/Fall	Feet	Inches
	c a r t & l	SW		*c a f & r*	W			

Weather Diary   a cloud came on about sunset and continued to rain moderately all night. rained at 3 PM

Lewis   passing the dividing ridge between the waters of the Columbia and Missouri rivers at $1/4$ of a miles.

Gass   We had a wet night, and a cloudy morning…we came to the dividing ridge between the waters of the Missouri and Columbia

Clark's Party

	Sunrise			4 p.m.			River	
Temp	Weather	Wind	Temp	Weather	Wind	Rise/Fall	Feet	Inches
	c a r	W		*f a r*	SW by W			

Weather Diary   Saw a blowing Snake. a violent rain from 4 to $1/2$ past 5 last evening & Some rain in the latter part of last night. a small Shower of rain at 4 this morning accompanied with wind from the S.S.W.

Clark   We arived at a Boiling Spring [Jackson Hot Spring]…this Spring contains a very considerable quantity of water and actually blubbers with heat for 20 paces below where it rises. It has every appearance of boiling, to hot for a man to endure his hand in it 3 seconds. A little sulferish. This extensive vally Surround with covered with snow is extreemly fertile covered esculent plants &c….I now take my leave of this butifull extensive vally which I call hot spring Vally. remarkable Cold night

Ordway   had several Showers of rain & Thunder in the course of this afternoon—

Note   *Clark's party arrives at Camp Fortunate at present-day Clark Reservoir, Montana.*

TUESDAY, JULY 8

Lewis's Party

	Sunrise			4 p.m.			River	
Temp	Weather	Wind	Temp	Weather	Wind	Rise/Fall	Feet	Inches
	f	SW		*f*	W			

Weather Diary   heavy white frost last night. very cold
Lewis   the grass generally about 9 inches high.
Gass   The morning was pleasant with some white frost.

Clark's Party

	Sunrise			4 p.m.			River	
Temp	Weather	Wind	Temp	Weather	Wind	Rise/Fall	Feet	Inches
	f a r	W		*f*	SW			

Weather Diary   a Small Shower of rain a little after dark a heavy rain and wind from SW at 4 P.M. yesterday a heavy Shower of rain accompanied with rain from the SW from 4 to 5 PM. Passed the boiling hot springs emerced 2 peces of raw meat in the Spring and in 25 Minits the Smallest pece was sufficiently cooked and in 32 the larger was also sufficiently cooked

Clark   opened the cache...I found every article Safe, except a little damp. The Country through which we passed to day was diversified high dry and uneaven Stonry open plains and low bottoms very boggy [Grasshopper Creek, Montana] with high mountains on the tops and North sides of which there was Snow

Ordway   a clear cold morning & hard frost. Cam to a boiling hot spring [Jackson Hot Spring] I drank some of the water found it well tasted but So hot [has been measured at 136 °F]

WEDNESDAY, JULY 9

Lewis's Party

	Sunrise			4 p.m.			River	
Temp	Weather	Wind	Temp	Weather	Wind	Rise/Fall	Feet	Inches
	c a r	NE		*r*	NE			

Weather Diary   rained slightly last night. Air cold. Rained constantly all day air extremly cold it began to rain about 8 AM and continued with but little intermission all day   in the evening late it abated and we obtained a view of the mountains we had just passed   they were covered with snow apparrently several feet deep which had fallen during this day.—

Lewis   Set out early and had not proceeded far before it began to rain. The air extreemly cold. Halted a few minutes in some old lodges untill it cased to rain in some measure. We then proceeded and it rained without intermission. Wet us to the skin. They day continued rainy and cold. The sun river is generally about 80 yds wide rapid...water clear.

Gass   A cloudy morning. We set out early to go down the river; but had not proceeded far before it began to rain, and we halted and took shelter. In an hours time the rain slackened, and we proceeded on; but had not gone far before it began to rain again, and the weather was very cold for the season...before noon. ...and lay by during the afternoon as the rain continued during the whole of it.

Clark's Party

	Sunrise			4 p.m.			River	
Temp	Weather	Wind	Temp	Weather	Wind	Rise/Fall	Feet	Inches
	c	SW		f	SW			

Weather Diary   Hard frost. Some ice this morning. last night was very Cold and wind hard from the NE all night. The river is 12 inches higher that it was last Summer when we made the deposit here and portage from this place. more Snow on the adjacent mountains than was at that time.

Clark   This day was windy and Cold. The wind dried our Canoes very much.

Ordway   a fair morning.

Note   *Clark's party starts down Jefferson (present-day Beaverhead) River heading toward Three Forks, Montana. On their way, they pass present-day Dillon, Montana.*

THURSDAY, JULY 10

Lewis's Party

	Sunrise			4 p.m.			River	
Temp	Weather	Wind	Temp	Weather	Wind	Rise/Fall	Feet	Inches
	f a r	NW		f	W			

Weather Diary   rain ceased a little after dark

Lewis   the ground is renderd so miry by the rain which fell yesterday that it is excessively fatiegueing to the horses to travel. We came 10 miles and halted for dinner the wind blowing down the river in the fore part of the day was unfavourable to the hunters.

They saw several gangs of Elk but they having the wind of them ran off. In the evening the wind set from the West and we fell in with a few elk. The most direct and best Course from the dividing ridge which divides the waters of the Columbia from those of the Missouri at the Gap. A fine road and about 45 miles, reducing the distance from Clark's river to 145 miles.

Gass   At dark last evening the weather cleared up, and was cold all night. This morning was clear and cold, and all the mountains in sight were covered with snow, which fell yesterday and last night. The road was very muddy after the rain.

Clarks' Party

Sunrise			4 p.m.			River		
Temp	Weather	Wind	Temp	Weather	Wind	Rise/Fall	Feet	Inches
	f	SE		*f*	SW			

Weather Diary   white frost this morning. ice ³/₄ of an inch thick on Standing water. grass killd by the frost. river falling proceviable. a large white frost last night. the air extreemlly Cold. Ice ³/₄ of an inch thick on Standing water.

Clark   last night was very cold and this morning everything was white with frost and the grass Stiff frozend. I had Some water exposed in a bason in which the ice was ³/₄ of an inch thick this morning. The Musquetors were troublesom all day and untill one hour after Sunset when it became Cool and they disappeared. I saw several large rattle Snakes in passing the rattle Snake Mountain   they were fierce.

Ordway   a Severe hard frost & Ice. Chilley and cold this morning

Note   *Lewi's party arrives at the upper portage camp at the White Bear Islands, near Great Falls, Montana.*

FRIDAY, JULY 11

Lewis's Party

Sunrise			4 p.m.			River		
Temp	Weather	Wind	Temp	Weather	Wind	Rise/Fall	Feet	Inches
	f	NW		*f*	NW			

Weather Diary   wind very hard in the latter part of the day

Lewis   the morning was fair and the plains looked beatifull   the grass much improved by the late rain. The air was pleasant and a vast assemblage of little birds which croud to the groves on the river sun most enchantingly. It is now the season at which the baffaloe begin to coppelate and the bulls keep a tremendious roaring we could hear them for many miles and there are such numbers of them that there is one continual roar. This evening….the wind blew very hard.

Gass This was a fine morning

Clark's Party

Sunrise			4 p.m.			River		
Temp	Weather	Wind	Temp	Weather	Wind	Rise/Fall	Feet	Inches
	f	SE		*f*	NNE			

Weather Diary frost this morning. goslins nearly grown fishing hawks have their young The yellow Current nearly ripe. a slight frost last night. the air Cool. the Musquetors retired a little after dark, and did not return untill about an hour after Sunrise.

Clark the wind rose and blew with great violence from the SW imediately off Some high mountains Covered with Snow. The violence of this wind retarded our progress very much and the river being emencly Crooked we had it imediately in our face nearly every bend. At 6 PM I passed Phalanthropy river which I proceved was very low. The wind Shifted about to the NE and blew very hard tho' much w[a]rmer than the forepart of the day. Wisdom river is very high and falling.

Ordway a fair morning. We took breakfast eairly and set off. The wind hard a head which is unfavourable to us. Wisdom river is verry high at this time.

SATURDAY, JULY 12

Lewis's Party

Sunrise			4 p.m.			River		
Temp	Weather	Wind	Temp	Weather	Wind	Rise/Fall	Feet	Inches
	f	NW		*f*	NW			

Weather Diary wind violent all last night and today untill 5 PM when it ceased in some measure

Lewis after 10 am….the wind blew so violently that I did not think it prudent to attempt passing the river— at 5 P.M. the wind abated and we transported our baggage and meat to the opposite shore in our canoes which we found answered even beyond our expectations. I think the river is somewhat higher than when we were here last spirng/summer. The present season has been much more moist than the preceeding one. The grass and weeds are much more luxouriant than they were when I left this place on the 13th of July 1805— the yellow Currants begining to ripen.

Gass Again a fine morning

Clark's Party

Sunrise			4 p.m.			River		
Temp	Weather	Wind	Temp	Weather	Wind	Rise/Fall	Feet	Inches
	f	SE		*f*	NW			

Weather Diary   wisdom river is high but falling. Prickly pears in blume

Clark   after completing the paddles &c and takeing Some Brackfast I set out   The current I find much Stronger below the forks than above and the river tolerably steight as low as panther Creek when it became much more Crooked   the Wind rose and blew hard off the Snowey mountains to the NW and rendered it very difficuelt to keep the canoes from running against the Shore. At 2 PM the Canoe in which I was in was driven by a Suden puff of wind under a log which projected over the water from the bank, and the man in the Stern Howard was caught in between the Canoe and the log and a little hurt after disingaging our selves from this log the canoes was driven imediately under a drift which projected over and a little above the Water, here the Canoe was very near turning over

Ordway   a clear morning. The canoe Capt. Clark was in got drove to Shore by the wind under Some tops of trees and was near being filled with water. Capt. Clark fired 2 guns as a Signal for help. I and the other canoes which was a head halted and went to their assistance. They soon got him safe off before 2 pm.

Note   *Clark's party splits. Clark and some men journey toward the Rochejhone (Yellowstone) River and pass present-day Bozeman, Montana. Meanwhile, Sergeant Ordway leads a smaller party with the canoes down the Missouri River toward Great Falls, Montana.*

SUNDAY, JULY 13

Lewis's Party

Sunrise			4 p.m.			River		
Temp	Weather	Wind	Temp	Weather	Wind	Rise/Fall	Feet	Inches
	f	NE		*f*	NE			

Lewis   had the cash opened. Found my bearskins entirely destroyed by the water, the river having risen so high that the water had penitrated. All my specimens of plants also lost. The Chart of the Missouri fortunately escaped. Musquetoes excessively troublesome insomuch that without the protection of my musquetoe bier I should have found it impossible to wright a moment.

Gass   The morning was pleasant

Clark's Party

	Sunrise			4 p.m.			River	
Temp	Weather	Wind	Temp	Weather	Wind	Rise/Fall	Feet	Inches
	f	*SSE*		*f*	*NE*			

Clark   The Country in the forks between Gallitins & Madisens rivers is a butifull leavel plain Covered with low grass—   The Current of the river is rapid and near the mouth contains Several islands

Ordway's Party

Ordway   a clear morning. I and 9 proceedd on down the river. The wind a head so we halted little before night. The Musquetoes more troublesome than ever we have seen them before.

MONDAY, JULY 14

Lewis's Party

	Sunrise			4 p.m.			River	
Temp	Weather	Wind	Temp	Weather	Wind	Rise/Fall	Feet	Inches
	f	*SW*		*f*	*SW*			

Lewis   had the meat cut thiner and exposed to dry in the sun.

Gass   There was a pleasant morning. We staid here also to day; and the musketoes continued to torment us until about noon, when a fine breeze of wind arose and drove them, for a while away.

Clark's Party

	Sunrise			4 p.m.			River	
Temp	Weather	Wind	Temp	Weather	Wind	Rise/Fall	Feet	Inches
	f	*NW*		*f*	*NW*			

Weather Diary   Saw a Tobacco worm shown me by York

Clark   Passed 3 Small Streams from the Mountains to my right. Some Snow on the mountains to the SE S SW. Marked my name & day & year on a Cotton tree. The

Main fork of Galletins River turn South and enter them mountains which are yet Covered with Snow.

Ordway's Party

Ordway   a fair morning. The wind rose hard a head   About noon we halted   the wind rose So high that we were unable to proceed. In the evening as the wind fell we mooved down the R. to a bottom and Camped.

Note   *Clark's party arrives at the Yellowstone River near present-day Livingston, Montana. Ordway's party passes Helena, the present-day capital of Montana.*

TUESDAY, JULY 15

Lewis's Party

	Sunrise			4 p.m.			River		
Temp	Weather	Wind	Temp	Weather	Wind	Rise/Fall	Feet	Inches
	f	*SW*		*f*	*E*			

Lewis   the musquetoes continue to infest us in such manner that we can scarcely exist; for my own part I am confined by them to my bier at least ³/₄ths of my time. My dog even howls with the torture he experiences from them, they are almost insupportable, they are so numerous that we frequently get them in our throats as we breath.

Gass   We had pleasant weather.

Clark's Party

	Sunrise			4 p.m.			River		
Temp	Weather	Wind	Temp	Weather	Wind	Rise/Fall	Feet	Inches
	f	*SE by E*		*f*	*NE*			

Weather Diary   Struck the river Rochejhone 120 yards wide water falling a little

Clark   river is about 120 yrd wide bold and deep [Yellowstone]. The water of a whiteish blue colour. A mountain which is ruged NW has Snow on parts of it. River is rugid and covered with Snow those on the West is also high but have now Snow. Much dead timber on its N side— Camped last night passing over a low dividing ridge to the head of a water Course which runs into the Rochejhone [Yellowstone River]. The River Rochejhone at which place I arrived at 2 PM. Shield River discharges itself into the Rochejhone…this river is 35 yards deep and affords a great quantity of water   it heads in those Snowey Mountain to the NW with Howards Creek. The Roche passes out of a high rugid mountain covered with Snow. This 2nd bottom over flows in high floods. But fiew flowers to be Seen in those plains. Low grass in the high plains

Ordway's Party

Ordway   a fair morning.

Note   *Lewis leaves Sergeant Gass in command of the upper portage camp at White Bear Islands, while he takes a small party to explore the Marias River basin from July 16–28. Ordway's party passes through the Gates of the Rockies.*

WEDNESDAY, JULY 16

Lewis's Party

Sunrise			4 p.m.			River		
Temp	Weather	Wind	Temp	Weather	Wind	Rise/Fall	Feet	Inches
	f	*SW*		*f*	*SW*			

Weather Diary   Saw the Cookkoo or rain corw and the redheaded woodpecker. The golden rye now heading. Both species of the prickly pare in blume.— the sunflower in blume.

Lewis   proceeded down the river to the handsome fall [Rainbow Falls]…where I halted about 2 hours and took a haisty sketch of these falls. here we encamped and the evening having the appearance of rain made our beds and slept under a shelving rock. these falls have abated much of their grandure since I first at them in June 1805, the water being much lower at present than it was at that moment, however they are still a sublimely grand object.

Clark's Party

Sunrise			4 p.m.			River		
Temp	Weather	Wind	Temp	Weather	Wind	Rise/Fall	Feet	Inches
	c	*NE*		*c*	*NE*			

Weather Diary   Saw the wild indigo & common sunflower

Clark   the current of the Rochejhone is too rapid to depend on Skinn canoes no other alternative for me but to proceed on down untill I can find a tree sufficently large &c. to make a Canoe—

Gass's Portage Party

Gass   There was a fine morning.

Ordway's Party

Ordway   a fair morning. The wind rose a head and blew so high about nono that obledged us to lay too near the gates of the rockey mountains. About 3 PM the wind abated a little and we proced. on thro the gates of the mn

THURSDAY, JULY 17

Lewis's Party

	Sunrise			4 p.m.			River	
Temp	Weather	Wind	Temp	Weather	Wind	Rise/Fall	Feet	Inches
	f a t l	SW		*f*	SW			

Weather Diary   wind voilent all day. Distant thunder last evening to the West.

Lewis   I steered my course through the wind and level plains which have somewhat the appearance of an ocean, not a tree nor a shrub to be seen. The land is not fertile, at least far less so, than the plains of the Columbia or those lower down this river, it is a light coloured soil. Rose river [Teton] is at this place fifty yards wide, the water which is only about 3 feet deep…is very terbid of a white colour. From the size of rose river at this place and [its] direction I have no doubt but it takes [its] source within the first range of the Rocky Mountains.

Clark's Party

	Sunrise			4 p.m.			River	
Temp	Weather	Wind	Temp	Weather	Wind	Rise/Fall	Feet	Inches
	f a r h t & l	SE		*f*	SW			

Weather Diary   Heavy showers of rain   Hard Thunder & Lightning last night   a heavy Shower of rain accompanied with hail Thunder and Lightning at 2 a.m. with hard wind from the SW   after the Shower was over it Cleared away and became fair.

Clark   The rain of last night wet us all. Cross a large Creek which heads in a high Snow toped Mountain to the NW imediately opposit to the enterance of the Creek one Somthing larger falls in from the high Snow mountains to the SW & South. All the mountains to the SW is covered with Snow [Absaroka Range and Beartooth Mountains]

Gass's Portage Party

Gass   We had a pleasant day, and high wind; which drives away the musquitoes and relieves us from those tormenting insects.

Ordway's Party

Ordway   a clear morning. The wind rose so high that Some of the canoes were near being filled. Halted at the creek above as the wind too high to pass these rapids with safety. Towards evening the wind abated a little So we passed down the rapids with Safety.

FRIDAY, JULY 18

Lewis's Party

	Sunrise			4 p.m.		River		
Temp	Weather	Wind	Temp	Weather	Wind	Rise/Fall	Feet	Inches
	f	SW		f	NE			

Lewis   many prickly pears now in blume

Clark's Party

	Sunrise			4 p.m.		River		
Temp	Weather	Wind	Temp	Weather	Wind	Rise/Fall	Feet	Inches
	f	SW		f	SE			

Weather Diary   yellow, purple & black Currents ripe and abundant

Clark   currents ripe…I think the purple Superior to any I have ever tasted. The river here is about 200 yard wide rapid as usial and the water gliding over corse gravel

Gass's Portage Party

Gass   There was another plesant day

Ordway's Party

Ordway   a clear cool windy morning. Proceeded down the gentle current   the musquetoes and small flyes are very troublesome. My face and eyes are Swelled by the poison of those insects which bite verry Severe indeed.

Note   *Clark establishes Yellowstone River Canoe Camp between July 19 and 24 near present-day Columbus, Montana. Ordway's party arrives at the upper portage camp at the White Bear Islands and begins to assist Sergeant Gass's party in moving the canoes and materials down to lower portage camp.*

SATURDAY, JULY 19

Lewis's Party

	Sunrise			4 p.m.		River		
Temp	Weather	Wind	Temp	Weather	Wind	Rise/Fall	Feet	Inches
	f	SE		*f*	NE			

Lewis   completed my observation of the sun's meridian Altitude we set out.

Clark's Party

	Sunrise			4 p.m.		River		
Temp	Weather	Wind	Temp	Weather	Wind	Rise/Fall	Feet	Inches
	f	NW		*f*	SE			

Weather Diary   Saw the 1st Grape vine of the dark purple kind   the grape nearly grown

Clark   we passed over two high points of Land from which I had a view of the rocky Mounts. to the W & S SE all Covered with Snow. It may be proper to observe that the emence Sworms of Grass hoppers have distroyed every Spring of Grass for maney miles on this Side of the river, and appear to be progressing upwards.

Gass–Ordway Portage Party

Gass   The weather continues pleasant

Ordway   a clear & pleasant morning.

SUNDAY, JULY 20

Lewis's Party

	Sunrise			4 p.m.		River		
Temp	Weather	Wind	Temp	Weather	Wind	Rise/Fall	Feet	Inches
	f	E		*f*	N			

Lewis the day proved excessively warm and we lay by four hours during the heat of it; there is scarcely any water at present in the plains and what there is, lies in small pools and is so strongly impregnated with the mineral salts that it is unfit for any purpose except the use of the buffaloe. Those animals appear to prefer this water to that of the river.

Clark's Party

	Sunrise			4 p.m.			River	
Temp	Weather	Wind	Temp	Weather	Wind	Rise/Fall	Feet	Inches
	f	NE		*f*	NE			

Weather Diary Sworms of grass hoppers have eaten the grass of the plains for many miles. The River Rachejhone falls about ½ an in in 24 hours and becomes much Clearer than above. The Grass hoppers are emencely noumerous and have distroyed every Species of grass from one to 10 Miles above on the river & a great distance back.

Gass–Ordway Portage Party

Gass We had a fine day
Ordway a clear warm morning.

MONDAY, JULY 21

Lewis's Party

	Sunrise			4 p.m.			River	
Temp	Weather	Wind	Temp	Weather	Wind	Rise/Fall	Feet	Inches
	f	N		*f*	NE			

Lewis the water of this stream is nearly clear [Cut Bank Creek]. From the appearance of this rock and the apparent hight of the bed of the steem I am induced to believe that there are falls in these rivers somewhere about their junction.

Clark's Party

	Sunrise			4 p.m.			River	
Temp	Weather	Wind	Temp	Weather	Wind	Rise/Fall	Feet	Inches
	f	NE		*c*	NE			

Weather Diary   river falls a little and the water is nearly Clear

Clark   This evening late a very black Cloud from the SE accompanied with Thunder and lightning with hard winds which Shifted about and was worm and disagreeable.

Gass–Ordway Portage Party

Gass   A plesant morning.

Ordway   a fair warm morning

Note   *Lewis's party establishes Camp Disappointment on Cut Bank Creek of the Marias River, west of present-day Cut Bank, Montana. They remain here until July 26.*

TUESDAY, JULY 22

Lewis's Party

Sunrise			4 p.m.			River		
Temp	Weather	Wind	Temp	Weather	Wind	Rise/Fall	Feet	Inches
	f	SE		*f*	NE			

Lewis   there being no wood we were compelled to make our fire with the buffaloe dung which I found answered the purpose very well. as I could see from hence very distinctly where the river entered the mountains and the bearing of this point being S of West I thought it unnecessary to proceed further and therefore encamped resolving to rest ourselves and horses a couple of days at this place and take the necessary observations…the rocky mountains to the SW of us appear but low from their base up yet are partially covered with snow nearly to their bases. There is no timber on those mountains within our view. the river appears to possess at least double the vollume of water which it had where we first arrived on it below; this no doubt proceeds from the avapporation caused by the sun and air and the absorbing of the earth in [its] passage through these open plains. I now have lost all hope of the waters of this river ever extending to N Latitude 50^0.

Clark's Party

Sunrise			4 p.m.			River		
Temp	Weather	Wind	Temp	Weather	Wind	Rise/Fall	Feet	Inches
	fatl&r	NE		*c*	NE			

Weather Diary   rained Slightly last evening about dark with hard winds   Thunder & lightning   a fiew drops of rain last night at dark. The Cloud appd. to hang to the SW, wind blew hard from different points from 5 to 8 PM which time it thundered and Lightened. The river by 11 am to day had risen 15 inches, and the water of a milky white Colour.

Clark The wind continued to blow very hard from the NE and a little before day lght was moderately Cool. The plains imediately out from Camp is So dry and hard that the track of a horse Cannot be Seen without close examination. The plains being so remarkably hard and dry as to render it impossible to see a track. grass is but Short and dry.

Gass–Ordway Portage Party

Gass We had a fine morning. Here a heavy shower of rain came on with thunder and lightning; and we remained at this place all night.

Ordway a fair morning.

WEDNESDAY, JULY 23

Lewis's Party

	Sunrise			4 p.m.			River	
Temp	Weather	Wind	Temp	Weather	Wind	Rise/Fall	Feet	Inches
	fat&l	SE		f	SW			

Weather Diary a distant thundercloud last evening to the west. Mountains covered with snow.

Lewis The clouds obscured the moon and put an end to further observations. We indeavoured to take some fish but took only one small trouht. Musquetoes uncommonly large and reather troublesome.

Clark's Party

	Sunrise			4 p.m.			River	
Temp	Weather	Wind	Temp	Weather	Wind	Rise/Fall	Feet	Inches
	f	NE		c	SE			

Weather Diary violent wind last night from SW   The river has fallen within the last 24 hours 7 inches. the wind was violent from the SW for about 3 hours last night from the hours of 1 to 3 am

Gass–Ordway Portage Party

Gass There was a pleasant morning after the rain

Ordway a hard Shower of rain hail and wind last evening.

Note   *Clark's party passes the area of present-day Billings, Montana.*

THURSDAY, JULY 24

Lewis's Party

	Sunrise			4 p.m.		River		
Temp	Weather	Wind	Temp	Weather	Wind	Rise/Fall	Feet	Inches
	cart&l	NW		*cartl*	NW			

Weather Diary   a violent gust of thunder Lighting last evening at 6 PM rain and wind all night untill this evening with some intervales.

Lewis   At 8 A.M. the sun made [its] appearance for a few minutes and I took [its] altitude but it shortly after clouded up again and continued to rain the ballance of the day. I was therefore unable to complete the observations I wished to take at this place. I determined to remain another day in the hope of [its] being fair. the air has become extreemly cold which in addition to the wind and rain renders our situation extreemly unpleasant.

Clark's Party

	Sunrise			4 p.m.		River		
Temp	Weather	Wind	Temp	Weather	Wind	Rise/Fall	Feet	Inches
	f	SW		*r*	SW			

Weather Diary   Violent wind last night. River falling a little. Since the last rise is had fallen 13 inches. River falling a little it is 6 feet lower than the highest appearance of [its] rise. Rained from 3 to 4 PM but slightly. The wind violent from the SW

Clark   river 300 yds. wide. Came to big horn river…is 150 yards wide at [its] Mouth…the water of a light Muddy Colour and much Colder that the Rochejhone. For me to mention or give an estimate of the differant Spcies of wild animals on this river particularly Buffalow, Elk Antelopes & Wolves would be increditable. I shall therefore be silent on the Subject further. Current rapid.

Gass–Ordway Portage Party

Gass   This was a cloudy morning. About 4 o'clock a very heavy shower of rain, accompanied with thunder and lightning, came on, and lasted about an hour and a half. After this we had a fine evening,

Ordway   a clear morning. Had a hard shower of rain which rendred the plains verry muddy

Note   *Clark's party stops and names Pompey's Tower [Pillar] after Sacagawea's son. This is now a National Historic Landmark. At this rock, the only remaining physical evidence of the Lewis and Clark Expedition remains: William Clark's name and the day inscribed in the sandstone.*

FRIDAY, JULY 25

Lewis's Party

	Sunrise			4 p.m.			River	
Temp	Weather	Wind	Temp	Weather	Wind	Rise/Fall	Feet	Inches
	c a r	*NW*		*c a r*	*NW*			

Weather Diary   rained and wind violent all day and night.

Lewis   The weather still continues cold cloudy and rainy, the wind also has blown all day with more than usual violence from the NW. I remained in camp with R. Fields to avail myself of every opportunity to make my observations should any offer, but it continued to rain and I did not see the sun through the whole course of the day. I determined that if tomorrow continued cloudy to set out as I now begin to be apprehensive that I shall not reach the United States within this season unless I make every exertion in my power which I shall certainly not omit when once I leave this place which I shall do with much reluctance without having obtained the necessary data to establish [its] longitude— as if the fates were against me my chronometer from some unknown cause stoped today, when I set her to going she went as usual.

Clark's Party

	Sunrise			4 p.m.			River	
Temp	Weather	Wind	Temp	Weather	Wind	Rise/Fall	Feet	Inches
	c	*E*		*c a r*	*SW*			

Weather Diary   rained from 3 to 4 PM yesterday but Slight. Rained several showers Several showers of rain with hard winds from the S and SW the fore part of the day. The brooks on each side are high and water muddye.

Clark   and had not proceeded far before a heavy shower of rain pored down on us, and the wind blew hard from the SW. The wind increased and the rain (began) continued to fall. I halted...covered with deerskins to keep off the rain, and a large fire made to dry ourselves. The rain continued moderately untill near twelve oClock when it Cleared away and become fair. The wind Contined high untill 2 PM. At 4 PM arived at a remarkable rock. This rock I shall Call Pompy's tower is 200 feet high. The nativs have ingraved on the face of this rock the figures of animals &c. near which I marked my name and the day of the month & year. From the top of this Tower I Could discover two Mountains & the Rocky Mts. covered with Snow SW. I proceeded on a Short distance and encamped, and earlyer than I intended on acout of a heavy cloud which was comeing up from the SSW. and some appearance of a Violent wind. About Sunset the wind blew hard from the W. and some little rain.

Gass–Ordway Portage Party

Gass   This was a fine morning with a very heavy dew. About 2 o'clock…we had another very heavy shower of rain accompanied with thunder and lightning. At 3 o'clock…it rained on us hard all the way, and the road was so muddy that the horses were not able to haul the loads

Ordway   hard rain comd. About noon and continued the remainder part of the day, but did not Stop us from our urgent labours. halted as much as we were able to help the horses as the place So amazeing muddy & bad. Rained very hard and we having no Shelter some of the men and myself turned over a canoe & lay under it others Set up by the fires. The water run under us and the ground was covrd with water. The portage River raises fast

Note   *The Gass–Ordway portage party leaves the lower portage camp site and heads down the Missouri River.*

SATURDAY, JULY 26

Lewis's Party

	Sunrise			4 p.m.		River		
Temp	Weather	Wind	Temp	Weather	Wind	Rise/Fall	Feet	Inches
	c a r	N		*f*	NW			

Weather Diary   wind violent   rain continues.

Lewis   The morning was cloudy and continued to rain as usual, tho' the cloud seemed somewhat thiner I therefore posponed seting out untill 9 A.M. in the hope that it would clear off but find the contrary result I had the horses caught and we set out biding a lasting adieu to this place which I know call camp *disappointment.* A small creek [Willow Creek] is shallow and rappid; has the appearance of overflowing [its] banks frequently and discharging vast torrants of water at certain seasons of the year. I halted and used my spye glass...which I discovered several indians....this was a very unpleasant sight.

Clark's Party

Sunrise			4 p.m.			River		
Temp	Weather	Wind	Temp	Weather	Wind	Rise/Fall	Feet	Inches
	c	SSW		f a r	NW			

Weather Diary   a slight shower this morning with hard wind from the SW the river falling, but very slowly 1 inch in 24 hs.

Clark   the Current of the river reagulilarly Swift.

Gass–Ordway Portage Party

Gass   The morning was cloudy. It rained very hard all night, which has made the plains so muddy, that it is with greatest difficulty we can get along with the canoe. A few drops of rain fell in the course of the day.

Ordway   a wet disagreeable morning. The portage River too high to waid but is falling fast. The truck wheels Sank in the mud nearly to the hub.

Note   *Lewis's party meets with an unfortunate tragedy. Two Blackfeet Indians are killed as they attempt to steal the party's rifles and horses. Lewis's party leaves with haste and rides all day and nearly all night toward the confluence of the Marias and Missouri rivers.*

SUNDAY, JULY 27

Lewis's Party

Sunrise			4 p.m.			River		
Temp	Weather	Wind	Temp	Weather	Wind	Rise/Fall	Feet	Inches
	f	NW		f	SW			

Lewis   the day proved warm but the late rains had supplied the little reservors in the plains with water and had put them in fine order for traveling our whole rout so far was as level as a bowling green with but little stone and few prickly pears. after refreshing ourselves we again set out by moonlight and traveled leasurely, heavy thunderclouds lowered arround us on every quarter but that from which the moon gave us light.

Clark's Party

Sunrise			4 p.m.			River		
Temp	Weather	Wind	Temp	Weather	Wind	Rise/Fall	Feet	Inches
	f	*NE*		*f*	*SW*			

Weather Diary   Saw a flight of gulls   a small rattle snake   several flocks of crows and black burds

Clark   I marked my name with red paint on a Cotton tree near my camp…the river is much wider from 4 to 600 yards much divided by Islands. When we pass the Big horn I take my leave of the view of the tremendious chain of Rocky Mountains white with Snow in view of which I have been Since the 1st of Nay last.

Gass–Ordway Portage Party

Gass   a fine clear pleasant morning
Ordway   a clear morning.

Note   *Lewis's party meets the Gass–Ordway portage party and together they proceed down the Missouri River through the area of present-day Missouri Breaks National Monument, Montana.*

MONDAY, JULY 28

Lewis's Party (now including the Gass–Ordway party)

Sunrise			4 p.m.			River		
Temp	Weather	Wind	Temp	Weather	Wind	Rise/Fall	Feet	Inches
	f a r t & l	*NE*		*c a f b r t & l*	*NE*			

Weather Diary   a thundershower last night from NW but little rain where we were. heavy hail storm at 3 PM   the prickly pear has now cast [its] blume

Lewis   The morning proved fair . After 1 PM…during the time we halted at the entrance of Maria's river we experienced a very heavy shower of rain and hail attended with violent thunder and lightning.

Gass   The morning was fine and pleasant. About one o'clock….about this time we started a heavy gust of rain and hail, accompanied with thunder and lightning came on and lasted about an hour, after which we had a cloudy wet afternoon

Ordway   about 9 AM we discovrd on a high bank a head Capt. Lewis & the three men who went with him on horse back comming towards us on N Side. Got his observations for the Lat. [at the Marias River] but the cloudy weather prevented him from gitting the Longitude &C. About 1 PM we arived at the forks of the Marriah....we Soon had a hard Shower of rain & large hail. Some larger than a musket Ball Thunder and high winds a head but we procd. Late in the evening we had a Shower of rain which lasted about a hour—

Clark's Party

	Sunrise			4 p.m.			River		
Temp	Weather	Wind	Temp	Weather	Wind	Rise/Fall	Feet	Inches
	c a f r	*NE*		*f*	*NW*			

Weather Diary   a fiew drops of rain this morning a little before day light. River still falling a little   Bratten [caught] a beaver   Labeech shot 2 last evenig. I saw a wild cat lying on a log over the water

TUESDAY, JULY 29

Lewis's Party

	Sunrise			4 p.m.			River		
Temp	Weather	Wind	Temp	Weather	Wind	Rise/Fall	Feet	Inches
	rart&l	*SW*		*c a r*	*NE*			

Weather Diary   heavy rain last night, continued with small intervales all night

Lewis   Shortly after dark last evening a violent storm came on from the NW attended with rain hail Thunder and lightning which continued the greater part of the night. Not having the means of making a shelter I lay in the water all night. The rain continued with but little intermission all day. I intend halting as soon as the weather proves fair in order to dry our baggage which much wants it. We set out early and the currant being strong we proceeded with great rapidity. At 11 AM we passed that very interesting part of the Missouri where the natural walls appear, particularly discribed in my outward bound journey. The river is now nearly as high as it bas been this season and is so thick with mud and sand that it is with difficulty I can drink it. Every little rivulet now discharges a torrant of water bringing down immece boddies of mud sand and filth from the plains and broken bluffs—

Gass   Early in a cloudy morning we commenced our voyage from the mouth of Maria's river; and the current of the Missouri being very swift, we went down rapidly. A considerable quantity of rain fell in the course of the day.

Ordway   cloudy and rain. About 11 AM we entered the high clay broken country   white clay hills and the white walls resembling ancient towns & buildings &C. We had a Shower of rain

Clark's Party

Sunrise			4 p.m.			River		
Temp	Weather	Wind	Temp	Weather	Wind	Rise/Fall	Feet	Inches
	cart&l	NE		f	N			

Weather Diary a fiew drops of rain (rain slightly with Thunder and lightning) accompanied with hard Claps of Thunder and Sharp lightning last night   wind hard from the NE

Clark a Slight rain last night with hard thunder and Sharp lightening accompanied with a violent NE wind. I set out early this morning   wind So hard a head that we made but little way. The Tongue River....so muddy and w[a]rm as to render it very disagreeable to drink...nearly milk w[a]rm very muddy. The river widens. I think it may be generally Calculated at from 500 to a half a mile wide in width

Note   *Lewis's party passes the confluence of the Missouri and Judith rivers in central Montana.*

WEDNESDAY, JULY 30

Sunrise			4 p.m.			River		
Temp	Weather	Wind	Temp	Weather	Wind	Rise/Fall	Feet	Inches
	r a r	NE		r	NE			

Weather Diary   rained almost without intermission

Lewis   The rain still continued this morning it was therefore unnecessary to remain as we could not dry our baggage. The currant being strong and the men anxious to get on they plyed their oars faithfully and we went at the rate of about seven miles an hour. The rain continued with but little intermission all day; the air is cold and extreemly disagreeable. Nothing extraordinary happened today.

Gass   We embarked early in a cloudy morning with some rain. The water of the river is thick and muddy, on account of the late falls of rain, which wash those clay hills very much. Heavy rain fell at intervals during the day.

Ordway   cloudy and wet. Rained all day.

Clark's Party

Sunrise			4 p.m.			River		
Temp	Weather	Wind	Temp	Weather	Wind	Rise/Fall	Feet	Inches
	fartl	NW		far	SE			

Weather Diary   Great number of Swallows, they have their young. Killed 1s black tail deer. Young gees beginning to fly a slight shower of rain accompanied with thunder and lightning. Several showers in the course of this day. It cleared away in the evening and became fair   river falling a little. Great quantities of Coal appear in the bluffs of either Side. Some appearance of Burnt hills at a distance from the river.

Clark   first appearance of Birnt hills. They have the apperanc of dischargeing emence torrents of water. The late rains which has fallen in the plains raised Suden & heavy Showers of rain must have fallen, Several of which I have seen dischargeing those waters. A violent Storm from the NW obliged us to land imediately below this rapid…above enterance of a river…I call it Yorks dry R. After the rain and wind passed over I proceeded on at 7 miles. Water was disagreeably muddy.

Note   *Lewis's party leaves the area of present-day Missouri Breaks National Monument, Montana.*

THURSDAY, JULY 31

Lewis's Party

	Sunrise			4 p.m.			River	
Temp	Weather	Wind	Temp	Weather	Wind	Rise/Fall	Feet	Inches
	c a r	NE		*r*	NW			

Weather Diary   rained almost without intermission

Lewis   the rain still continuing. The river is still rising and excessively muddy more so I think than I ever saw it. We experienced some very heavy showers of rain today.

Gass   We set out early, though it continued at intervals to rain hard. At noon we halted to dine, and had then a very heavy shower of rain. Though the afternoon was wet and disagreeable, we came 70 miles to day.

Ordway   cloudy and rain. Had several showers of rain. The river verry muddy owing to the heavy rains washing those Clayey hills.

Clark's Party

	Sunrise			4 p.m.			River	
Temp	Weather	Wind	Temp	Weather	Wind	Rise/Fall	Feet	Inches
	f	NW		*c a r*	NE			

Weather Diary   rained only a fiew drops last night. A small showers to day. Wind hard [from] the NE   the wind blew hard and it was showery all day tho not much rain. The clouds came up from the W and NW frequently in course of the day.

Clark   Showers all this day.

Note   *In early August, Lewis's party passes by the confluence of the Missouri and Musselshell rivers in central Montana. Clark's party is proceeding down the Yellowstone River and nears the area of present-day Glendive, Montana. On August 1, Clark began to record data on the rise and fall of the Yellowstone River. The data is not for a 24-hour period unless they are encamped. In most cases, when Clark's party stopped for the evening, he followed Lewis's habit from previous journal entries, placing a mark and measuring the next morning. Thus many observations were only 8- to 10-hour-long periods.*

FRIDAY, AUGUST 1

Lewis's Party

Sunrise			4 p.m.			River		
Temp	Weather	Wind	Temp	Weather	Wind	Rise/Fall	Feet	Inches
	r a r	*NE*		*r a r*	*NW*			

Lewis   The rain still continuing I set out early as usual   after 1 PM...as the rain still continued with but little intermission and appearances seemed unfavorable to [its] becomeing fair shortly, I deteremined to halt at this place. Shortly after we landed the rain ceased tho' it still continued cloudy all this evening.

Gass   We embarked early in a wet disagreeable morning   The afternoon was cloudy with some rain

Ordway   hard rain. Here we delayed this afternoon to dry our deer Skins Mount. Sheep skins &C. which were near Spoiling as the weather has been some time wet.

Clark's Party

Sunrise			4 p.m.			Yellowstone River		
Temp	Weather	Wind	Temp	Weather	Wind	Rise/Fall	Feet	Inches
	c a r	*NW*		*r*	*N*	*R*		5 ¹/₂

Weather Diary   rained last night and all day to day at intervals

Clark   We Set out early as usial the wind was high and ahead which caused the water to be a little rough and delayed us very much   aded to this we had Showers of rain repeetedly all day at the intermition of only a fiew minits between them. My Situation a very disagreeable one. In an open Canoes wet and without a possibility of keeping my Self dry. The brooks have all Some water in them from the rains which has fallen. The water is excessively muddy. This gangue of Buffalow was entirely across and as thick as they could Swim. The Chanel on the Side of the Island the went into the river was crouded with those animals for ½ an hour. [...a discussion of meat...] nearly Spoiled from the wet weather.

SATURDAY, AUGUST 2

Lewis's Party

	Sunrise			4 p.m.			River	
Temp	Weather	Wind	Temp	Weather	Wind	Rise/Fall	Feet	Inches
	f a r	NW		*f*	NW			

Weather Diary   it became fair soon after dark last evening and continued so.—

Lewis   The morning proved fair and I determined to remain all day and dry the baggage and give the men an opportunity to dry and air their skins and furr. The day proved warm fair and favourable for our purpose. The river fell 18 inches since yesterday evening. We are all extreemly anxious to reach the entrance of the Yellowstone river where we expect to join Capt. Clark and party.

Gass   This was a fine clear morning

Ordway   a fair morning. The day warm

Clark's Party

	Sunrise			4 p.m.			Yellowstone River	
Temp	Weather	Wind	Temp	Weather	Wind	Rise/Fall	Feet	Inches
	c a r	N		*f a r*	N	R		3

Weather Diary   rained a little last night and Several Showers this morng

Clark   river wide. The river in this days decent is less rapid crouded with Islds and muddy bars and is generally about one mile in wedth.

Note   *Clark's party passes the confluence of the Yellowstone and Missouri rivers and crosses the present-day Montana–North Dakota state line.*

SUNDAY, AUGUST 3

Lewis's Party

	Sunrise			4 p.m.		River		
Temp	Weather	Wind	Temp	Weather	Wind	Rise/Fall	Feet	Inches
	f	*SE*		*f*	*SE*			

Gass   We had a fine morning

Ordway   a fair morning.

Clark's Party

	Sunrise			4 p.m.		Yellowstone River*		
Temp	Weather	Wind	Temp	Weather	Wind	Rise/Fall	Feet	Inches
	f	*SW*		*f*	*SW*	*R*		2 $^1/_4$

Weather Diary   Musquetors troublesom. I arive at the Missouri. heavy dew.

Clark   last night the Musquetors was so troublesom that no one of the party Slept half the night. For my part I did not Sleep one hour. Those tormenting insects found their way into My beare and tormented me the whole night. They are not less noumerous or troublesom this morning. At 8 AM I arived at the Junction of the Rochejhone with the Missouri. A large Buck Elk which I shot & had his flesh dryed in the Sun for a Store down the river. Had the canoes unloaded and every article exposed to dry & Sun. Maney of our things were wet. The current of this river may be estimated at 4 miles and ½ pr. hour from the Rocky Mts. as low as Clarks Fork, at 3 ½ Miles pr. hour from thence as low as the Bighorn…3 [mph] as the Tongue…2 ³/₄ [mph] as Wolf rapid and at 2 ½ [mph] from thence to its enterance into the Missouri.

Note   *Lewis's party passes by the confluence of the Missouri and Milk rivers, southeast of present-day Glasgow, Montana.*

---

*River observations are now on the Missouri River following the same technique noted just before the August 1 entry.

MONDAY, AUGUST 4

Lewis's Party

	Sunrise			4 p.m.			River	
Temp	Weather	Wind	Temp	Weather	Wind	Rise/Fall	Feet	Inches
	f	SE		*f*	SE			

Lewis   Set out at 4 AM this morning. At ½ after eleven O'Ck. passed the entrance of big dry river; found the water in this river about 60 yds. wide tho' shallow. It runs with a boald currant. At 3 PM we arrived a the entrance of Milk river where we halted a few minutes. This stream is full at present and [its] water is much the colour of that of the Missouri

Gass   This was another pleasant day. At five o'clock we passed the mouth of Milk River, which was very high and the current strong.

Clark's Party

	Sunrise			4 p.m.			Missouri River	
Temp	Weather	Wind	Temp	Weather	Wind	Rise/Fall	Feet	Inches
		NW		*f*	NE	F		6 ½

Weather Diary   Rochejhone falling much faster than the Missouri

Clark   Musquetors excessively troublesom...those insects is on the Sand Bars in the river and even those Situations are only clear of them when the Wind Should happen to blow which it did to day for a fiew hours in the middle of the day. The torments of those Misquetors. The Child of Shabono has been So much bitten by the Musquetor that his face is much puffed up & Swelled.

TUESDAY, AUGUST 5

Lewis's Party

	Sunrise			4 p.m.			River	
Temp	Weather	Wind	Temp	Weather	Wind	Rise/Fall	Feet	Inches
	c a f	NW		*f*	SE			

Lewis   the geese cannot fly at present; I saw a solitary [pelican] the other day in the same situation. This happens from their shedding or casting the fathers of the wings at this season.

Gass   Last night was cloudy and thunder was heard at a distance. This morning was also cloudy. The forenoon had become clear and pleasant, and at noon we got under way. At sunset we encamped and at dark a violent gust of rain and wind came on with thunder and lightning, which lasted about an hour; after which we had a fine clear night.

Ordway   a fair morning. A little after dark came up a Thunder Shower of wind and rain and nearly filled our canoes, so that we had to unload them. The sand flew so that we could Scarsely See & cut our faces by the force of the wind—

Clark's Party

	Sunrise			4 p.m.			Missouri River	
Temp	Weather	Wind	Temp	Weather	Wind	Rise/Fall	Feet	Inches
	f	NE		f	NE	F		7

Weather Diary   Musquetors excessively troublsom   both rivers falling.

Clark   the Misquetors was So numerous that I could not keep them off my gun long enough to take Sight and by thair means missed. At 10 AM the wind rose with a gentle breeze from the NW which in Some measure thinned the Misquetors. Camped…our Situation was exposed to a light breeze of wind which continued all the forepart of the night from the SW and blew away the misquetors.

Note   *Clark's party passes the area of present-day Williston, North Dakota.*

WEDNESDAY, AUGUST 6

Lewis's Party

	Sunrise			4 p.m.			River	
Temp	Weather	Wind	Temp	Weather	Wind	Rise/Fall	Feet	Inches
	fart&l	NE		f	NE			

Weather Diary   a violent gust of Thunder Lightning wind and hail last night.

Lewis   A little after dark last evening a violent storm arrose to the NE and shortly after came on attended with violent Thunder lightning and some hail; the rain fell in a mere torrant and the wind blew so violently that it was with difficulty I could have the small canoes unloaded before they filled with water; they sustained no injury. Our situation was open and exposed to the storm. I obtained a few hours of broken rest; the wind and rain continued almost all night and the air became very cold. We set out early this morning...decended 10 miles below Porcupine river [Poplar River, Montana]...when the wind became so violent that I laid by untill 4 PM. The wind then abaited in some measure we again resumed our voyage

Gass   a fine morning, but high wind. At 12 o'clock the wind blew so violent that it became dangerous to go on and we halted...left after three hours

Ordway   a fair morning. The wind rose high   so halted.

Clark's Party

	Sunrise			4 p.m.			Missouri River		
Temp	Weather	Wind	Temp	Weather	Wind	Rise/Fall	Feet	Inches
	cart&l	*SW*		*f*	*NE*	F		2 ½

Weather Diary   rained hard last night with Thunder Lightning & hard wind from S.W. Killed a white Bear & Bighorn

Clark   I rose very wet. About 11 PM last night the wind become very hard for a fiew minits Suckceeded by Sharp lightning and hard Claps of Thunder and rained for about 2 hours very hard after which it continued Cloudy the balance of the night. Wind hard from the NW I halted. Bear plunged into river...we all fired into him without killing him, the wind so high that we could not pursue him. I have observed buffalow floating down which I suppose must have been drounded in Crossing above. The wind blew hard all the after part of the day.

Note   *Lewis's party passes the confluence of the Yellowstone and Missouri rivers and crosses the present-day Montana–North Dakota state line.*

THURSDAY, AUGUST 7

Lewis's Party

	Sunrise			4 p.m.			River		
Temp	Weather	Wind	Temp	Weather	Wind	Rise/Fall	Feet	Inches
	r a r	*NE*		*c a r*	*NE*			

Weather Diary   rained form 12 last night untill 10 AM today—.

Lewis It began to rain about midnight and continued with but little intermission untill 10 AM today. The air was cold and extreemly unpleasant. The currant favoured our progress being more rapid than yesterday

Gass The morning was cloudy, and we set out early, after a very heavy shower of rain which fell before day light. About 4 o'clock arrived at the mouth of Yellow Stone river.

Ordway a Showery wet morning. About 4 PM we arived at the mouth of the River Roshjone [Yellowstone]. The wind his this evening.

Clark's Party

Sunrise			4 p.m.			Missouri River		
Temp	Weather	Wind	Temp	Weather	Wind	Rise/Fall	Feet	Inches
	r	NE		c a r	N	F		2 ½

Weather Diary Commenced raining at daylight and continued at intervals all day. air Cool.

Clark Some hard rain this morning after daylight which wet us all. I formed a Sort of Camped and delayed untill 11 AM when it Stoped raining for a short time. Proceeded…the rain Continued at intervales all day tho' not hard   at 6 PM…Campd…soon after we landed the wind blew very hard for about 2 hours, when it lulled a little. The air was exceedingly Clear and Cold and not a misquetor to be Seen, which is a joyfull circumstance to the Party.

Note   *Lewis's party stays near present-day Williston, North Dakota, from August 8–10.*

FRIDAY, AUGUST 8

Lewis's Party

Sunrise			4 p.m.			River		
Temp	Weather	Wind	Temp	Weather	Wind	Rise/Fall	Feet	Inches
	f	NE		f	NE			

Weather Diary   wind hard but not so much so as to detain us.—

Lewis I set our early; the wind [hard] from the NE but by the force of the oars and currant we traveled at a good rate. This evening…the air is cold yet the Musquetoes continue to be troublesome—

Gass   We had a fine cool morning with some white frost
Ordway   a fair morning.

Clark's Party

Sunrise			4 p.m.			Missouri River		
Temp	Weather	Wind	Temp	Weather	Wind	Rise/Fall	Feet	Inches
	f	N		f	NW	F		2

Weather Diary   air cool. Sergt. Pryor arrive in Skin Canoes.

Clark   Sergt. N. Pryor informed me that the Second night after he parted with me [Pryor and party left with horses on July 24, 1806]…about 4 PM he halted to let the horses graze during which time a heavy Shower of rain raised the Creek so high that Several horses which had Stragled across the Chanel of this Creek was obliged to Swim back [July 26, 1806].

SATURDAY, AUGUST 9

Lewis's Party

Sunrise			4 p.m.			River		
Temp	Weather	Wind	Temp	Weather	Wind	Rise/Fall	Feet	Inches
	f	NE		f	SE			

Weather Diary   heavy dew last night. air cold.

Lewis   The day proved fair and favourable for our purposes.
Gass   This was another fine day
Ordway   a cool windy morning.

Clark's Party

Sunrise			4 p.m.			Missouri River		
Temp	Weather	Wind	Temp	Weather	Wind	Rise/Fall	Feet	Inches
	f	NE		f	NE	F		1 1/4

Weather Diary   a heavy dew. air cool and clear   found red goose berries and a dark purple current & Service's

Clark   a heavy dew this morning.

Sunday, August 10

Lewis's Party

Sunrise			4 p.m.			River		
Temp	Weather	Wind	Temp	Weather	Wind	Rise/Fall	Feet	Inches
	f	NE		c a r	NE			

Weather Diary   a slight shower about 3 PM   wind hard.

Lewis   The morning was somewhat cloudy I therefore apprehended rain however it shortly after became fair. At 4 in the evening it clouded up and began to rain the wind has blown very hard all day but did not prove so much so this evening as absolutely to detain us.

Gass   We had a fine morning   In the afternoon some drops of rain fell; and the musquitoes here were very bad indeed.

Ordway   a cool windy morning. Camped...the musquetoes troublesome indeed. We could not all this night git a moment of quiet rest for them—

Clark's Party

Sunrise			4 p.m.			Missouri River		
Temp	Weather	Wind	Temp	Weather	Wind	Rise/Fall	Feet	Inches
	f	E		c	E	F		3/4

Weather Diary   found a Species of Cherry resembling the read Heart cherry of our country

Clark   wind blew hard from the East all day. In the after part of the day it was (cloudy) & a fiew drops of rain. I finished a Copy of my Sketches of the River Rochejhone.

Monday, August 11

Lewis's Party

Sunrise			4 p.m.			River		
Temp	Weather	Wind	Temp	Weather	Wind	Rise/Fall	Feet	Inches
	f	NE		f	NW			

Weather Diary   air cool this evening   wind hard.

Lewis   It being my wish to arrive at the birnt hills [Crow Hills, North Dakota] by noon in order to take the latitude of that place as it is the most northern point of the

Missouri…when I arrived here it was about 20 minutes after noon and of course the observation for the sun's meridian Altitude was lost.

Gass   The morning was pleasant

Ordway   a fair morning. High winds.

Clark's Party

Sunrise			4 p.m.			Missouri River		
Temp	Weather	Wind	Temp	Weather	Wind	Rise/Fall	Feet	Inches
	f	*NW*		*f*	*NW*	*F*		*2*

Weather Diary   sarvis berries in abundance & ripe.

Clark   at 10 AM landed on a Sand bar…and during brackfast…and my delay at this place which was 2 hours had the Elk meat exposed to the Sun.

Note   *Lewis's party meets Clark's party in west-central North Dakota. The reunited expedition proceeds together down the Missouri River. Regarding the separate weather tables, according to Moulton, Lewis "apparently ceased keeping it after his reunion with Clark on August 12, as with his other journal-keeping, due to the wound he received on the eleventh."*

Tuesday, August 12

Lewis's Party

Sunrise			4 p.m.			River		
Temp	Weather	Wind	Temp	Weather	Wind	Rise/Fall	Feet	Inches
	f	*NW*						

Weather Diary   wind violent last night.

Lewis   At 1 PM I overtook Capt. Clark and party and had the pleasure of finding them all well. As wrighting in my present situation is extreemly painfull to me I shall desist untill I recover and leave to my frind Capt. C. the continuation of our journal. This cherry…is now ripe…I have never seen it in blume.*

Gass   The morning was pleasant. Clarke's party…found the Yellow Stone river a pleasant and navigable stream, with a rich soil along it; but timber scarce.

Ordway   a fair morning. A little rain this evening &C—

---

*Here ends Captain Meriwether Lewis's entries into the Lewis and Clark Expedition Journals. He never does resume writing for the rest of the journey. And only after arriving in St. Louis does he write about the expedition's completion and findings to President Thomas Jefferson.

Clark's Party

Sunrise			4 p.m.			Missouri River		
Temp	Weather	Wind	Temp	Weather	Wind	Rise/Fall	Feet	Inches
	f	SW		c	SW	F		2 ¼

Weather Diary   Capt. Lewis overtake me with the party

Clark   encamped. The wind blew very hard from the SW and Some rain.

Note   *The expedition passes by the confluence of the Missouri and Little Missouri rivers in North Dakota.*

WEDNESDAY, AUGUST 13

Sunrise			4 p.m.			Missouri River		
Temp	Weather	Wind	Temp	Weather	Wind	Rise/Fall	Feet	Inches
	f a r	SW		f	SW	F		2 ½

Weather Diary   a fiew drops of rain last night at 8 P.M. with hard S W wind

Clark   the last night was very Cold with a Stiff breeze from the NW. All hands were on board and we Set out at Sunrize and proceeded on very well with a Stiff breeze astern the greater part of the day. haveing came by the assistance of the wind, the current and our oars 86 miles. The air is cool &c.

Gass   After a stormy night of wind and rain we set out early in a fine morning

Ordway   a fair morning. A fair breeze from the NW

Note   *The expedition arrives at Fort Mandan near present-day Washburn, North Dakota, and stays here August 17. Here they leave their interpreter, Charbonneau; his wife, Sacagawea; and their son, Jean-Baptiste; as well as Private John Colter.*

THURSDAY, AUGUST 14

Sunrise			4 p.m.			Missouri River		
Temp	Weather	Wind	Temp	Weather	Wind	Rise/Fall	Feet	Inches
	f	NE		f	SW	F		3 ½

Weather Diary   Mandan Corn [is ripe] now full and beginning to harden

Clark   I proceeded on to the black cats village on the NE side of the Missouri where I intended to Encamp but the Sand blew in Such a manner that we deturmined not to continue on that side. After the Council I directed the Canoes to cross the river to a brook opposit where we Should be under the wind and in a plain where we would be Clear of musquetors

Gass   The morning of this day was pleasant

Ordway   a fair morning.

FRIDAY, AUGUST 15

	Sunrise			4 p.m.		Missouri River		
Temp	Weather	Wind	Temp	Weather	Wind	Rise/Fall	Feet	Inches
	f	*NW*		*f*	*NW*	*F*		*2*

Clark   The evening is Cool and windy.

Gass   We had a fine clear pleasant morning

Ordway   a clear pleasant morning.

SATURDAY, AUGUST 16

	Sunrise			4 p.m.		Missouri River		
Temp	Weather	Wind	Temp	Weather	Wind	Rise/Fall	Feet	Inches
	f	*NW*		*f*	*NW*	*F*		*3 1/2*

Weather Diary   Northern lights Seen last night which was in Streaks

Clark   a cool morning.

Gass   There was a fine cool day; and we yet remained here, waiting an answer from the natives.

Ordway   a clear cool morning.

Note   *The expedition leaves the Mandan villages and starts down the Missouri again.*

SUNDAY, AUGUST 17

	Sunrise			4 p.m.		Missouri River		
Temp	Weather	Wind	Temp	Weather	Wind	Rise/Fall	Feet	Inches
	c	*SE*		*c*	*SE*			

Weather Diary   leave the Mandans

Clark   a cool morning. We proceeded on to the old Ricara village the SE wind was so hard and the waves So high that we were obliged to Come too & Camp on the SW side near the old Village.

Gass   There were some flying clouds this morning, and the weather was cold for the season. We proceeded on at two o'clock; the wind was high, and river rough

Ordway   we set out and procd. On the wind a head.

Note   *The expedition passes the area of Bismarck, present-day capital of North Dakota.*

MONDAY, AUGUST 18

	Sunrise			4 p.m.		Missouri River		
Temp	Weather	Wind	Temp	Weather	Wind	Rise/Fall	Feet	Inches
	c a r	SE		*f*	SE	F		1 ½

Weather Diary   rained moderately last night in forpart of the night.

Clark   moderate rain last night, the wind of this morning from the SE as to cause the water to be So rough that we Could not proceed on untill 8 AM at which time it fell a little & we proceeded on tho' the waves were yet high and the wind Strong. At 2 PM....wind Still high and from the Same point. The winds blew hard from the SE all day which retarded our process very much

Gass   We set out early in a cloudy morning, and the wind high.

Ordway   the wind high and a little rain. The wind continued

TUESDAY, AUGUST 19

	Sunrise			4 p.m.		Missouri River		
Temp	Weather	Wind	Temp	Weather	Wind	Rise/Fall	Feet	Inches
	t l & r	SE		*c*	SE	F		¾

Weather Diary   Comenced raining at 5 A.M. and Continued with a hard wind until [blank]

Clark   Some rain last night and this morning the wind rose and blew with great Violence untill 4 PM and as our camp was on a Sand bar we were very much distressd with the blows of San. At 4 PM the wind Seased to blow with that violence which it had done all day we Set out and proceeded on down. The wind rose and become very strong from the SE and a great appearance of rain. Stretched it over Some Stickes, under this piece of leather I Slept dry, it is the only covering which I have had Suffecient to keep off the rain Since I left the Columbia. It began to rain moderately soon after night.

Gass   This was a cloudy windy morning; and the water so rough, that our small canoes could not safely ride the waves. At 3 o'clock in the afternoon the wind ceased,

Ordway   a Showery morning. Thunder and high wind So it detained us. About 4 pm the wind fell a little and we procd. On. Windy & cold—

Note   **The expedition crosses the present-day North Dakota–South Dakota state line.**

WEDNESDAY, AUGUST 20

	Sunrise			4 p.m.			Missouri River	
Temp	Weather	Wind	Temp	Weather	Wind	Rise/Fall	Feet	Inches
	catl&r	SW			f	NW	F		1 1/4

Clark   a violent hard rain about day light this morning. All wet except myself and the indians. We embarked a little after Sun rise wind moderate and ahead. the wind blew hard all day which caused the waves to rise high and flack over into the Small Canoes in Such a manner as to employ one hand in throwing the water out. The plains begin to Change their appearance the grass is turning of a yellow colour. I observe a great alteration in the Corrent course and appearance of this pt. of the Missouri. In places where there was Sand bars in the fall 1804 at this time the main Current passes, and where the current then passed is now a Sand bar— San bars which were then naked are now covered with willow Several feet high. the enteranc of some of the Rivers & Creeks changed owing to the mud thrown into them, and a layor of mud over Some of the bottoms of 8 inches thick.

Gass   We embarked early after a heavy gust of wind and rain, and proceeded on very well. The forenoon was cloudy, without rain; and in the afternoon the weather became clear and pleasant

Ordway   the after part of the day pleasant.

THURSDAY, AUGUST 21

	Sunrise			4 p.m.			Missouri River	
Temp	Weather	Wind	Temp	Weather	Wind	Rise/Fall	Feet	Inches
	f	SE		f	NW	F		2 1/2

Weather Diary   rained a little in the course of the night. at day a violent hard Shower for 1/2 an hour

Clark   The wind rose and blew from the NW at half past 11 AM. The Sun being very hot the Chyenne Chief envited us to his Lodge which was pitched in the plain at no great distance from the River.

Gass   had a fine morning

Ordway   a fair morning.

Note   *The expedition passes present-day Mobridge, South Dakota.*

Friday, August 22

Sunrise			4 p.m.			Missouri River		
Temp	Weather	Wind	Temp	Weather	Wind	Rise/Fall	Feet	Inches
	c a r	SW		*f*	SE	F		4

Weather Diary   rained the greater part of last night. grape and plums ripe. The rains which have fallen in this month is most Commonly from flying Clouds which pass in different directions, those Clouds are always accompanied with hard winds and Sometimes accompanied with thunder and lightning— The river has been falling moderately Since the third of the month. the rains which has fallen has no impression of the river than Causing it to be more muddy and probably prevents its falling fast.—

Clark   rained all the last night   every person and all our bedding wet, the Morning cloudy. Below the ricaras the river widens and the Sand bars are emencely noumerous much less timber in the bottoms than above—

Gass   There was a cloudy wet morning, after a night of hard rain. At little after noon...the weather became clear

Ordway   hard Thunder Shower all last night. I about 10 AM cleared off fair and we set out

Saturday, August 23

Sunrise			4 p.m.			Missouri River		
Temp	Weather	Wind	Temp	Weather	Wind	Rise/Fall	Feet	Inches
	c	SE		*f*	NW	F		1 ½

Weather Diary   rained at 10 A.M. & 4 PM   hard wind

Clark   We set out very early, the wind rose & became very hard, we passed the Sar-war-kar-na-har river [Moreau River, South Dakota] at 10 AM and half past eleven the wind became so high and the water So rough that we were obliged to put to Shore and Continue untill 3 PM when we had a Small Shower of rain after which the wind lay, and we proceeded on. At 4 PM a Cloud from the NW with a violent rain for about half an hour   after the rain we again proceeded on.

Gass   We set out early in a fine morning, but the wind was high; and we went on very well till near noon, when the wind blew so hard that we had to halt, and were detained about four hours. Later....we had a very heavy shower of rain which detained us another hour.

Ordway   a little rain & Thunder. About 11 AM the wind rose so high that it detained us about 3 hours. Had light Showers of rain all day.

Sunday, August 24

	Sunrise			4 p.m.			Missouri River	
Temp	Weather	Wind	Temp	Weather	Wind	Rise/Fall	Feet	Inches
	f	NE		*f*	NW	F		2

Weather Diary   wind blew hard all day   grapes in abundance

Clark   a fair morning. Proceeded on untill 2 PM when the wind blew So hard from the NW that we could not proceed   came too on the SW Side where we continued untill 5 PM when the wind lay a little and we proceeded on.

Gass   We had a fine morning, and went on very well till noon, when the wind rose, and blew so strong that we were obliged to halt. Having lain by three hours we again proceeded, but did not go far before we were obliged on account of the wind, again to stop, and encamp for the night.

Ordway   a clear pleasant morning. About noon the wind rose high from SW which detained us about 3 hours   then procd. on though the work against us

Monday, August 25

	Sunrise			4 p.m.			Missouri River	
Temp	Weather	Wind	Temp	Weather	Wind	Rise/Fall	Feet	Inches
	f	SW		*f*	NW	F		1 1/2

Clark   a cool clear morning a Stiff breeze ahead. The Chyenne discharges but little water which is much the colour of the missouri not So muddy. This day proved a fine Still day

Gass   The morning was again pleasant.

Ordway   a clear pleasant morning.

Note   *The expedition passes Pierre, present-day capital of South Dakota.*

Tuesday, August 26

	Sunrise			4 p.m.			Missouri River	
Temp	Weather	Wind	Temp	Weather	Wind	Rise/Fall	Feet	Inches
	f	SE		*f*	SE	F		3/4

Weather Diary   Heavy dew this morning. Saw a pilecan

Clark   a heavy dew this morning. After 5 PM...we had a Stiff breeze from the SE which continued to blow the greater part of the night dry and pleasant. We made 60 miles to day with the wind ahead greater part of the day—

Gass   had a pleasant morning

Ordway   a fair morning.

Note   *The expedition passes around the Big Bend (known also as the Grand De Tour) of the Missouri in central South Dakota.*

WEDNESDAY, AUGUST 27

	Sunrise			4 p.m.			Missouri River		
Temp	Weather	Wind	Temp	Weather	Wind	Rise/Fall	Feet	Inches
	f	SE		*f*	SE	F		1 1/4

Weather Diary   first Turkeys at Tylor River above the big bend

Clark   a Stiff breeze a head from the East. At 1 PM we halted in the big bend [Grand Detour of the Missouri, South Dakota]

Gass   again had a pleasant day

Ordway   a fair morning.

Note   *The expedition camps just south of present-day Chamberlain, South Dakota.*

THURSDAY, AUGUST 28

	Sunrise			4 p.m.			Missouri River		
Temp	Weather	Wind	Temp	Weather	Wind	Rise/Fall	Feet	Inches
	f	SE		*f*	NW	F		1 1/4

Gass   We had another pleasant day

Ordway   a fair morning.

Note   *On August 29, the expedition records the rise and fall of the Missouri River for the last time, and records no more river data.*

FRIDAY, AUGUST 29

	Sunrise			4 p.m.		Missouri River		
Temp	Weather	Wind	Temp	Weather	Wind	Rise/Fall	Feet	Inches
	c	NW		f a r	SE	F		1/2

Weather Diary   Some rain this morning only a fiew drops and at 10 A.M.

Clark   a cloudy morning. Willard and Labiech waded white river a fiew miles above its enterance and inform me that they found it 2 feet water and 200 yards wide. The water of this river at this time nearly as white as milk. I assended to the high Country and from an eminance, I had a view of the plains for a great distance. From this eminance I had a view of a greater number of buffalow than I had ever Seen before at one time. I must have Seen near 20,000 of those animals feeding on this plain.

Gass   The morning was cloudy

Ordway   a little rain.

SATURDAY, AUGUST 30

	Sunrise			4 p.m.		River		
Temp	Weather	Wind	Temp	Weather	Wind	Rise/Fall	Feet	Inches
	c a r	SE		f	SE			

Weather Diary   a fiew drops of rain last night   I saw the Tetons

Clark   our encampment of this evening was a very disagreeable one, bleak exposed to the winds, and the Sand wet.

Gass   We had a pleasant morning

Note   *The expedition now travels parallel to the present-day South Dakota–Nebraska state line.*

SUNDAY, AUGUST 31

	Sunrise			4 p.m.		River		
Temp	Weather	Wind	Temp	Weather	Wind	Rise/Fall	Feet	Inches
	c a r t & l & w	SE		c a r	SE			

Weather Diary   rained most of last night with T. Li & a hard wind from the S.W. some rain to day

Clark   all wet and disagreeable this morning. A half past 11 PM last night the wind Shifted about to the NW and it began to rain with hard Claps of thunder and lightning the Clouds passd over and the wind Shifted about to the SW & blew with great violence So much So that all hand were obliged to hold the Canoes & Perogue to prevent their being blown off from the Sand bar. However a Suden Squal of wind broke the cables of the two small Canoes and with Some dificuelty they were got to Shore. Soon after the 2 canoes in which Sergt. Pryor and the indians go in broke loose...and were blown quite across the river to the NE Shore where fortunately they arived Safe. the wind Slackened a little and by 2 AM Segt with [party] returned safe, the wind continud to blow and it rained untill day light all wet and disagreeable. the morning Cloudy and wind down the river   After 4 PM...the Sun Shone with a number of flying Clouds.

Gass   There was a cloudy morning, after a disagreeable night of wind and hard rain. We set our early

Ordway   we had hard Showers of rain all last night and verry high winds caused one of our canoes broke loose and I took another canoes and to take it back and with Some difficulty goot it back to Camp. Verry disagreeable night.

Note   *The expedition camps near the future historical site of Fort Yankon, just below the present-day Gavin Dam along the Nebraska–South Dakota state line.*

MONDAY, SEPTEMBER 1

	Sunrise			4 p.m.			River	
Temp	Weather	Wind	Temp	Weather	Wind	Rise/Fall	Feet	Inches
	fog	SE		*f a r*	SE			

Weather Diary   a thick fog untill 8 A.M. A fiew drops of rain about 1 P.M.

Clark   Musquitors very troublesom last night. we set out at the usial hour and had not proceeded on far before the fog became So thick that we were oblige to come too and delay half an hour for the fog to pass off which it did in Some measure and we again proceeded on. At 9 AM we passed the enterance of River Quiequur [Niobrara River, Nebraska] which had the Same appearance it had when we passed up water rapid and of a milky white colour. the musquitors excessively troublesom untill about 10 PM when the SW wind became Strong and blew the most of them off. we came 52 miles to day only with a head wind.

Gass   a fine pleasant day. After 10 o'clock proceeded on with an unfavorable wind.

Ordway   a fair morning. About 9 AM we passd. the mo. of Rapid Water River [Niobrara].

Tuesday, September 2

	Sunrise			4 p.m.			River	
Temp	Weather	Wind	Temp	Weather	Wind	Rise/Fall	Feet	Inches
	f	SE		*f*	SE			

Weather Diary Hard wind all day. Saw the prarie fowl common in the Illinois plains. Saw Linn and Slipery elm

Clark The wind was hard a head & continued to increas which obliged us to lay be nearly all day. The wind Still high and water rough we did not Set out untill near Sun Set. We proceeded to a Sand abar…Come to on account of the wind and Encamped on a Sand bar, the woods being the harbor of the Musquetors and the party without means of Screaning themselves from those tormenting insects. on the Sand bars the wind which generaly blows moderately at night blows off these pests and we Sleep Soundly. The wind Continued to blow hard from the Same point SE untill 3 PM.

Gass a fine morning, but high wind. After noon….the wind blew so violent that we had to encamp for the night.

Ordway a fair morning. About 11 AM the wind rose So high a head that it detained us untill towards evening.

Note *The expedition passes present-day Vermillion, South Dakota.*

Wednesday, September 3

	Sunrise			4 p.m.			River	
Temp	Weather	Wind	Temp	Weather	Wind	Rise/Fall	Feet	Inches
	f	SW		*f*	SW			

Weather Diary a Stiff breeze form the S.E. untill 12 at night when it changed to S.W. and blew hard all night

Clark Wind Continued to blow very hard this morning. It Shifted last night to the SW and blew the Sand over us in Such a manner as to render the after part of the night very disagreeable. The wind luled a little and we Set out and proceeded on with the wind a head. After 4 pm…soon after we Landed a violent Storm of Thunder Lightning and rain from the NW which was violent with hard Claps of thunder and Sharp Lightning which continued untill 10 PM after which the wind blew hard. I set up late and partook of the tent of Mr. Aires which was dry. Mr. Aires unfortunately had his boat Sunk on the 25 of July last by a violent Storm of wind and hail by which accident he lost the most of his usefull articles as he informed us. ....the river much crowded with Sand bars, which are very differently Situated from what they were when we went up.

Gass   a pleasant morning. At sunset a violent gust of wind and rain, with thunder and lightning came on and lasted two hours.

Ordway   the day warm & Sultry. A verry hard Storm of wind and hard rain this evening.

Note   *The expedition camps near present-day Sioux City, Iowa, and crosses the border from South Dakota into present-day Nebraska.*

THURSDAY, SEPTEMBER 4

Sunrise			4 p.m.			River		
Temp	Weather	Wind	Temp	Weather	Wind	Rise/Fall	Feet	Inches
	far t&l	SE		*f*	SE			

Weather Diary   at 6 P.M. a violent Storm of Thunder Lightng and rain untill 10 P.M. when it ceased to rain and blew hard from N W untill 3 A.M.

Clark   I rose at the usial hour found all the party as wet as rain could make them.

Gass   There was a cloudy morning.

Ordway   a fair morning, but the hard rain and Thunder continued the greater part of last night.

FRIDAY, SEPTEMBER 5

Sunrise			4 p.m.			River		
Temp	Weather	Wind	Temp	Weather	Wind	Rise/Fall	Feet	Inches
	f	SE		*c*	SW			

Clark   The Musquetors being So excessively tormenting that the party was all on board and we Set out at day light and proceeded on very well. The river...becoms much narrower more Crooked and the Current more rapid and Crouded with Snags and Sawyers.

Gass   This was a fine morning.
Ordway   a fair morning.

SATURDAY, SEPTEMBER 6

Sunrise			4 p.m.			River		
Temp	Weather	Wind	Temp	Weather	Wind	Rise/Fall	Feet	Inches
	c	*SE*		*f*	*SE*			

Weather Diary   head the whipper will   Common to the u states at Soldiers river.

Clark   proceeded on wind-hard a head. The evening proved Cloudy and the wind blew hard.

Gass   a fine morning

Ordway   a fair morning.

SUNDAY, SEPTEMBER 7

Sunrise			4 p.m.			River		
Temp	Weather	Wind	Temp	Weather	Wind	Rise/Fall	Feet	Inches
	f	*SE*		*f*	*SE*			

Weather Diary   Saw the whiperwill and heard the common hooting owl   Musquetors very troublesom. killed 3 Elk.

Clark   we proceeded on with a Stiff Breeze ahead. Note the evaporation on this portion of the Missouri has been noticed as we assended this river, and it now appears to be greater than it was at that time. I am obliged to replenish my ink Stand every day with fresh ink at least $9/10$ of which must evaperate. We all Set out at 4 PM wind ahead as usial. Found the Musquetors excessively tormenting not withstanding a Stiff breeze from the SE   a little after dark the wind increased the Musquetors dispersed

Gass   We had a pleasant morning

Ordway   a pleasant morning. Abt. 10 AM set out…the wind So high that we could Scarsely proced. About 2 oClock PM we overtook the party who had halted to hunt as the wind was So high. Towards evening the wind abated So that we procd. on untill after Sunset.

Note   *The expedition camps near present-day Omaha, Nebraska.*

MONDAY, SEPTEMBER 8

Sunrise			4 p.m.			River		
Temp	Weather	Wind	Temp	Weather	Wind	Rise/Fall	Feet	Inches
	f	*SE*		*f*	*SE*			

Weather Diary   warmest day we have experienced in this year. passed River platt—

Clark   The Missouri at this place does not appear to Contain (as much) more water than it did 1000 Miles above this place, the evaporation must be emence; in the last 1000 miles this river receives the water 20 rivers and maney Creeks   Several of the Rivers large and the Size of this river or the quantity of water does not appear to increas any—

Gass   We again had a plesant morning

Ordway   a fair morning.

Note   *The expedition passes by the confluence of the Platte and Missouri rivers in present-day Nebraska. They also cross the present-day Iowa–Missouri state line.*

TUESDAY, SEPTEMBER 9

Sunrise			4 p.m.			River		
Temp	Weather	Wind	Temp	Weather	Wind	Rise/Fall	Feet	Inches
	f	*SE*		*f*	*SE*			

Clark   passed the enterance of the great river Platt which is at this time which is a this time low the water nearly clear the current turbelant as usial;   our party appears extreamly anxious to get on, and every day appears produce new anxieties in them to get to their Country and friends. The Musquetors are yet troublesome, tho' not So much So as they were above the River platt. the Climate is every day perceptably w[a]rmer and air more Sultery than I have experienced for a long time. the nights are now so w[a]rm that I sleep Comfortable under a thin blanket, a fiew days past 2 was not more than sufficient.

Gass   passed the mouth of the great river Platte; went on very well all day

Ordway   a fair morning.

WEDNESDAY, SEPTEMBER 10

Sunrise			4 p.m.			River		
Temp	Weather	Wind	Temp	Weather	Wind	Rise/Fall	Feet	Inches
	f	*SE*		*f*	*SE*			

Clark   we Set our very early this morning and proceeded on very well with wind moderately a head. We find the river in this timbered Country narrow and more moveing Sands and a much greater quantity of Sawyers or Snags than above. Great caution and much attention is required to Stear Clear of all those dificuelties in this low State of the water.

Gass   We had a pleasant morning

Ordway   a fair morning.

Note   *The expedition crosses the present-day Nebraska–Kansas state line.*

THURSDAY, SEPTEMBER 11

Sunrise			4 p.m.			River		
Temp	Weather	Wind	Temp	Weather	Wind	Rise/Fall	Feet	Inches
	c a r	*SE*		*f a r*	*SE*			

Weather Diary   a fiew drops of rain only a little before day and Some rain at 2 P M

Clark   a heavy Cloud and wind from the W detained us untill after sunrise. The [mosquitoes] are no longer troublesom on the river, from what cause they are noumerous above and not So on this part of the river I cannot account.

Gass   had a cloudy morning, and slight showers of rain during the forenoon

Ordway   a Showery morning.

Note   *The expedition camps near present-day St. Joseph, Missouri.*

FRIDAY, SEPTEMBER 12

Sunrise			4 p.m.			River		
Temp	Weather	Wind	Temp	Weather	Wind	Rise/Fall	Feet	Inches
	f	*SE*		*c a r*	*SE*			

Weather Diary   Heavy dew this morning and fog   Some rain from 12 to 4 P M

Clark   a thick fog a little before day which blew off at day light. A heavy Dew this morning. We set out at Sunrise the usial hour and proceeded on very well about 7 miles met 2 perogues....The wind blew a head Soon after we passed those perogues. The evening proveing to be wet and Cloudy we Concluded to continue all night

Gass   The morning was fine

Ordway   a foggy morning. We had Small Showers of rain this evening.

SATURDAY, SEPTEMBER 13

	Sunrise			4 p.m.			River	
Temp	Weather	Wind	Temp	Weather	Wind	Rise/Fall	Feet	Inches
	f	SE		*f*	SE			

Clark   a little after Sunrise we set out   the wind hard a head from the SE   At 8 AM….the wind being too high for us to proceed in Safty through the eme[n]city of Snags…we concluded to lye by. At 11 AM we proceeded on. The day disagreeably w[a]rm.

Gass   We had a pleasant morning after some rain that fell yesterday, and again proceeded on early with unfavourable wind.

Ordway   a fair morning. The wind being high and as we were out of meat we detained along at different places to hunt. Camped having made but a Short distance this day—

Note   The expedition passes present-day Atchison, Kansas, and Independence Creek.

SUNDAY, SEPTEMBER 14

	Sunrise			4 p.m.			River	
Temp	Weather	Wind	Temp	Weather	Wind	Rise/Fall	Feet	Inches
	f	SE		*c*	SE			

Clark   Set out early and proceeded on very well. Our party received a dram and Sung Songs untill 11 oClock at night in the greatest harmoney.

Gass   a fine morning

Ordway   a fair morning.

Note   The expedition passes by the confluence of the Kansas and Missouri rivers, at present-day Kansas City, Missouri.

MONDAY, SEPTEMBER 15

	Sunrise			4 p.m.			River	
Temp	Weather	Wind	Temp	Weather	Wind	Rise/Fall	Feet	Inches
	f	SE		*f*	SE			

Weather Diary   day very w[a]rm Smokey and w[a]rm

Clark   we set out early with a Stiff Breeze a head. At 11 AM passed the enterance of the Kanzas river which was very low. As the winds were unfavourable the greater part of the day we only decended 49 miles and encamped. The weather disagreeably w[a]rm and if it was not for the constant winds which blow from the S and SE we Should be almost Suficated Comeing out of a northern Country open and Cool between the Latd. Of 46^0 and 49^0 North in which we had been for nearly two years, rapidly decending into a woody Country in a w[a]rmer Climate between the Latds. 38^0 and 39^0 North is probably the Cause of our experiencing the heat much more Senceable than those who have Continued within the parralel of Latitude.

Gass   The morning was pleasant

Ordway   a fair morning. We set off at eight and procd. on   the wind a head as usal.

TUESDAY, SEPTEMBER 16

	Sunrise			4 p.m.			River	
Temp	Weather	Wind	Temp	Weather	Wind	Rise/Fall	Feet	Inches
	f	SE		*f*	SE			

Weather Diary   this day very Sultry and much the hotest which we have experienced

Clark   the Day proved excessively w[a]rm and disagreeable, so much so that the men rowed but little.

Gass   This was another pleasant day

Ordway   a fair morning. The day verry warm indeed.

WEDNESDAY, SEPTEMBER 17

	Sunrise			4 p.m.			River	
Temp	Weather	Wind	Temp	Weather	Wind	Rise/Fall	Feet	Inches
	f	SE		*f*	SE			

Weather Diary   day w[a]rm, but fiew musquetors

Clark   passed the Island of the little Osage Vilage [near Malta Bend, Missouri] which is considered by the navigater of this river to be the worst place in it. The current passing with great velocity against the banks which cause them to fall &c. This day proved [warm].

Gass   had a pleasant day, but very warm. About 11 o'clock we passed through a bad part of the river, where it was so filled with sawyers that we could hardly find room to pass through safe.

Ordway   a fair morning. We passed through a verry bad part of the river which was filled So thick with log Standing on end & Sawyers that we only found room to pass through.

THURSDAY, SEPTEMBER 18

	Sunrise			4 p.m.			River	
Temp	Weather	Wind	Temp	Weather	Wind	Rise/Fall	Feet	Inches
	f	SE		c	SE			

Clark the weather we found excessively hot as usial. We find the current of this part of the Missouri much more jentle than it was as we assended, the water is now low and where it is much confin'd it is rapid. one of our party J. Potts complains very much of one of his eyes which is burnt by the Sun from exposeing his face without a cover from the Sun. Shannon also complains of his face & eyes &c.

Ordway a clear morning.

Note *The expedition passes the area of present-day Jefferson City, state capitol of Missouri.*

FRIDAY, SEPTEMBER 19

	Sunrise			4 p.m.			River	
Temp	Weather	Wind	Temp	Weather	Wind	Rise/Fall	Feet	Inches
	f	SE		f	SE			

Weather Diary Saw a green Snake as high up as Salt Rivr on the missouri. the limestone bluffs commence below Salt river on S. Side

Clark a very singular disorder is takeing place amongst our party that of the Sore eyes. Three of the party have their eyes inflamed and Sweled in Such a manner as to render them extreamly painfull, particularly when exposed to the light, the eye ball is much inflaimed and the lid appears burnt with the Sun, the cause of this complaint of the eye I can't [account?] for. From [its] Sudden appearance I am willing to believe it may be owing to the reflection of the sun on the water.

Gass a fine day....being so anxious to reach St. Louis, where, without any important occurrence, we arrived on the 23rd, and were received with great kindness and marks of friendship by the inhabitants, after an absence of two years, four months and ten days.

Ordway a fair morning.

SATURDAY, SEPTEMBER 20

	Sunrise			4 p.m.		River		
Temp	Weather	Wind	Temp	Weather	Wind	Rise/Fall	Feet	Inches
	f	*NE*		*f*	*SE*			

Clark   The Osage river very low and discharges but a Small quantity of water at this time for so large a river We Saw Some cows on the bank which was a joyfull Sight to the party and Caused a Shout to be raised for joy. We Came in Sight of the little french Village called Charriton. We landed and were very politely received...as it was like to rain we accepted of a bed in one of their tents.

Ordway   As Several of the part have Sore eyes [Clark speculates that they are sun burned from viewing sun on the water. Moulton (ed.) later speculates it was caused by bacteria.] & unable to work, our officers leave 2 small canoes

Note   *The expedition arrives at St. Charles, Missouri.*

SUNDAY, SEPTEMBER 21

	Sunrise			4 p.m.		River		
Temp	Weather	Wind	Temp	Weather	Wind	Rise/Fall	Feet	Inches
	c a r	*SE*		*c*	*SE*			

Weather Diary   a Slight Shower of rain a little before day light this morning

Clark   at 4 PM we arived in Sight of St. Charles, the party rejoiced at the Sigh.

Ordway   late in the evening hard rain commend. and continued hard during the night.

MONDAY, SEPTEMBER 22

	Sunrise			4 p.m.		River		
Temp	Weather	Wind	Temp	Weather	Wind	Rise/Fall	Feet	Inches
	ratl&r	*S*		*c a r*	*S*			

Weather Diary   at St. Charles the raine commenced about 9 P.M and was moderate untill 4 A.M. when it increased and rained without intermition untill 10 A.M: Some Thunder and lightning about daylight. it Continued Cloudy with Small Showers of rain all day. we arived at the Mississippi

Clark   This morning being very wet and the rain Still continueing hard, and our party being all Sheltered in the houses of those hospitable people, we did not [think?] proper to proceed on untill after the rain was over   at 10 AM it seased raining and we Colected our party and Set out and proceeded on down to the Contonemt. We were honored with a Salute of Guns and a harty welcom—

Ordway   the hard rain continued this morning untill about 11 Oclock AM at which time the party was collected and we Set out & procd. on   Some Rain this evening.

Note   *The expedition arrives in St. Louis after completing a journey of more than 8,000 miles.*

TUESDAY, SEPTEMBER 23

	Sunrise			4 p.m.			River	
Temp	Weather	Wind	Temp	Weather	Wind	Rise/Fall	Feet	Inches
	c & r	NE		c a r	NE			

Weather Diary   at St. Louis Several light Showers in the course of this day. we arrived at St Louis at 12 oClock

Clark   set out decended to the Mississippi and down that river to St. Louis at which place we arived about 12 oClock. We Suffered the party to fire off their pieces as a Salute to the Town.

Ordway   a wet disagreeable morning. Soon arived at the Mouth of the Missouri entered the Mississippi River and landed at River deboise where we wintered in 1804. About 12 oClock we arived in Site of St. Louis. Drew out the canoes then the party all considerable much rejoiced that we have the Expedition Completed and now we look for boarding in Town and wait for our Settlement and then we entend to return to our native homes to See our parents once more as we have been So long from them.—   Finis.

WEDNESDAY, SEPTEMBER 24

	Sunrise			4 p.m.			River	
Temp	Weather	Wind	Temp	Weather	Wind	Rise/Fall	Feet	Inches
	r			c a r				

Weather Diary   rained moderately this morning and continued Cloudy with moderate rain at intervales all day

Clark   I sleped but litte last night. However we rose early and Commencd wrighting our letters. Capt. Lewis wrote one to the presidend and I wrote Govr. Harrison & my friends in Kentucky. Capt. Lewis in opening his trunk found all his papers wet, and Some Seeds spoiled.

THURSDAY, SEPTEMBER 25

	Sunrise			4 p.m.			River	
Temp	Weather	Wind	Temp	Weather	Wind	Rise/Fall	Feet	Inches
	c	NE		*f*				

Clark   had all of our Skins &c. Suned and Stored away in a Storeroom

Note   *On September 26, three days after arriving in St. Louis, Captain William Clark records his last daily journal entry into the Daily Narrative Journals of the Lewis and Clark Expedition.*

FRIDAY, SEPTEMBER 26

	Sunrise			4 p.m.			River	
Temp	Weather	Wind	Temp	Weather	Wind	Rise/Fall	Feet	Inches
	f	SE		*f*	SE			

Weather Diary   fair and w[a]rm

Clark   a fine morning   we commenced wrightin &c.

SATURDAY, SEPTEMBER 27

	Sunrise			4 p.m.			River	
Temp	Weather	Wind	Temp	Weather	Wind	Rise/Fall	Feet	Inches
	f	NE		*f*	SE			

Weather Diary   emencely w[a]rm

SUNDAY, SEPTEMBER 28

	Sunrise			4 p.m.			River	
Temp	Weather	Wind	Temp	Weather	Wind	Rise/Fall	Feet	Inches
	f	SE		f	SE			

Weather Diary   emencely w[a]rm

MONDAY, SEPTEMBER 29

	Sunrise			4 p.m.			River	
Temp	Weather	Wind	Temp	Weather	Wind	Rise/Fall	Feet	Inches
	f	S		f	SE			

Weather Diary   emencely w[a]rm

Note   *Tuesday, September 30 marks the last entry in the Weather Diary that pertains to weather or climate.*

TUESDAY, SEPTEMBER 30

	Sunrise			4 p.m.			River	
Temp	Weather	Wind	Temp	Weather	Wind	Rise/Fall	Feet	Inches
	f	SE		f	E			

Weather Diary   emencely w[a]rm

Lewis and Clark Trail Pictorial

One of only a few stretches of free-flowing water in the lower Missouri River Basin. Photograph taken near Lisbon, Missouri. (See June 7, 1804.)

A replica keelboat built from specifications and drawings in the journals. This was the largest vessel to accompany the expedition up the Missouri River. Photograph taken at Lewis and Clark State Park, Onawa, Nebraska. (See August 10, 1804.)

Swirling currents, falling and sloughing banks, and tree snags were just some of the daily problems confronting the expedition. Photograph taken near Yankton, South Dakota. (See August 26, 1804.)

Fearful of the Teton Sioux tribes, the expedition spent several nights in central South Dakota in the middle of the river with their boats shored up on unstable sand bars. Photograph taken north of Pierre, South Dakota. (See September 19, 1804.)

Low water during the late fall made travel difficult up the Missouri as they neared their winter quarters at Fort Mandan. Photograph taken near Washburn, North Dakota. (See October 17, 1804.)

A replica of Fort Mandan, the winter quarters 1804-05. Photograph taken near Washburn, North Dakota. (See December 24, 1804.)

An earthen lodge replica at Knife River Indian Villages National Historic Site near Stanton, North Dakota. (See November 17, 1804.)

Bull boats used by the Mandan and Hidatsa to cross the Missouri River were made of animal skins and small tree branches. Photograph of replica at Jefferson National Expansion Memorial. (See March 28, 1805.)

Vast prairies, large buffalo and antelope herds, and sweeping winds greeted the expedition in 1805 as they pushed upstream from winter quarters. Photograph taken near New Town, North Dakota. (See June 26, 1805.)

Wide open spaces and the lack of trees combined with spring winds and associated dust and sandstorms caused the corps to slow their ascent during the spring of 1805. Photograph taken near Price, North Dakota. (See April 20, 1805.)

Yellow bluffs and rocks characterize the Yellowstone River, which was named by the local tribes. Clark and a small party passed here on their return trip in July 1806. Photograph taken east of Fairview, Montana. (See July 31, 1806.)

The only known remaining evidence on the land from the expedition is William Clark's signature etched into the sandstone on Pompey's Pillar. The journals mention this kind of activity occurring at several locations along the trail, but this is the only etching left. (See July, 25 1806.)

A view up the Yellowstone River toward Bozeman Pass, which Sacagawea pointed out to Clark as the correct way east through the last of the Rocky Mountains. Photograph taken near Ballantine, Montana. (See July 19, 1806.)

While standing on this hill near modern-day Fort Peck, Montana, Clark was able to view the confluence of the Missouri and Milk rivers ("The River that Scolds all Others"), and see nearly 50 miles in every direction. (See May 8, 1805.)

One day the corps came upon a river full of feathers. Fearful of an Indian attack, they prepared. To their surprise and relief, as they rounded a bend in the river they came upon a flock of pelicans shedding their feathers for the season. Photograph taken near Great Falls, Montana. (See August 8, 1804.)

Passing through central Montana, the expedition entered the Missouri Breaks region (site of today's Upper Missouri River Breaks National Monument). Clark considered this area the vast deserts of North America. Lewis referred to the canyons as visionary enchantments. Photograph taken near Portage, Montana. (See May 26, 1805.)

Decision Point—the confluence of the Missouri and Maria's rivers. Spring floodwaters caused both rivers to look nearly identical in size, and scouting trips were made up each river to determine which was the Missouri. Lewis and Clark wisely chose the less muddy channel. Photograph taken at Loma, Montana, with Bear Paw Mountains in the distance. (See June 3, 1805.)

Steep canyon walls near the "Great Falls of the Missouri." Photograph taken near Great Falls, Montana. (See May 31, 1805.)

Today, Ryan Dam holds back the mighty Missouri at the "Great Falls of the Missouri." Lewis heard the falls from more than 7 miles away and discovered the "sublimely grand specticle" on June 13. It confirmed the choice he and Clark made on which river to follow. Photograph taken near Great Falls, Montana. (See June 13, 1805.)

Looking downriver from Rainbow Falls, Crooked Falls is visible near the water backed up from Morony Dam. The expedition had portage around these falls. Photograph taken near Great Falls, Montana. (See June 14, 1805.)

White Bear Islands (named after the many grizzly bears in the area) was the site of the upper portage camp. The expedition left supplies, as well as Lewis's failed cast-iron boat, here for the return trip, and headed into the mountains on July 15, 1805. Photograph taken at Great Falls, Montana. (See July 9, 1805.)

Square Butte (named Fort Mountain by Lewis and Clark) dominates the plains between Great Falls and the Rocky Mountains. Lewis used it to navigate back to the Sun (Medicine) River on his return to the White Bear Islands. Photograph taken near Cascade, Montana. (See July 15, 1805.)

The Rocky Mountains' glistening snowpack contained large amounts of snow during the late summers of 1805 and 1806 due to the little ice age and numerous winter storms. Photograph taken near Wisdom, Montana. (See July 4, 1805.)

After leaving the plains and traveling into the mountains, the corps entered steep canyon walls, which Lewis named the "Gates of the Rocky Mountains." Photograph taken north of Helena, Montana. (See July 19, 1805.)

The confluence of the Madison and Jefferson rivers—the captains chose to follow the westernmost river and named it after President Thomas Jefferson. Photograph taken near Three Forks, Montana. (See July 30, 1805.)

Jefferson's River split again near today's Twin Bridges, Montana, into the Big Hole (Wisdom) River and Beaverhead River (pictured). (See August 5, 1805.)

Sacagawea gave hope to the weary expedition members when she recognized the Beaver's Head, a landmark used by her fellow Shoshone-Lemhi when hunting. Photograph taken near Dillon, Montana. (See August 8, 1805.)

Lewis and his small reconnoitering party would have seen this rock on their way up Horse Prairie Creek in search of the Shoshone-Lemhi and the end of the Missouri River drainage. Photograph taken near Grant, Montana. (See August 11, 1805.)

The western end of the Missouri River, known as Trail Creek. The expedition drank from the springs here, which produce some of the first waters of the Mighty Missouri. (The furthermost point on the Missouri follows the Red Rock River southeast and begins as Hell Roaring Creek on the western slopes of Sawtell Peak, Idaho.) Photograph taken near Lemhi Pass, Montana. (See August 12, 1805.)

At the Continental Divide, looking west from Lemhi Pass into Idaho. Photograph taken near Tendoy, Idaho. (See August 12, 1805.)

The expedition found the waters of the Salmon (Lewis's) River too rapid and the walls too steep to navigate, so Lewis and Clark hired Shoshone-Lemhi guide "Old Toby" to lead them through the mountains. Sacagawea is honored with an Interpretive Center in her homeland near Salmon, Idaho. (See August 23, 1805.)

Entering the Bitterroot River valley via the Lost Trail Pass, one of the expedition's pack horses slipped on a path slick with rain and snow. Their last thermometer was broken. Photograph taken near Darby, Montana. (See September 3, 1805.)

The view of the Bitterroot Mountains over the Lolo Pass prompted Sergeant Gass to exclaim that these were "the most terrible mountains I ever beheld." A snowstorm with six to eight inches of snow on September 16 greatly impeded their progress. Photograph taken near Lolo, Montana. (See September 16, 1805.)

Nearly staved to death, the corps finally used one of their colts for food on the upper stretches of the Lochsa River. They named a nearby stream Colt Killed Creek (near today's Powell Ranger Station). (See September 15, 1805.)

Along the Clearwater River near Kamiah, Idaho, the expedition waited in "long camp" during May and June 1806 while the higher mountain snows melted. (See May 1806.)

The Nez Perce Indian Tribe assisted the Corps in building canoes for their journey downriver at "Canoe Camp" near present-day Orofino, Idaho. Dworshak Dam blocks the canyon on the north fork of the Clearwater River in the distance. (See September 26, 1805.)

Snake River rapids flow through towering basalt canyons. The corps started down river in their dugout ponderosa pine canoes near this location. Photograph taken near Lewiston, Idaho. (See October 10, 1805.)

Just downriver from the confluence of the Snake and Columbia rivers, the channel widens to nearly a mile. Members noted numerous dead salmon in this area. Photograph taken near Wallula, Washington. (See October, 17 1805.)

Large islands in the middle of the Columbia River provided resting points between the cascades and shutes. Photograph taken near Boardman, Oregon. (See October 18, 1805.)

The corps moved through five sets of large cascades. The first, known as the "Great Falls of the Columbia," or Celilo Falls is now covered by the John Day Dam, but the terrain is much the same. Photograph taken near Wishram, Washington. (See October 22, 1805.)

Correctly identified by Clark in the journals, Mt. Hood loomed to the southwest like a beacon. When the corps saw this large volcano, they knew from previous explorers' findings along the Columbia River a decade earlier that they were back on the map. Photograph taken near The Dalles, Oregon. (October 19, 1805.)

The Columbia River just below the last set of rapids known as the Cascade Locks. Beacon Rock, in the distance, was reached by Lt. William Broughton in 1792 as the furthermost point explored upriver during Captain Vancouver's sailing voyage. The Lewis and Clark Expedition was elated to see this landmark and note the tidal changes in the river. Photograph taken near Skamania, Washington. (See November 1, 1805.)

The Columbia River becomes a large lake below the Cascades, and the large estuary can reach more than 10 miles wide at points. The corps stayed closed the north shore on their journey downriver due to dense fog and rainy conditions. Photograph taken near Cathlamet, Washington. (See November 4, 1805.)

"Ocean in View, O' the Joy," wrote William Clark when they reached the Pacific Ocean. This photograph shows the North Head Lighthouse at the mouth of the Columbia River. (See November 7, 1805.)

At the mouth of the Columbia River, looking north toward Cape Disappointment. Photograph taken at Point Adams Spit, Oregon. (See November 16, 1805.)

A replica of Fort Clatsop stands where the expedition spent the winter of 1805-06. Photograph taken near Astoria, Oregon. (See December 7, 1805.)

While wintering at Fort Clatsop, members heard of a large whale washed ashore near present day Cannon Beach. Clark, Sacagawea, and several members trekked through rain-swollen rivers and streams, only to find that most of the blubber had already been removed by the natives. They traded to obtain a small ration. Clark and a few members climbed a high ridge at the ocean's edge, which Lewis named "Clark's Mountain" (Tillamook Head). Photograph of preserved whale skeleton at Long Beach, Washington. (See January 10, 1806.)

APPENDIX A

**President Thomas Jefferson's
Confidential Letter to Congress**

Confidential. [January 18, 1803]

Gentlemen of the Senate, and of the House of Representatives:

As the continuance of the act for establishing trading houses with the Indian tribes will be under the consideration of the Legislature at its present session, I think it my duty to communicate the views which have guided me in the execution of that act, in order that you may decide on the policy of continuing it, in the present or any other form, or discontinue it altogether, if that shall, on the whole, seem most for the public good.

The Indian tribes residing within the limits of the United States, have, for a considerable time, been growing more and more uneasy at the constant diminution of the territory they occupy, although effected by their own voluntary sales: and the policy has long been gaining strength with them, of refusing absolutely all further sale, on any conditions; insomuch that, at this time, it hazards their friendship, and excites dangerous jealousies and perturbations in their minds to make any overture for the purchase of the smallest portions of their land. A very few tribes only are not yet obstinately in these dispositions. In order peaceably to counteract this policy of theirs, and to provide an extension of territory which the rapid increase of our numbers will call for, two measures are deemed expedient.

First: to encourage them to abandon hunting, to apply to the raising stock, to agriculture and domestic manufacture, and thereby prove to themselves that less land and labor will maintain them in this, better than in their former mode of living. The extensive forests necessary in the hunting life, will then become useless, and they will see advantage in exchanging them for the means of improving their farms, and of increasing their domestic comforts.

Secondly: to multiply trading houses among them, and place within their reach those things which will contribute more to their domestic comfort, than the possession of extensive, but uncultivated wilds. Experience and reflection will develop to them the wisdom of exchanging what they can spare and we want, for what we can spare and they want. In leading them to agriculture, to manufactures, and civilization; in bringing together their and our settlements, and in preparing them ultimately to participate in the benefits of our governments, I trust and believe we are acting for their greatest good. At these trading houses we have pursued the principles of the act of Congress, which directs that the commerce shall be carried on liberally, and requires only that the capital stock shall not be diminished. We consequently undersell private traders, foreign and domestic, drive them from the competition; and thus, with the good will of the Indians, rid ourselves of a description of men who are constantly endeavoring to excite in the Indian mind suspicions, fears, and irritations towards us. A letter now enclosed, shows the effect of our competition on the operations of the traders, while the Indians, perceiving the advantage of purchasing from us, are soliciting generally, our establishment of trading houses among them. In one quarter this is particularly interesting. The Legislature, reflecting on the late occurrences on the Mississippi, must be sensible how desirable it is to possess a respectable breadth of country on that river, from our Southern limit to the Illinois at least; so that we may present as firm a front on that as on our Eastern border. We possess what is below the Yazoo, and can probably acquire a certain breadth from the Illinois and Wabash to the Ohio; but between the Ohio and Yazoo, the country all belongs to the Chickasaws, friendly tribe within our limits, but the most decided against the alienation of lands. The portion of their country most important for us is exactly that which they do not inhabit. Their settlements are not on the Mississippi, but in the interior country. They have lately shown a desire to become agricultural; and this leads to the desire of buying implements and comforts. In the strengthening and gratifying of these wants, I see the only prospect of planting on the Mississippi itself, the means of its own safety. Duty has required me to submit these views to the judgment of the Legislature; but as their disclosure might embarrass and defeat their effect, they are committed to the special confidence of the two Houses.

While the extension of the public commerce among the Indian tribes, may deprive of that source of profit such of our citizens as are engaged in it, it might be worthy the attention of Congress, in their care of individual as well as of the general interest, to point, in another direction, the enterprise of these citizens, as profitably for themselves, and more usefully for the public. The river Missouri, and the Indians inhabiting it, are not as well known as is rendered desirable by their connexion with the Mississippi, and consequently with us. It is, however, understood, that the country on that river is inhabited by numerous tribes, who furnish great supplies of furs and peltry to the trade of another nation, carried on in a high latitude, through an infinite number of portages and lakes, shut up by ice through a long season. The commerce on that line could bear no competition with that of the Missouri, traversing a moderate climate, offering according to the best accounts, a continued navigation from its source, and possibly with a single portage, from the Western Ocean, and finding to the Atlantic a choice of channels through the Illinois or Wabash, the lakes and Hudson, through the Ohio and Susquehanna, or Potomac or James rivers, and through the Tennessee and Savannah, rivers. An intelligent officer, with ten or twelve chosen men, fit for the enterprise, and willing to undertake it, taken from our posts, where they may be spared without inconvenience, might explore the whole line, even to the Western Ocean, have conferences with the natives on the subject of commercial intercourse, get admission among them for our traders, as others are admitted, agree on convenient deposits for an interchange of articles, and return with the information acquired, in the course of two summers. Their arms and accoutrements, some instruments of observation, and light and cheap presents for the Indians, would be all the apparatus they could carry, and with an expectation of a soldier's portion of land on their return, would constitute the whole expense. Their pay would be going on, whether here or there. While other civilized nations have encountered great expense to enlarge the boundaries of knowledge by undertaking voyages of discovery, and for other literary purposes, in various parts and directions, our nation seems to owe to the same object, as well as to its own interests, to explore this, the only line of easy communication across the continent, and so directly traversing our own part of it. The interests of commerce place the principal object within the constitutional powers and care of Congress, and that it should incidentally advance the geographical knowledge of our own continent, cannot be but an additional gratification. The nation claiming the territory, regarding this as a literary pursuit, which is in the habit of permitting within its dominions, would not be disposed to view it with jealousy, even if the expiring state of its interests there did not render it a matter of indifference. The appropriation of $2,500, "for the purpose of extending the external commerce of the United States," while understood and considered by the Executive as giving the legislative sanction, would cover the undertaking from notice, and prevent the obstructions which interested individuals might otherwise previously prepare in its way.

TH: JEFFERSON
Jan. 18. 1803.
(Richardson 1897, 340–342; Jackson 1978, 10–13)

APPENDIX B

Lewis's Expedition Requirements List

Mathematical Instruments

1	Hadley's Quadrant
1	Mariner's Compas & 2 pole chain
1	Sett of plotting instruments
3	Thermometers
1	Cheap portable Microscope
1	Pocket Compass
1	brass Scale one foot in length
6	Magnetic needles in small straight silver or brass cases opening on the side with hinges.
1	Instrument for measuring made of tape with feet & inches mark'd on it,...
2	Hydrometers
1	Theodolite
1	Sett of planespheres
2	Artificial Horizons
1	Patent log
6	papers of Ink powder
4	Metal Pens brass or silver
1	Set of Small Slates & pencils
2	Creyons
	Sealing wax one bundle
1	Miller's edition of Lineus in 2 Vol:
	Books
	Maps
	Charts
	Blank Vocabularies
	Writing paper
1	Pair large brass money scales with two setts of weights...

Arms & Accouterments

15	Rifle
15	Powder Horns & pouches complete
15	Pairs of Bullet Moulds
15	do. Of Wipers or Gun worms
15	Ball Screws
24	Pipe Tomahawks
24	large knives
	Extra parts of Locks & tools for repairing arms
15	Gun Slings
500	best Flints

Ammunition

200	Lbs. Best rifle powder
400	lbs. Lead

Clothing

15	3 pt. Blankets
15	Watch Coats with Hoods & belts
15	Woolen Overalls
15	Rifle Frocks of waterproof Cloth if possible
30	Pairs of Socks or half Stockings
20	Fatigue Frocks or hinting shirts
30	Shirts of Strong linnen
30	yds. Common flannel.

Camp Equipage

6	Copper kettles (1 of 5 Gallons, 1 of 3, 2 of 2, & 2 of 1)
35	falling Axes.
4	Drawing Knives, short & strong
2	Augers of the patent kind...
1	Small permanent Vice
1	Hand Vice
36	Gimblets assorted
24	Files do.
12	Chisels do.
10	Nails do.
2	Steel plate hand saws
2	Vials of Phosforus
1	do. Of Phosforus made of allum & sugar
4	Groce fishing Hooks assorted

12	Bunches of Drum Line
2	Foot Adzes
12	Bunches of Small cord
2	Pick Axes
3	Coils of rope
2	Spades
12	Bunches Small fishing line assorted
1	lb. Turkey or Oil Stone
1	Iron Mill for Grinding Corn
20	yds. Oil linnen for wrapping & securing Articles
10	yds do. do. Of thicker quality for covering and lining boxes. &c
40	yds Do. Do. To form two half faced Tents or Shelters…
4	Tin blowing Trumpets
2	hand or spiral spring Steelyards
20	yds Strong Oznaburgs
24	Iron Spoons
24	Pint Tin Cups (without handles)
30	Steels for striking or making fire
100	Flints for do. do. do.
2	Frows
6	Saddlers large Needles
6	Do. Large Awls
	Muscatoe Curtains
2	patent chamber lamps & wicks
15	Oil Cloth Bags for securing provision
1	Sea Grass Hammock

Provisions and Means of Subsistence

150	lbs. Portable Soup.
3	bushels of Allum or Rock Salt
	Spicies assorted
6	Kegs of 5 Gallons each making 30 Gallons of rectified spirits such as is used for the Indian trade
6	Kegs bound with iron Hoops

Indian Presents

5	lbs. White Wampum

5	lbs. White Glass Beads mostly small
20	lbs. Red Do. Do. Assorted
5	lbs. Yellow or Orange Do. Do. Assorted
30	Calico Shirts
12	Pieces of East India muslin Hanckerchiefs striped or check'd with brilliant Colours.
12	Red Silk Hanckerchiefs
144	Small cheap looking Glasses
100	Burning Glasses
4	Vials of Phosforus
288	Steels for striking fire
144	Small cheap Scizors
20	Pair large Do.
12	Groces Needles Assorted No. 1 to 8 Common points
12	Groces Do. Assorted with points for sewing leather
288	Common brass thimbles — part W. office
10	lbs. Sewing Thread assorted
24	Hanks Sewing Silk
8	lbs. Red Lead
2	lbs. Vermillion — at War Office
288	Knives Small such as are generally used for the Indian trade, with fix'd blades & handles inlaid with brass
36	Large knives
36	Pipe Tomahawks - at H. Ferry
12	lbs. Brass wire Assorted
12	lbs. Iron do. Do. generally large
6	Belts of narrow Ribbons colours assorted
50	lbs. Spun Tobacco.
20	Small falling axes to be obtained in Tennessee
40	fish Griggs such as the Indians use with a single barbed point — at Harper's ferry
3	Groce fishing Hooks assorted
3	Groce Mockerson awls assorted

50	lbs. Powder secured in a Keg covered with oil Cloth
24	Belts of Worsted feiret or Gartering Colours brilliant and Assorted
15	Sheets of Copper Cut into strips of an inch in width & a foot long
20	Sheets of Tin
12	lbs. Strips of Sheet iron 1 In. wide 1 foot long
1	Pc. Red Cloth second quality
1	Nest of 8 or 9 small copper kettles
100	Block-tin rings cheap kind ornamented with Colour'd Glass or Mock-Stone
2	Groces of brass Curtain Rings & sufficently large for the Finger
1	Groce Cast Iron Combs
18	Cheap brass Combs
24	Blankets.
12	Arm Bands Silver at War Office
12	Wrist do. do. Do.
36	Ear Trinkets Do. Part do.
6	Groces Drops of Do. Part Do.
4	doz Rings for Fingers of do.
4	Groces Broaches of do.
12	Small Medals do.

Means of Transportation

1	Keeled Boat light strong at least 60 feet in length her burthen equal to 8 Tons
1	Iron frame of Canoe 40 feet long
1	Large Wooden Canoe
12	Spikes for Setting-Poles
4	Boat Hooks & points Complete
2	Chains & Pad-Locks for confining the Boat & Canoes &c.

Medicine

15	lbs. Best powder's Bark
10	lbs. Epsom or Glauber Salts
4	oz. Calomel

12	oz. Opium
_	oz. Tarter emetic
8	oz. Borax
4	oz. Powder'd Ipecacuana
8	oz. Powder Jalap
8	oz. Powdered Rhubarb
6	Best lancets
2	oz. White Vitriol
4	oz. Lacteaum Saturni
4	Pewter Penis syringes
1	Flour of Sulphur
3	Clyster pipes
4	oz. Turlingtons Balsam
2	lbs. Yellow Bascilicum
2	Sticks of Symple Diachylon
1	lb. Blistering Ointments
2	lbs. Nitre
2	lbs. Coperas

Materials for making up the Various Articles into portable Packs

30	Sheep skins taken off the Animal as perfectly whole aspossible, without being split on the belly as usual and dress'd only with lime to free them from the wool; or otherwise about the same quantity of Oil Cloth bags well painted
	Raw hide for pack strings
	Dress'd letter for Hoppus-Straps
	Other packing

Do. = ditto

&c. = etcetera

Oznaburgs = strong cloth

Worsted [ferret] = woven wool tape, used for embellishment and trade

Hoppus = might possible refer to an Indian term for knapsack

(Jackson 1978, 69–75)

APPENDIX C

President Jefferson's Expedition Instructions to Lewis

[20 June 1803]

To Meriwether Lewis, esquire, Captain of the 1st regiment of infantry of the United States of America.

Your situation as Secretary of the President of the United States has made you acquainted with the objects of my confidential message of Jan. 18, 1803, to the legislature. You have seen the act they passed, which, tho' expressed in general terms, was meant to sanction those objects, and you are appointed to carry them into execution.

Instruments for ascertaining by celestial observations the geography of the country thro' which you will pass, have been already provided. Light articles for barter, & presents among the Indians, arms for your attendants, say for from 10 to 12 men, boats, tents, & other travelling apparatus, with ammunition, medicine, surgical instruments & provision you will have prepared with such aids as the Secretary at War can yield in his department; & from him also you will receive authority to engage among our troops, by voluntary agreement, the number of attendants above mentioned, over whom you, as their commanding officer are invested with all the powers the laws give in such a case.

As your movements while within the limits of the U.S. will be better directed by occasional communications, adapted to circumstances as they arise, they will not be noticed here. What follows will respect your proceedings after your departure from the U.S.

Your mission has been communicated to the Ministers here from France, Spain, & Great Britain, and through them to their governments: and such assurances given them as to [its] objects as we trust will satisfy them. The country *of Louisiana* having been ceded by Spain to France, *and possession by this time probably given*, the passport you have from the Minister of France, the representative of the present sovereign of the country, will be a protection with all its subjects: and that from the Minister of England will entitle you to the friendly aid of any traders of that allegiance with whom you may happen to meet.

The object of your mission is to explore the Missouri river, & such principal stream of it, as, by [its] course & communication with the water of the Pacific ocean may offer the most direct & practicable water communication across this continent, for the purposes of commerce.

Beginning at the mouth of the Missouri, you will take *careful* observations of latitude and longitude at all remarkable points on the river, & especially at the mouths of rivers, at rapids, at islands & other places & objects distinguished by such natural marks & characters of a durable kind, as that they may with certainty be recognized hereafter. The courses of the river between these points of observation may be supplied by the compass, the log-line & by time, corrected by the observations themselves. The variations of the compass too, in different places should be noticed.

The interesting points of the portage between the heads of the Missouri & the water offering the best communication with the Pacific ocean should be fixed by observation, & the course of that water to the ocean, in the same manner as that of the Missouri.

Your observations are to be taken with great pains & accuracy to be entered distinctly, & intelligibly for others as well as yourself, to comprehend all the elements necessary, with the aid of the usual tables to fix the latitude & longitude of the places at which they were taken,

& are to be rendered to the war office, for the purpose of having the calculations made concurrently by proper persons within the U.S. Several copies of these as well as of your other notes, should be made at leisure times, & put into the care of the most trustworthy of your attendants, to guard by multiplying them against the accidental losses to which they will be exposed. A further guard would be that one of these copies be written on the paper of the birch, as less liable to injury from damp than common paper.

The commerce which may be carried on with the people inhabiting the line you will pursue, renders a knolege of these people important. You will therefore endeavor to make yourself acquainted, as far as a diligent pursuit of your journey shall admit, with the names of the nations & their numbers;

> the extent & limits of their possessions;
> their relations with other tribes or nations;
> their language, traditions, monuments;
> their ordinary occupations in agriculture, fishing, hunting, war, arts, & the implements for these;
> their food, clothing, & domestic accommodations;
> the diseases prevalent among them, & the remedies they use;
> moral and physical circumstance which distinguish them from the tribes they know;
> peculiarities in their laws, customs & dispositions;
> and articles of commerce they may need or furnish, & to what extent.

And considering the interest which every nation has in extending & strengthening the authority of reason & justice among the people around them, it will be useful to acquire what knolege you can of the state of morality, religion & information among them, as it may better enable those who endeavor to civilize & instruct them, to adapt their measures to the existing notions & practises of those on whom they are to operate.

Other objects worthy of notice will be

> the soil & face of the country, [its] growth & vegetable productions, especially those not of the U.S.
> the animals of the country generally, & especially those not known in the U.S. the remains & accounts of any which may be deemed rare or extinct;
> the mineral productions of every kind; but more particularly metals, limestone, pit coal & saltpetre; salines & mineral waters, noting the temperature of the last & such circumstances as may indicate their character;
> volcanic appearances;
> climate as characterized by the thermometer, by the proportion of rainy, cloudy & clear days, by lightening, hail, snow, ice, by the access & recess of frost, by the winds, prevailing at different seasons, the dates at which particular plants put forth or lose their flowers, or leaf, times of appearance of particular birds, reptiles or insects.

Altho' your route will be along the channel of the Missouri, yet you will endeavor to inform yourself, by inquiry, of the character and extent of the country watered by its branches, & especially on [its] Southern side. The North river or Rio Bravo which runs into the gulph of Mexico, and the North river, or Rio colorado which runs into the gulph of California, are understood to be the principal streams heading opposite to the waters of the Missouri, and running Southwardly. Whether the dividing grounds between the Missouri & them are mountains or flatlands, what are their distance from the Missouri, the character of the intermediate country, & the people inhabiting it, are worthy of particular enquiry. The Northern waters of the Missouri are less to be enquired after, because they have been ascertained to a considerable degree, and are still in a course of ascertainment by English traders & travellers. But if you can learn anything certain of the most Northern source of the Mississippi, & of [its] position relative to the lake of the woods, it will be interesting to us.

Two copies of your notes at least & as many more as leisure will admit, should be made & confided to the care of the most trusty individuals of your attendants. Some account too of the path of the Canadian traders from the Mississippi, at the mouth of the Ouisconsing river, to where it strikes the Missouri, and of the soil and rivers in [its] course, is desirable.

In all your intercourse with the natives treat them in the most friendly & conciliatory manner which their own conduct will admit; allay all jealousies as to the object of your journey, satisfy them of [its] innocence, make them acquainted with the position, extent, character, peaceable & commercial dispositions of the U.S., of our wish to be neighborly, friendly & useful to them, & of our dispositions to a commercial intercourse with them; confer with them on the points most convenient as mutual emporiums, & the articles of most desirable interchange for them & us. If a few of their influential chiefs, within practicable distance, wish to visit us, arrange such a visit with them, and furnish them with authority to call on our officers, on their entering the U.S. to have them conveyed to this place at the public expense. If any of them should wish to have some of their young people brought up with us, & taught such arts as may be useful to them, we will receive, instruct & take care of them. Such a mission, whether of influential chiefs, or of young people, would give some security to your own party. Carry with you some matter of the kine pox, inform those of them with whom you may be, of [its] efficacy as a preservative from the small pox; and instruct & encourage them in the use of it. This may be especially done wherever you may winter.

As it is impossible for us to foresee in what manner you will be received by those people, whether with hospitality or hostility, so is it impossible to prescribe the exact degree of perseverance with which you are to pursue your journey. We value too much the lives of citizens to offer them to probably destruction. Your numbers will be sufficient to secure you against the unauthorised opposition of individuals, or of small parties: but if a superior force, authorised or not authorised, by a nation, should be arrayed against your further passage, & inflexibly determined to arrest it, you must decline [its] further pursuit, and return. In the loss of yourselves, we should lose also the information you will have acquired. By returning safely with that, you may enable us to renew the essay with better calculated means. To your own discretion therefore must be left the degree of danger you may risk, & the point at which you should decline, only saying we wish you to err on the side of your safety, & to bring back your party safe, even if it be with less information.

As far up the Missouri as the white settlements extend, an intercourse will probably be found to exist between them and the Spanish posts at St. Louis, opposite Cahokia, or Ste. Genevieve opposite Kaskaskia. From still farther up the river, the traders may furnish a conveyance for letters. Beyond that you may perhaps be able to engage Indians to bring letters for the government to Cahokia or Kaskaskia, on promising that they shall there receive such special compensation as you shall have stipulated with them. Avail yourself of these means to communicate to us, at seasonable intervals, a copy of your journal, notes & observations of every kind, putting into cypher whatever might do injury if betrayed.

Should you reach the Pacific ocean, inform yourself of the circumstances which may decide whether the furs of those parts may not be collected as advantageously at the head of the Missouri (convenient as is supposed to the waters of the Colorado & Oregon or Columbia) as at Nootka sound or any other point of that coast; & that trade be consequently conducted through the Missouri & U.S. more beneficially than by the circumnavigation now practised.

On your arrival on that coast, endeavor to learn if there be any port within your reach frequented by the sea-vessels of any nation, and to send two of your trusty people back by sea, in such way as *they shall judge* shall appear practicable, with a copy of your notes. And should you be of opinion that the return of your party by the way they went will be eminently dangerous, then ship the whole, & return by sea by way of Cape Horn or the Cape of Good Hope, as you shall be able. As you will be without money, clothes or provisions, you must endeavor to use the credit of the U.S. to obtain them; for which purpose open letters of credit shall be furnished you authorizing you to draw on the Executive of the U.S. or any of its officers in any part of the world, in which draughts can be disposed of, and to apply with our recommendations to the consuls, agents, merchants or citizens of any nation with which we have intercourse, assuring them in our name that any aids they may furnish you shall be honorably repaid, and on demand. Our consuls Thomas Howes at Batavia in Java, William Buchanan of the Isles of France and Bourbon, & John Elmslie at the Cape of Good Hope will be able to supply your necessities by draughts on us.

Should you find it safe to return by the way you go, after sending two of your party round by sea, or with your whole party, if no conveyance by sea can be found, do so; making such observations on your return as may serve to supply, correct or confirm those made on your outward journey.

On re-entering the U.S. and reaching a place of safety, discharge any of your attendants who may desire & deserve it: procuring for them immediate paiment of all arrears of pay & cloathing which may have incurred since their departure and assure them that they shall be recommended to the liberality of the legislature for the grant of a souldier's portion of land each, as proposed in my message to Congress: & repair yourself with your papers to the seat of government *to which I have only to add my sincere prayer for your safe return.*

To provide, on the accident of your death, against anarchy, dispersion & the consequent danger to your party, and total failure of the enterprise, you are hereby authorised, by any instrument signed & written in your own hand, to name the person among them who shall succeed to the command on your decease, & by like instruments to change the nomination from time to time, as further experience of the characters accompanying you shall point out superior fitness: and all the powers & authorities given to yourself are, in the event of your death, transferred to & vested in the successor so named, with further power to him, & his successors in like manner to name each his successor, who, on the death of his predecessor shall be invested with all the powers & authorities given to yourself.

Given under my hand at the city of Washington, this 20th. day of June 1803.

Th Jefferson, Pr. U.S. of America.
(Jackson 1978, 61–66)

APPENDIX D

Lewis's Letter to President Jefferson Upon Return to St. Louis

To Thomas Jefferson, President of the United States

Sir   St. Louis, September 23rd 1806

It is with pleasure that I announce to you the safe arrival of myself and party at 12 OClk. today at this place with out papers and baggage. In obedience to your orders we have penitrated the Continent of North America to the Pacific Ocean, and sufficiently explored the interior of the country to affirm with confidence that we have discovered the most practicable rout which does exist across the continent by means of the navigable branches of the Missouri and Columbia Rivers. Such is that by way of the Missouri to the foot of the rapids five miles below the great falls of the Missouri of that river a distance of 2575 miles, thence by land passing the Rocky Mountains to a navigable part of the Kooskooske 340; with the Kooskooske 73 mls. a South Easterly branch of the Columbia 154 miles and the latter river 413 mls. to the Pacific Ocean; making the total distance from the confluence of the Missouri and Mississippi to the discharge of the Columbia into the Pacific Ocean 3555 miles. The navigation of the Missouri may be deemed safe and good; [its] difficulties arrise from [its] falling banks, timber imbeded in the mud of [its] channel, [its] sand bars and steady rapidity of [its] current, all which may be overcome with a great degree of certainty by taking the necessary precautions. The passage by land of 340 miles from the Missouri to the Kooskooske is the most formidable part of the tract proposed across the Continent; of this distance 200 miles is along a good road, and 140 over tremendious mountains which for 60 mls. are covered with eternal snows; however a passage over these mountains is practicable from the latter part of June to the last of September, and the cheep rate at which horses are to be obtained from the Indians of the Rocky Mountains and West of them, reduces the expences of transportation over this portage to a mere trifle. The navigation of the Kooskooske, the South East branch of the Columbia itself is safe and good from the 1st of April to the middle of August, by making three portages on the latter; the first of which in decending is that of 1200 paces at the great falls of the Columbia, 261 mls. from the Ocean, the second of two miles at the long narrows six miles below the falls, and the 3rd also of 2 miles a the great rapids 65 miles still lower down. The tides flow up the Columbia 183 miles, or within seven miles of the great rapids, thus far large sloops might ascend in safety, and vessels of 300 tons burthen could with equal safety reach the entrance of the river Multnomah, a large Southern branch of the Columbia, which taking [its] rise on the confines of Mexico with the Callarado and Apostles river, discharges itself into the Columbia 125 miles from [its] mouth. From the head of tide water to the foot of the long narrows the Columbia could be most advantageously navigated with large batteauxs, and from thence upwards by perogues. The Missouri possesses sufficient debth of water as far as is specifyed for boats of 15 tons burthern, but those of smaller capacity are to be prefered.

We view this passage across the Continent as affording immence advantages to the fur trade, but fear that the advantages which it offers as a communication for the productions of the East Indies to the United States and thence to Europe will never be found equal on an extensive scale to that by way of the Cap of Good hope; still we believe that many articles not bulky brittle nor of a very perishable nature may be conveyed to the United States by this rout with more facility and at less expence than by that at present practiced.

The Missouri and all [its] branches from the Chyenne upwards abound more in beaver and Common Otter, than any other streams on earth, particularly that proportion of them lying within the Rocky Mountains. The furs of all this immence tract of country including such as may be collected on the upper portion of the River St. Peters, Red river and the Assinniboin with the immence country watered by the Columbia, may be conveyed to the mouth of the Columbia by the 1st of August in each year and from thence be shiped to, and arrive in Canton earlier than the furs at present shiped from Montreal annually arrive in London. The British N. West Company of Canada were they permitted by the United States might also convey their furs collected in the Athabaske, on the Saskashawan, and South and West of Lake Winnipic by that rout within the period before mentioned. Thus the productions [of] nine tenths of the most valuable fur country of America could be conveyed by the rout proposed to the East Indies.

In the infancy of the trade across the continent, or during the period that the trading establishments shall be confined to the Missouri and [its] branches, the men employed in this trade will be compelled to convey the furs collected in that quarter as low on the Columbia as tide water, in which case they could not return to the falls of the Missouri untill about the 1st of October, which would be so late in the season that there would be considerable danger of the river being obstructed by ice before they could reach this place and consequently that the comodities brought from the East indies would be detained untill the following spring; but this difficulty will at once vanish when establishments are also made on the Columbia, and a sufficient number of men employed at them to convey annually the productions of the East indies to the upper establishment on the Kooskooske, and there exchange them with the men of the Missouri for their furs, in the begining of July. By this means the furs not only of the Missouri but those also of the Columbia may be shiped to the East indies by the season before mentioned, and the comodities of the East indies arrive at St. Louis or the mouth of the Ohio by the last of September in each year.

Although the Columbia dose not as much as the Missouri abound in beaver and Otter, yet it is by no means despicable in this rispect, and would furnish a valuable fur trade distinct from any other consideration in addition to the otter and beaver which it could furnish. There might be collected considerable quantities of the skins of three species of bear affording a great variety of colours and of superior delicacy, those also of the tyger cat, several species of fox, martin and several others of an inferior class of furs, besides the valuable Sea Otter of the coast.

If the government will only aid, even in a very limited manner, the enterprise of her Citizens I am fully convinced that we shal shortly derive the benifits of a most lucrative trade from this source, and that in the course of ten or twelve years a tour across the Continent by the rout mentioned will be undertaken by individuals with as little concern as a voyage across the Atlantic is a present.

The British N. West Company of Canada has for several years, carried on a partial trade with the Minnetares Ahwayhaways and Mandans on the Missouri from their establishments on the Assinniboin at the entrance of Mouse river; at present I have good reason for believing that they intend shortly to form an establishment near those nations with a view to engroce the fur trade of the Missouri. The known enterprize and resources of this Company, latterly strengthened by an union with their powerfull rival the X.Y. Company renders them formidable in that distant part of the continent to all other traders; and in my opinion if we are to regard the trade of the Missouri as an object of importance to the United States; the strides of this Company towards the Missouri cannot be too vigilantly watched nor too firmly and speedily opposed by our government. The embarrasments under which the navigation of the Missouri at present labours from the unfriendly dispositions of the kancez, the several bands of Tetons, Assinniboins and those tribes that resort to the British establishments on the Saskashawan is also a subject which requries the earliest attention of our government. As I shall shortly be with you I have deemed it unnecessary here to detail the several ideas which have presented themselves to my mind on those subjects, more especially when I consider that a thorough knowledge of the geography of the country is absolutely necessary to their being understood and leasure has not yet permited us to make but one general map of the country which I am unwilling to wrisk by the Mail.

As a sketch of the most prominent features of our perigrination since we left the Mandans may not be uninterseting, I shall indeavour to give it to you by way of letter from this place, where I shall necessarily be detained several days in order to settle with and discharge the men who accompanyed me on the voyage as well as to prepare for my rout to the City of Washington.

We left Fort Clatsop where we wintered near the entrance of the Columbia on the 27th of March last, and arrived at the foot of the Rocky mountains on the 10th of May where we were detained untill the 24th of June in consequence of the snow which rendered a passage over those Mountains impractiable untill that moment; had it not been for this detention I should ere this have joined you at Montichello. In my last communication to you from the Mandans I mentioned my intention of sending back a canoe with a small party from the Rocky Mountains; but on our arrival at the great falls of the Missouri on the 14th of June 1805, in view of that formidable snowey barrier, the discourageing difficulties which we had to encounter in making a portage of eighteen miles of our canoes and baggage around those falls were such that my friend Capt. Clark and myself conceived in inexpedient to reduce the party, lest by doing so we should lessen the ardor of those who remained and thus hazard the fate of the expedition, and therefore declined that measure, thinking it better that the government as well as our friends should for a moment fell some anxiety for our fate than to wrisk to much; experience has since proved the justice of our decision, for we have more than once owed our lives and the fate of the expedition to our number which consisted of 31 men.

I have brought with me several skins of the Sea Otter, two skins of the native sheep of America, five skins and skelitons complete of the Bighorn or mountain ram, and a skin of the Mule deer beside the skins of several other quadrupeds and b[ir]ds nat[iv]es of the countries through which we have passed. I have also preserved a pretty extensive collection of plants, and collected nine other vocabularies.

I have prevailed on the great Chief of the Mandan nation to accompany me to Washington; he is now with my frind and colligue Capt. Clark at this place, in good health and sperits, and very anxious to proceede.

With rispect to the exertions and services rendered by that esteemable man Capt. William Clark in the course of late voyage I cannot say too much; if sir any credit be due for the success of that ardous enterprize in which we have been mutually engaged, he is equally with myself entitled to your consideration and that of our common country.

The anxiety which I feel in returning once more to the bosom of my friends is a sufficient guarantee that no time will be unnecessarily expended in this quarter.

I have detained the post several hours for the purpose of making you this haisty communication. I hope that while I am pardoned for this detention of the mail, the situation in which I have been compelled to write will sufficiently apologize for having been this lcaonic.

The rout by which I purpose traveling fro hence to Washington is by way of Cahokia, Vincennes, Louisvill Ky., the Crab orchard, Abington, Fincastle, Stanton and Chorlottsville. Any letters directed to me at Louisville ten days after the receipt of this will most probably meet me at that place. I am very anxious to learn the state of my friends in Albermarle particularly whether my mother is yet living. I am with every sentiment of esteem Your Obt. And very Humble servent.

Meriwether Leiws Capt.
1st. U.S. Regt. Infty.

N.B. The whole of the party who accompanyed me f[ro]m the Mandans have returned in good health, which is not, I assure you, to me one of the least pleasing considerations of the Voyage.

M. L.
(Jackson 1978, 319–324)

APPENDIX E

Clark's Letter to his Brother Upon Return to St. Louis

[This letter was the first published in the United States describing the expedition's return in the Frankfort, Kentucky *Palladium* newspaper on October 9, 1806. It was copied and spread throughout the nation as rapid as the means of the day allowed.]

Dear Brother   St. Louis 23rd September 1806

We arrived at this place at 12 oClock today from the Pacific Ocian where we remained dureing the last winter near the entrance of the Columbia river. This station we left on the 27th of March last and should have reached St. Louis early in August had we not been detained by snow which bared our passage across the Rocky Mountains untill the 24th of June. In returning through those mountains we devided ourselves into several parties, disgressing from the rout by which we went out in order the more effectually to explore the Country and discover the most practicable rout which does exist across the Continent by way of the Missouri and Columbia rivers, in this we were completely successfull and have therefore no hesitation in declaring that such as nature has permited it we have discovered the best rout which does exist across the continent of North America in that direction. Such is that, by way of the Missouri to the foot of the rapids below the great falls of that river a distance of 2575 miles thence by land passing the Rocky Mountains to a navigable part of the Kooskooske 340. and with the Kooskooske 73 mls. Lewis's river 154 miles and the Columbia 413 miles to the Pacific Ocian makeing the total distance from the confluence of the Missouri and Mississippi to the discharge of the Columbia into the Pacific Ocean 3555 miles. The navegation of the Missouri may be deemed good; [its] dificulties arise from [its] falling banks, timber embeded in the mud of [its] channel, [its] sand bars and steady rapidity of [its] current all which may be over come with a great degree of certainty by useing the necessary precautions. The passage by land of 340 miles from the Falls of the Missouri to the Kooskooske is the most formidable part of the tract proposed across the Continent. Of this distance 200 miles is along a good road, and 140 over tremendious Mountains which for 60 miles are covered with eternal snows. A passage over these mountains is however practicable from the latter part of June to the last of September and the cheap rate at which horses are to be obtained from the Indians of the Rocky Mountains and West of them reduce the expences of transportation over this portage to a mere trifle. The navigation of the Kooskooske, the Lewis's river and the Columbia is safe and good from the 1st of April to the middle of August by makeing these portages on the latter river. The first of which in decending is that of 1200 paces at the Falls of the Columbia 261 miles up that river, the second of 2 miles at the long narrows 6 miles below the falls and the third also of 2 miles a the great rapids 65 miles still lower down. The tide flows up the Columbia 183 miles and within 7 miles of the great rapids. Large sloops may with safety ascend as high as tide water and Vessels of 300 tons burthen reach the entrance of the Multnomah River a large Southern branch of the Columbia, which takes [its] rise on the confines of New Mexico with the Callarado and Apostles river, dischargeing itself into the Columbia 125 miles from [its] entrance into the Pacific Ocian. I consider this tract across the continent emence advantage to the fur trade, as all the furs collected in 9/10ths of the most valuable furr country in America may be conveyed to the mouth of the Columbia and shiped from thence to East indies by the 1st of August in each year, and will of course reach Canton earlier than the furs which are annually exported from Montrall arive in Great Britain.

In our outward bound voyage we ascended to the foot of the rapids below the great falls of the Missouri where we arived on the 14th of June 1805. Not haveing met with any of the nativs of the Rocky Mountains we were of course ignorant of the passes by land which existed through those mountains to the Columbia river, and had we even known the rout we were destitute of horses which would have been indispensibly necessary to enable us to transport the requisit quantity of amunition and other stores to ensure the success of the remaining part of our voyage down the Columbia; we therefore deturmined to navigate the Missouri as far as it was practicable, or unless we met with some of the natives from whom we could obtain horses and information of the Country. Accordingly we undertook a most laborious portage at the falls of the Missouri of 18 miles which we effected with our Canoes and baggage by the 3rd of July. From hence ascending the Missouri we penetrated the Rocky Mountain at the distance of 71 miles above the upper part of the portage and penetrated as far as the three forks of that river a distance of 181 miles further; here the Missouri devides into three nearly equal branches at the Same point. The two largest branches are so nearly of the same dignity that we did not conceive that either of them could with propriety retain the name of the Missouri and therefore called these three streems Jefferson's Madisons and Gallitin's rivers. The confluence of those rivers is 2848 miles from the mouth of the Missouri by the meanders of that river. We arived at the three forks of the Missouri the 27th of July. Not haveing yet been so fortunate as to meet with the nativs altho' I had previously made several excurtions for that purpose, we were compelled still to continue our rout by water. The most northerly of the three forks, that to which we had given the name of Jeffersons river was deemed the most proper for our purpose and we accordingly ascended it 248 miles to the upper forks and [its] extreem navigable point, makeing the total distance to which we had navigated the waters of the Missouri 3096 miles of which 429 lay within the Rocky Mountains. On the morning of the 17th of August 1805 I arrived a the forks of the Jeffersons river where I met Capt. Lewis who had previously penitrated with a party of three men to the waters of the Columbia discovered a band of the Shoshone Nation and had found means to induce thirty five of their chiefs and warriors to accompany him to that place. From these people we learned that the river on which they resided was not navagable and that a passage through the mountains in that direction was impracticable; being unwilling to confide in the unfavorable account of the natives it was concerted between Capt. Lewis and myself that one of us should go forward immediately with a small party and explore the river while the other in the intirem would lay up the Canoes at that place and engage the nativs with their horses to assist in transporting our stores and baggage to their camp. Accordingly I set out the next day passed the deviding mountains between the waters of the Missouri and Columbia and decended the river which I since call the East fork of Lewis's river about 70 miles. Finding that the Indians account of the country in the direction of this river was correct I returned and joined Capt. Lewis on the 29th of August at the Shoshone camp excessively fatigued as you may suppose, haveing passed mountains almost inexcessable and compelled to subsist on berries dureing the greater part of my rout. We now purchased 27 horses of these indians and hired a guide who assured us that he could in 15 days take us to a large river in an open country west of these mountains by a rout some Distance to the North of the river on which they lived, and that by which the nativs west of the mountains visited the plains of the Missouri for the purpose of hunting the buffalow. Every preperation being made we set forward with our guide on the 31st of August through those tremendious mountains, in which we continued untill the 22nd of September before we reached the lower country beyond them; on our way we met with the Ootelachshoot a band of the Tuchapahs from whome we obtained axcession of seven horses and exchanged eight or ten others. This proved of infinite service to us as we were compelled to subsist on horse beef about Eight days before we reached the Kooskooske. Dureing our passage over those mountains we suffered every thing which hunger cold and fatigue could impose; nor did our difficulties with respect to provision cease on our arival at the Kooskooske for although the Pallotepallors a noumerous nation inhabiting that country were extremely hositable and for a fiew trifling articles furnish us with an abundance of roots and dryed salmon the food to which they were accustomed we found that we could not subsist on those articles and almost all of us grew sick on eating them. We were obliged therefore to have recourse to the flesh of horses and dogs as food to supply the dificiency of our guns which produced but little meat as game was scarce in the vicinity of our camp on the Kooskooske where we were compelled to remain in order to construct our perogues to decend the river. At this season the salmon are meagre and form but indifferent food. While we remained here I was my self sick for several days and my friend Capt. Lewis suffered a serious indesposition. Haveing completed four Perogues and a small canoe we gave our horses in charge to the Palletepallors untill we returned and on the 7th of October reimbarked for the Pacific Ocian. We decended by the rout I have already mentioned. The water of the river being low at this season we experienced much dificuelty in decending, we found it obstructed by a great number of dificuelt and dangerous rapids in passing of which our perogues several times filled and the men escaped narrowly with their lives. However the dificuelty does not exist in high water which happens within the periods which I have perviously mentioned.

We found the nativs extremely noumerous and generally friendly though we have on several occasion owed our lives and the fate of the expedition to our number which consisted of 31 men. On the 17th of November we reached the Ocian where various considerations induced us to spend the winter. We therefore searched for an eligible situation for that purpose and selected a spot on the South side of the little river called by the nativs the Netul which discharges itself at a small bar on the South Side of the Columbia and 14 miles within point Adams. Here we constructed some log houses and defended them with a common stockade work; this place we called Fort Clatsop after a nation of that name who were our nearest neighbours. In this country we found an abundance of Elk on which we subsisted principally during the last winter. We left Fort Clatsop on the 27th of March. On our homeward bound voyage being much better acquinted with the Country we were enabled to take such precautions as in a great measure secured us from the want of provision at any time, and greatly lessoned our fatigues, when compared with those to which we were compelled to submit, in our Outward bound journey. We have not lost a man sence we left the Mandans a circumstance which I assure you is a pleasing consideration for me. As I shall shortly be with you and the post is now waiting I deem it unnecessary here to attempt minutely to detail the Occurencies of the last 18 month.

I am &c. Yr. affectunate brother,

Wm. Clark
(Jackson 1978, 325–329)

BIBLIOGRAPHY

Adams, G., ed., 1965: *Early Voyages in the Pacific Northwest, 1813–1818, Peter Corney.* Ye Galleon Press, 238 pp.

Aegerter, M., and S. Russell, 2002: *Hike Lewis and Clark's Idaho.* University of Idaho Press, 223 pp.

Ahrens, C. D., 1985: *Meteorology Today.* West Publishing Company, 523 pp.

Allen, J. L., 1975: *Lewis and Clark and the Image of the American Northwest.* Dover, 412 pp.

————, 1991: *Jedediah Smith and the Mountain Men of the American West.* Chelsea House Publishers, 119 pp.

Ambrose, S. E., 1978: Snow conditions on the Lolo Trail—Some comparisons. *We Proceeded On*, February, 12–14.

————, 1996: *Undaunted Courage: Meriwether Lewis, Thomas Jefferson, and the Opening of the American West.* Simon and Schuster, 511 pp.

————, 1998: *Lewis & Clark: Voyage of Discovery.* National Geographic Society, 255 pp.

Appleman, R. E., 1975: *Lewis & Clark's Transcontinental Exploration 1804–1806.* U.S. Department of the Interior, 429 pp.

Arrington, L., 1994: *History of Idaho.* University of Idaho Press, 406 pp.

Bagley, J., 2000: *The First Known Man in Yellowstone.* Old Faithful Eye-Witness Publishing, 300 pp.

Bakeless, J., 1947: *Lewis and Clark: Partners in Discovery.* W. Morrow, 498 pp.

————, ed., 1964: *The Journals of Lewis and Clark.* Reprint. Penguin Putnam, 382 pp.

Baron, W. R., 1995: Historical climate records from the northeastern United States, 1640–1900.

Climate since A.D. 1500, R. S. Bradley and P. D. Jones, eds., Routledge, 74–91.

Beckham, S., 2003: *The Literature of the Lewis and Clark Expedition, a Bibliography and Essays.* Lewis and Clark College, 315 pp.

————, and R. Reynolds, 2002: *Lewis & Clark From the Rockies to the Pacific.* Graphic Arts Center Publishing, 144 pp.

Bedini, S. A., 1986: *Early American Scientific Instruments and Their Makers.* Landmark Enterprises, 189 pp.

————, 2002: *Jefferson and Science.* University of North Carolina Press, 126 pp.

Bergon, F., ed., 1989: *The Journals of Lewis and Clark.* Viking Penguin Nature Library, 464 pp.

Biddle, N., 1814: *The Journals of the Expedition under the Command of Captains Lewis and Clark.* 2 vols. Reprint. The Heritage Press, 547 pp.

Blume, J., 1999: Lewis and Clark, Geology and North Dakota. *North Dakota Geological Survey Newsl.*, **26** (2), 16–20.

Botkin, D. B, 1995: *Our Natural History: The Lessons of Lewis and Clark.* Berkley Publishing Group, 300 pp.

Bradley, R. S., and P. D. Jones, 1993: "Little Ice Age" summer temperature variations: Their nature and relevance to recent global warming trends. *Holocene*, 3, 387–396.

Brooks, N., 1935: *First Across the Continent: The Story of the Lewis and Clark Expedition.* Charles Scribner's Sons, 361 pp.

Bruun, E., and J. Crosby, 1999: *Our Nation's Archive: The History of the United States in Documents.* Workman, 886 pp.

Burnette, D., 2002: Meteorological reconstruction of the Lewis and Clark Expedition. Masters thesis, Emporia State College, 105 pp.

Burns, K., 1997: Lewis & Clark: The Journey of the Corps of Discovery. 240 minutes color video, American Lives Film Project. PBS Home Video, B#3499.

Burroughs, R. D., 1995: *The Natural History of the Lewis and Clark Expedition.* Michigan State University Press, 340 pp.

Carpenter, A., 1991: *The Encyclopedia of The Far West.* Facts on File, 544 pp.

Catchpole, A. J. W., 1995: Hudson's Bay Company ships' log-books as sources of sea ice data.

Climate since A.D. 1500, R. S. Bradley and P. D. Jones, eds., Routledge, 17–39.

Chittenden, H. M., 1986: *The American Fur Trade of the Far West.* 2 vols. University of Nebraska Press, 994 pp.

Chuinard, E. G., 1998: *Only One Man Died, the Medical Aspects of the Lewis and Clark Expedition.* Arthur H. Clark Company, 444 pp.

Clarke, C. G., 1970: *The Men of the Lewis and Clark Expedition.* University of Nebraska Press, 351 pp.

Cohen, D., 2002: *Mapping the West: America's Westward Movement 1524–1890.* Rizzoli International Publications, 208 pp.

Cordes, K., 1999: *America's National Historic Trails.* University of Oklahoma Press, 369 pp.

Coues, E., Ed., 1893: *The History of the Lewis and Clark Expedition.* 4 vols. Reprint, 1987. Dover, 1364 pp.

_____, ed., 1965: *The Expeditions of Zebulon Montgomery Pike in the Years, A New Edition.* 2 vols. Reprint. Ross and Haines, 955 pp.

Cutright, P. R., 1976: *A History of the Lewis and Clark Journals.* University of Oklahoma Press, 311 pp.

_____, 1986: Rest, rest, perturbed spirit. *We Proceeded On*, March, 7–16.

_____, 2001: *Contributions of Philadelphia to Lewis and Clark History.* Lewis and Clark Trail Heritage Foundation Chapter, 50 pp.

_____, 2003: *Lewis and Clark: Pioneering Naturalists.* University of Nebraska Press, 538 pp.

Dear, E., 2000: *The Grand Expedition of Lewis and Clark as seen by C. M. Russell.* 2[d] ed. C. M. Russell Museum, 32 pp.

DeVoto, B., ed., 1953: *The Journals of Lewis and Clark.* Houghton Mifflin, 514 pp.

_____, 1998: *Across the Wide Missouri.* Reprint. Houghton Mifflin Company, 450 pp.

_____, 2002: *The Year of Decision—1846.* Reprint. Truman Talley Books, 537 pp.

Dippie, B., et al., 2002: *George Catlin and His Indian Gallery.* Smithsonian Books, 294 pp.

Druckenbrod, D. L., et al., 2003: Late-eighteenth-century precipitation reconstruction from James Madison's Montpelier plantation. *Bull. Amer. Meteor. Soc.*, **84**, 57–71.

Duncan, D., 1987: *Out West, A Journey Through Lewis & Clark's America.* University of Nebraska Press, 436 pp.

_____, and K. Burns, 1999: *Lewis and Clark: An Illustrated History.* Alfred A. Knopf, 250 pp.

Eastman, G., and M. Eastman, 2002: *Bitterroot Crossing: Lewis and Clark Across the Lolo Trail.* University of Idaho Library, 82 pp.

Encyclopedia Britannica, 1968: *The Annals of America, Volume 4, 1797–1820: Domestic Expansion and Foreign Entanglements.* William Benton, 158–189.

Evenson, T., et al., 2000: *The Lewis and Clark Cookbook.* Whisper Waters, 190 pp.

Fanselow, J., 2000: *Traveling the Lewis and Clark Trail.* 2[d] ed. Falcon Press, 321 pp.

Fifer, B., 2003: *Lewis and Clark Expedition Illustrated Glossary*. Farcountry Press, 80 pp.

Fifer, B., and V. Soderberg, 1998: *Along the Trail with Lewis and Clark*. Montana Magazine, 206 pp.

_____, and _____, 2001: *Along the Trail with Lewis and Clark*. 2d ed. Montana Magazine, 216 pp.

_____, and _____, 2002: *2002–2003 Along the Trail with Lewis and Clark Travel Planner and Guide*. Montana Magazine, 121 pp.

Fisher, V., 1962: *Suicide or Murder? The Strange Death of Governor Meriwether Lewis*. Sage Books, 288 pp.

Flores, D. L., 1984: *Southern Counterpart to Lewis & Clark, the Freeman & Custis Expedition of 1806*. University of Oklahoma Press, 386 pp.

Forrester, F. H., 1957: *1001 Questions Answered About the Weather*. Dodd, Mead and Company, 419 pp.

Frisinger, W. H., 1983: *The History of Meteorology: to 1800*. American Meteorological Society, 148 pp.

Gass, P., 1807: *A Journal of the Voyages and Travels of a Corps of Discovery*. Zadok Cramer, 262 pp.

Gilbert, B., 1979: *The Trailblazers*. Time Life Books, 236 pp.

Gilman, C., 2003: *Lewis and Clark Across the Divide*. Smithsonian Books, 424 pp.

Glaser R., et al., 1999: Seasonal temperature and precipitation fluctuations in selected parts of Europe during the sixteenth century. *Climate Change,* **43,** 169–200.

Glickman, T. S., ed., 2000: *Glossary of Meteorology*. 2d ed. American Meteorological Society, 855 pp.

Goetzmann, W., 1984: *Karl Bodmer's America*. Joslyn Art Museum, 376 pp.

_____, and G. Williams, 1992: *The Atlas of North American Exploration*. University of Oklahoma Press, 224 pp.

Graetz, R., and S. Graetz, 2001: *Lewis and Clark's Montana Trail*. Northern Rockies Publishing, 120 pp.

Gragg, R., 2003: *Lewis and Clark on the Trail of Discovery, A Museum in a Book*. Rutledge Hill Press, 48 pp.

Gulick, B., 1994: *Chief Joseph County: Land of the Nez Perce*. Caxton Printers, 316 pp.

Hafen, L., 1995: *Fur Traders, Trappers and Mountain Men of the Upper Missouri*. University of Nebraska Press, 138 pp.

Harris, B., 1993: *John Colter, His Years in the Rockies*. University of Nebraska Press, 180 pp.

Hayes, D., 2000: *Historical Atlas of the Pacific Northwest*. Sasquatch Books, 208 pp.

_____, 2001: *First Crossing*. Sasquatch Books, 320 pp.

Herbert, J., 2000: *Lewis and Clark for Kids*. Chicago Press Review, 143 pp.

Hill, W., 2000: *The Oregon Trail: Yesterday and Today*. Caxton Press, 197 pp.

_____, 2004: *The Lewis and Clark Trail: Yesterday and Today*. Caxton Press, 288 pp.

Hoganson, J., and E. Murphy, 2003: *Geology of the Lewis and Clark Trail in North Dakota*. Mountain Press, 247 pp.

Hollowy, D., 1974: *Lewis and Clark and the Crossing of North America*. Saturday Review Press, 224 pp.

Holmberg, J., 2002: *Dear Brother, Letters of William Clark to Jonathan Clark*. Yale University Press, 322 pp.

Howay, F., 1990: *Voyages of the "Columbia" to the Northwest Coast, 1787–1790 and 1790–1793*. Reprint. Oregon Historical Society Press, 518 pp.

Hunsuker, J. B., 2001: *Sacagawea Speaks, Beyond the Shining Mountains with Lewis & Clark*. TwoDot Book, 151 pp.

Hunt, S., ed., 1999: *Illustrated Atlas of Native American History.* Chartwell Books, 272 pp.

Ifland, P: 1998: *Taking the Stars: Celestial Navigation from Argonauts to Astronauts.* Mariners' Museum, 222 pp.

Ingram, M. J., D. J. Underhill, and T. M. L. Wigley, 1978: Historical climatology. *Nature,* **276,** 329–334.

Irving, W., ed., 2001: *The Adventures of Captain Bonneville.* The Narrative Press, 401 pp.

Jackson, D., 1978: *Letters of the Lewis and Clark Expedition with Related Documents, 1783–1854.* 2 vols. University of Illinois Press, 806 pp.

_____, 2002: *Thomas Jefferson & the Rocky Mountains: Exploring the West from Monticello.* University of Oklahoma Press, 339 pp.

Jones, L., 2004: *William Clark and the Shaping of the West.* Hill and Wang, 394 pp.

Knapp, P., 2004: Window of opportunity: The climatic conditions of the Lewis and Clark Expedition of 1804–1806. *Bull. Amer. Meteor. Soc.,* 85, 1289–1303.

Laliberte, A., and W. Ripple, 2003: Wildlife encounters by Lewis and Clark: A spatial analysis of interactions between Native Americans and wildlife. *BioScience,* 53, 994–1003.

Lamb, W. K., ed., 1960: *Simon Fraser: Letters and Journals, 1806–1808.* Pioneer Books, 292 pp.

Lange, R. E., 1979: The Expedition and the inclement weather of November – December 1805. *We Proceeded On,* November, 14–16.

Large, A. J., 1986: "…it thundered and lightened": The weather observations of Lewis and Clark. *We Proceeded On,* May, 6–10.

_____, 1991: Expedition aftermath: The Jawbone journals. *We Proceeded On,* February, 12–23.

Lavender, D., 1958: *Land of Giants: The Drive to the Pacific Northwest 1750–1950.* Doubleday and Company, 468 pp.

_____, 1998: *The Way to the Western Sea: Lewis and Clark across the Continent.* University of Nebraska Press, 444 pp.

Least Heat-Moon, W., 1999: *River-Horse Across America by Boat.* Penguin, 506 pp.

Ludlam, D. M., 1966: *Early American Winters, 1604–1820.* American Meteorological Society, 285 pp.

MacGregor, G., 2003: *Lewis and Clark Revisited: A Photographer's Trail.* University of Washington Press, 199 pp.

Mackenzie, A., 1801: *Voyages from Montreal Through the Continent of North America, In the Years 1789 and 1793.* Reprint, 1966. University Microfilms, 412 pp.

McKenney, T., 1972: *Sketches of a Tour to the Lakes.* Imprint Society, 414 pp.

Meadows, S., and J. Prewitt, 2003: *Lewis and Clark for Dummies.* Wiley Publishing, 382 pp.

Middleton, W. E. K, 1969: *Invention of the Meteorological Instruments.* John Hopkins University Press, 377 pp.

_____, 2003a: *The History of the Barometer.* John Hopkins University Press, 480 pp.

_____, 2003b: *A History of the Thermometer and Its Use in Meteorology.* John Hopkins University Press, 268 pp.

Miller, G., 2004: *Lewis and Clark's Northwest Journey: Disagreeable Weather.* Frank Amato Publications, 80 pp.

Moore, R., and M. Haynes, 2003: *Lewis and Clark Tailor Made, Trail Worn.* Farcountry Press, 288 pp.

Moulton, G. E., Ed., 1986: *The Journals of Lewis and Clark Expedition.* Vol.1, *Atlas of the Lewis & Clark Expedition.* University of Nebraska Press, 126 pp.

_____, 1986: *The Journals of Lewis and Clark Expedition.* Vol. 2, *August 30, 1803 – August 24, 1804.* University of Nebraska Press, 612 pp.

_____, 1987a: *The Journals of Lewis and Clark Expedition.*Vol.3, *August 25, 1804–April 6, 1805.* University of Nebraska Press, 544 pp.

_____, 1987b: *The Journals of Lewis and Clark Expedition.* Vol. 4, *April 7–July 27, 1805.* University of Nebraska Press, 464 pp.

_____, 1988: *The Journals of Lewis and Clark Expedition.* Vol.5, *July 28–November 1, 1805.* University of Nebraska Press, 415 pp.

_____, 1990: *The Journals of Lewis and Clark Expedition.* Vol. 6, *November 2, 1805–March 22, 1806.* University of Nebraska Press, 531 pp.

_____, 1991: *The Journals of Lewis and Clark Expedition.* Vol. 7, *March 23–June 9, 1806.* University of Nebraska Press, 383 pp.

_____, 1993: *The Journals of Lewis and Clark Expedition.* Vol. 8, *June 10–September 26, 1806.* University of Nebraska Press, 456 pp.

_____, 1995: *The Journals of Lewis and Clark Expedition.* Vol. 9, *The Journals of John Ordway, May 14, 1804–September 23, 1806 , and Charles Floyd, May 14–August 18, 1804.* University of Nebraska Press, 419 pp.

_____, 1996: *The Journals of Lewis and Clark Expedition.* Vol. 10, *The Journal of Patrick Gass, May 14, 1804–September 23, 1806.* University of Nebraska Press, 300 pp.

_____, 1997: *The Journals of Lewis and Clark Expedition.* Vol. 11, *The Journals of Joseph Whitehouse, May 14, 1804–April 2, 1806.* University of Nebraska Press, 459 pp.

_____, 1999: *The Journals of Lewis and Clark Expedition.* Vol. 12, *Herbarium of the Lewis & Clark Expedition.* University of Nebraska Press, 211 pp.

_____, 2001: *The Journals of Lewis and Clark Expedition.* Vol. 13, *Comprehensive Index.* University of Nebraska Press, 174 pp.

_____, 2003: *An American Epic of Discovery: The Lewis and Clark Journals.* University of Nebraska Press, 413 pp.

Mussulman, J., 2004: *Discovering Lewis and Clark from the Air.* Mountain Press Publishing Company, 261 pp.

Nell, D. F., and J. Taylor, 1996: *Lewis and Clark in the Three Rivers Valleys.* The Patrice Press, 284 pp.

_____, and A. Demetriades, 2002: The utmost reaches of the Missouri. *We Proceeded On*, November, 29–33.

Nathan, T., 2000: Meteorological Aspects of the Lewis and Clark Expedition. *Proc. 14th Conference on Biometeorology and Aerobiology*, University of California, Davis.

_____, 2003: *Lewis and Clark's Contributions to Meteorological Science.* Lewis and Clark Heritage Foundation, Annual Conference.

Nichols, R., and P. Halley, 1995: *Stephen Long and American Frontier Exploration.* University of Oklahoma Press, 276 pp.

Osgood, E. S., 1964: *The Field Notes of Captain William Clark, 1803–1805.* Yale University Press, 335 pp.

Patient, D. H., 2003: *Plants on the Trail with Lewis and Clark.* Houghton Mifflin, 112 pp.

Paton, B. C., 2001: *Lewis and Clark: Doctors in the Wilderness.* Fulcrum Publishing, 240 pp.

Peck, D. J., 2002: *Or Perish in the Attempt, Wilderness Medicine in the Lewis and Clark Expedition.* Farcountry Press, 351 pp.

Peterson, D. A., 1998: *Early Pictures of the Falls, A Lewis and Clark Portrait in Time.* Lewis and Clark Trail Heritage Foundation, 40 pp.

Pfister, C., 1995: Monthly temperature and precipitation in central Europe 1525–1979: Quantifying documentary evidence on weather and its effects. *Climate since A.D. 1500,* R. S. Bradley and P. D. Jones, eds., Routledge, 118–142.

Plamondon, M., II., 1991: The Instruments of Lewis and Clark. *We Proceeded On,* February, 7–11.

_____, 2000: *Lewis and Clark Trail Maps, A Cartographic Reconstruction, Volume I.* Washington State University Press, 191 pp.

_____, 2001: *Lewis and Clark Trail Maps, A Cartographic Reconstruction, Volume II.* Washington State University Press, 221 pp.

_____, 2004: *Lewis and Clark Trail Maps, A Cartographic Reconstruction, Volume III.* Washington State University Press, 256 pp.

Preston, R. S., 2000: The accuracy of the astronomical observations of Lewis and Clark. *Proc. Amer. Philos. Soc.,* **144,** 168–191.

Preston, V. L., 2004: Weather, water, and climate of the Lewis and Clark Expedition (1803–1806). Preprints, *14th Conf. on Applied Meteorology,* Seattle, WA, Amer. Meteor. Soc., P1.10.

Prucha, F., 1994: *Indian Peace Medals in American History.* University of Oklahoma Press, 186 pp.

Quaife, M. M., 1916: *The Journals of Captain Meriwether Lewis and Sergeant John Ordway Kept on the Expedition of Western Exploration, 1803–1806.* State Historical Society of Wisconsin, 444 pp.

Quinn, W. H., and V. T. Neal, 1995: The historical record of El Nino events. *Climate since A.D. 1500,* R. S. Bradley and P. D. Jones, eds., Routledge, 623–648.

_____, _____, and S.E. Antunez de Mayolo, 1987: El Nino occurrences over the past four and a half centuries. *J. Geophys. Res.,* **92** (C1), 14 449–14 461.

Richardson, J. D., 1897: *A Compilation of the Messages and Papers of the Presidents, Volume 1.* Bureau of National Literature, 446 pp.

Rhody, K., 1996: *Rendezvous, Reliving the Fur Trade Era of 1825–1840.* Sierra Press, 77 pp.

Ronda, J. P., 1984: *Lewis and Clark Among the Indians.* University of Nebraska Press, 310 pp.

_____, 2000: *Jefferson's West, A Journey with Lewis and Clark.* Thomas Jefferson Foundation, 80 pp.

_____, 2001: *Finding the West, Explorations with Lewis and Clark.* University of New Mexico Press, 138 pp.

_____, 2003: *Beyond Lewis and Clark: The Army Explores the West.* Washington State Historical Society, 106 pp.

Salisbury, A., and J. Salisbury, 1950: *Two Captains West.* Superior Publishing Company, 235 pp.

Saindon, R., 2003: *Explorations into the World of Lewis and Clark.* 3 vols. Digital Scanning, 1493 pp.

Schmidt, T., 1998: *Guide to the Lewis and Clark Trail.* National Geographic, 192 pp.

_____, and J. Schmidt, 1999: *The Saga of Lewis and Clark, into the Uncharted West.* Tehabi Books, 210 pp.

Solomon, S., and J. Daniel, 2004: Lewis and Clark: Pioneering meteorological observers in the American West. *Bull. Amer. Meteor. Soc.,* **85,** 1273–1288.

Space, R.S., 2001: *The Lolo Trail, a History and a Guide to the Trail of Lewis and Clark.* 2d ed. Historic Montana Publishing, 136 pp.

Steffen, J., 1977: *William Clark: Jeffersonian Man of the Frontier.* University of Oklahoma Press, 196 pp.

Stegner, P., 2002: *Winning the Wild West: The Epic Saga of the American Frontier 1800–1899.* The Free Press, 400 pp.

Strong, E., and R. Strong, 1995: *Seeking Western Waters: The Lewis and Clark Trail from the Rockies to the Pacific*. Oregon Historical Society Press, 383 pp.

Thomas, G., 2000: *Lewis and Clark Trail, the Photo Journal*. Snowy Mountain Publishing, 122 pp.

Thorp, D., 1998: *An American Journey: Lewis and Clark*. Metro Books, 160 pp.

Thwaites, R. G., ed., 1904a: *Original Journals of the Lewis and Clark Expedition, Volume 1.*
Dodd, Mead and Company, 374 pp.

_____, 1904b: *Original Journals of the Lewis and Clark Expedition, Volume 2*. Dodd, Mead and Company, 386 pp.

_____, 1904c: *Original Journals of the Lewis and Clark Expedition, Volume 3*. Dodd, Mead and Company, 363 pp.

_____, 1904d: *Original Journals of the Lewis and Clark Expedition, Volume 4*. Dodd, Mead and Company, 372 pp.

_____, 1904e: *Original Journals of the Lewis and Clark Expedition, Volume 5*. Dodd, Mead and Company, 395 pp.

_____, 1904f: *Original Journals of the Lewis and Clark Expedition, Volume 6*. Dodd, Mead and Company, 280 pp.

_____, 1904g: *Original Journals of the Lewis and Clark Expedition, Volume 7*. Dodd, Mead and Company, 534 pp.

_____, 1904h: *Original Journals of the Lewis and Clark Expedition, Atlas*. Dodd, Mead and Company, 54 pp.

Tubbs, S. A., 2003: *The Lewis and Clark Companion: An Encyclopedic Guide to the Voyage of Discovery*. Henry Holt, 345 pp.

Ward, G. C., 1996: *The West: An Illustrated History*. Little, Brown, 445 pp.

Wells, G., and D. Anzinger, 2001: *Lewis and Clark Meet Oregon's Forest: Lessons from Dynamic Nature*. Oregon State University Press, 224 pp.

Wheeler, O. D., 1904: *The Trail of Lewis and Clark, 1804–1904*. 2 vols. G.P. Putnam's Sons, 796 pp.

White, T., and D. Carrol: *America Looks West: Lewis and Clark on the Missouri*. Nebraskaland Magazine, 130 pp.

Whitnah, D.R., 1961: *A History of the United States Weather Bureau*. University of Illinois Press, 267 pp.

Wood, W. R., 2003: *Prologue to Lewis and Clark: The Mackay and Evans Expedition*. University of Oklahoma Press, 234 pp.

_____, and T. D. Thiessen, 1985: *Early Fur Trade on the Northern Plains*. University of Oklahoma Press, 353 pp.

Woodger, E., and B. Toropov, 2004: *Encyclopedia of the Lewis and Clark Expedition*. Checkmark Books, 438 pp.